T0255306

CATCH A
FALLING STAR

DONALD D. CLAYTON

iUniverse, Inc.
New York   Bloomington

iUniverse books may be ordered through booksellers or by contacting:

iUniverse
1663 Liberty Drive
Bloomington, IN 47403
www.iuniverse.com
1-800-Authors (1-800-288-4677)

Because of the dynamic nature of the Internet, any Web addresses or links contained in this book may have changed since publication and may no longer be valid. The views expressed in this work are solely those of the author and do not necessarily reflect the views of the publisher, and the publisher hereby disclaims any responsibility for them.

ISBN: 978-1-4401-6103-2 (sc)
ISBN: 978-1-4401-6104-9 (ebook)
ISBN: 978-1-4401-6102-5 (dj)

Printed in the United States of America

iUniverse rev. date: 10/23/2009

INTRODUCTION

This autobiography shares my life and the astrophysics that gave it lifelong meaning. I have three motives for undertaking this large effort.

Firstly, I hope to contribute to the genre of scientific biography. I love writing as a craft. My goal for my prose is to present not only my scientific life as accurately as possible but also insight into human aspects of its struggles. I see my story as a classical American story, having emerged from the melting pot of European immigration to the Midwest at the middle of the 19th century.

Secondly, I wish the reader to have an engrossing introduction to the origin of the atoms of the chemical elements. Discovering many aspects of that quest has been my life work. The atoms are the building blocks of our natural world. It has been my lifelong goal to convey interesting knowledge about our universe. My writing has always emphasized that goal. By sharing my scientific discoveries with the reader I hope to share the essence of that sublime science. The creation of our material world holds fascination for mankind.

Thirdly, I hope to contribute meaningfully to the history of science. I have striven to contribute primary material for its study. The possibility of achieving this goal derives from my intimate relationships with other pioneers of the science of nucleosynthesis in stars and observable consequences for astronomy. I have donated much of my photo collection to the APS Center for the History of Physics precisely because of my intent to contribute research materials for the history of science.

Having these three goals has made the writing of uniform prose difficult. For that reason, some lay readers may find the detailed scientific issues to be heavy going, and simply scan through them. Other readers interested in the scientific history of the origin of the chemical elements may find details of my personal life less compelling. I am content if each reader finds a meaningful experience in sharing my life quest, but I am unconcerned if I have provided more than some readers may want to know.

Psychologists have learned that we rewrite our own history continuously in the light of our self image. I accept that limitation; but I also try to diminish its force by heavy reliance on my diaries and on my mother's diaries, as well as on photo albums. These not only accurately reveal when and where events occurred but also place them in the context of my feelings at the time. I have researched my material carefully.

I report only occasional dialogue because I do not want to create it long after the fact. I strive throughout for historical truth. Some dialogue that I include is half a century old, allowing one to wonder about its accuracy. For that reason I limit dialogue to those especially meaningful times that I can never forget. Although the exact words may not be right, the truth of its occurrence and its meaning is certain.

Selected photographs are printed in the text. I call attention to a separate contribution to the history of science made available in my photo archive for the history of nuclear astrophysics. If the reader chooses to open this web site, http://www.astro.clemson.edu/NucleoArchive/index.html, he will find a lively accompaniment to my text. Click the *Photo List*. The photos are listed by year and include a carefully researched caption. They are referred to within this autobiography in the format <1962 Clayton Caltech PhD>. Higher resolution versions of my photographs can be obtained from the Emilio Segre Archive at the APS Center for the History of Physics.

Many of these photo citations are placed in end notes. The

main purpose of these notes is presentation of more technical and bibliographic material. These are primarily for scholars, and are not essential to the reading of the autobiography.

My professional web page is
http://www.astro.clemson.edu/NucleoArchive/index.html

I dedicate this book to every reader who places high value on understanding the history of our universe and its matter. May my words and my story touch your mind and your heart. And I dedicate it also to the great influence of four who are gone. I pay tribute in memoriam to William A. Fowler and to Fred Hoyle. Each was a deep personal friend and a very great scientist. They introduced me to the scientific path from which I have never strayed. Frank McDonald at SMU gave me the opportunity to become a physicist and to enter Caltech. My life as lived would have been unthinkable save for those three. My mother, Avis Kembery Clayton, daughter of German and English immigrants to Iowa, is never far removed from what I attempt in this work, and her historical contributions are large. I thank my wife, Nancy McBride Clayton, for her encouraging criticism of my text and for bringing happiness.

Donald D. Clayton
Clemson SC
June 20, 2009

PROLOGUE.
IN THE BEGINNING

In the beginning there was no earth. But long before the earth and planets formed, the chemical elements existed. Their atoms were created not at once but over time, over very long time. The creation of the elements that we inherited on our earth began about thirteen billion years ago, an age about three times greater than the age of our earth. The most abundant of our mineral-forming chemical elements were created by thermonuclear explosions near the centers of exploding stars. Stunningly, the chemical elements and their specific abundances are the products of natural history within the universe. The atoms of our own bodies, except for hydrogen in the water, were created by thermonuclear reactions within stars that lived out their lives before our solar system existed.

The natural creation of the atoms comprising the earth continued until almost five billion years ago, when our own atoms, floating initially within the gas and dust in the interstellar spaces between the stars, withdrew from that common pool of interstellar gas. Our atoms sequestered themselves at that time within a dense, cold, dark cloud of gas that was nearing a time of fundamental crisis. That cloud grew slowly more massive as more interstellar atoms joined it, becoming part of that unbelievably cold cloud. The weight of the cloud upon its inner self eventually exceeded what its internal pressure could support. The center could not hold. Squeezed by self weight into smaller and smaller size, gravity necessarily, according to its law, grew stronger and stronger, accelerating the compression. In a cosmic twinkling of an eye, the center gave way, collapsed, allowing overlying layers to rain down upon it until the whole dense cloud had collapsed. To what? The heat generated by the squeeze of gravity, soon caused a dense central ball of gas to glow, first infrared, then red, then orange, then yellow. Our sun was being born, even as more gas and dust fell onto it. Its central gases became so hot that the more energetic collisions in the heat tore the atoms apart, separating nuclei from their clouds of neutralizing electrons, increasing the pressure by increasing the number of free electrons. After 100,000 years that internal pressure was so large that it became sufficient to support the weight of the overlying gas. That support stopped the collapse, and the sun settled down for the next 4.6 billion years until the present day.

The entirety of that dense cloud was not physically able to join the sun, however. Fundamental laws of nature forbade it. The cloud had been rotating very slowly prior to its collapse, but its rotational speed necessarily accelerated as the cloud collapsed. Those fragments with the largest rotational momentum could not make it into the sun. Those fragments, in response to the pull of gravity toward the sun, could only settle into a flat disk of matter orbiting about the sun, held in their orbits by gravitational attraction toward the central sun. Rotational momentum prevented closer approach. Here, in this slightly warmed disk of orbiting interstellar debris, the planetary bodies were born from the concentrated gas and rubble.

Because the atoms of the orbiting disk were much older than the planets that were then forming from them, those atoms had existed in the prior interstellar gas and dust in a great variety of chemical forms reflecting their varied environmental situations. The chemical associations of these atoms retained an isotopic memory of their presolar histories. *Cosmic chemical memory* persisted in the solar disk of gas and rubble. For each chemical element many separate histories were mixed together, but still present, in every gram of solar disk. The free gaseous atoms retained no trace of memory, such memory being forbidden according to modern understanding of quantum mechanics. But other atoms were bonded together into molecules that documented the circumstances and locations of their chemical bondings. Their memory lay in the specific isotopes to which they were bonded. Some other atoms were contained within solid particles of small size, called dust, each of which retained some memory of the time and place in an interstellar cloud wherein those dust grains had grown larger by accumulating atoms and molecules. Their memory was of the isotopic compositions of their parent clouds, which was imprinted onto the dust. Still other atoms were building blocks of tiny minerals, hard cosmic gemlike rocks that had been fused by the heat of the dying stars into minerals that condensed while the hot gas was flowing away from the stars. These were *stardust*, solid pieces of individual presolar stars. The stardust persisted later in the disk rubble of the new sun. These stardust rocks were incorporated into the planetary bodies. They are solids older than the sun and planets. These various chemical forms of each chemical element were the fingerprints of cosmic chemical memory.

4,575,000,000 years later

About 4,575 million years later, on the third planet outward from the sun, in just that century when mankind had first evolved to be able to consider these questions scientifically, a farm boy from the Iowa cornfields would discover and understand key aspects of this cosmic history.

CHAPTER 1.
THE ADAIR COUNTY FARMS

I loved the farm. I have loved it throughout my life, intensely and perhaps beyond reason. I loved the roosters, their wake up calls in the gray dawn. I loved the hen house, gathering their eggs, and the little "peep peep" of tiny yellow fuzzballs. I loved milking the cows, how they already started home at 5:00 with full udders. I loved the baby pigs and feared the big hogs. Even more I loved the fields of oats when they were ready to cut, golden and heavy. I loved the alfalfa and clover and the smell of hay, fresh cut, drying, or of last year's thrown down to the cows. I loved how the horses crunched on their oats. As I lay in the oat bin and opened and chewed them, I embraced life. The love was primitive, an expression of subconscious well being. I loved the corn rows so tall that I could not reach their top ears. I loved their dried color, yellow, red, brown, kernels arranged by nature's whim on the cob. I was fascinated by the shelled corn, letting it rain through my fingers, and I savored its rough and chewy taste, tough but sweet. I wondered if the horses tasted it as I did.

I loved mowing times--hay, oats, wheat and barley. I loved the relentless rhythm of their coming due. I loved the windmill pumping water in a good breeze. I loved the sheep, especially the lambs' worried bleating. I loved the traditional farmer's machinery, and the horse team, its bridles, reins, halters, and fastening the neck

1

yoke to the wagon tongue. I admired the tractors, the combines, the manure spreaders, the planters, the harvesters and combines, and the hay mowers. I loved to help my grandparents, all four of them, and my aunt and uncle with their work. I marveled at their cream separator, and in their keeping of one part, trading the cream for goods in town, and slopping the hogs with the remainder. I loved the long power belts that drove the oat threshing machine, cycling with one lengthy twist from the tractor to the thresher. I loved loading the hay into the barn, how the horse team would walk away with the rope until the big fork was up and in. I watched it trip, fall and settle into the big red-barn haymow. I loved gathering dried corncobs to fuel the large iron stove.

I loved it all, and still do.

The memories of farming that I carry are of a world now past. Unforseen by me as a boy and by society, farming was changing. I had thought it immutable. For me it was families, my families, living on 160 acres, more or less, growing all those many things, some to feed to the diverse animals, each having their own function for the family, some to trade in town for cash or credit, and some to eat. Farming to me was hundreds of such families living around a town in which manufactured goods could be purchased. I carried this concept of farming with me through schooling, through marriages, through a career in scientific research, and it blooms within me even now as I write. It is a lasting part of the American mythology, even though it vanished while I was not watching. Politicians paid it lip service even while they helped its destruction with embrace of a new political religion—low-cost efficiency. Farmers today are businessmen, no matter how much they too may love the land; and their crops are more like manufactured items than byproduct of family subsistence on the land. Farmers today have the same bottom line, profit, as any industry. They manufacture a product to sell. With cash proceeds they buy their food at supermarket chains just as do citizenry of Chicago or Houston. Their machinery no longer symbolizes the tools of living, as I imagined, but rather capital investment, often huge, with which any manufacturing plant is saddled in its war with economic efficiency.

The images of farming in the old way remained fresh in my mind throughout my life. They imprinted upon a receptive center of my brain, something similar to the newly born animal imprinting its mother. I find no other way to understand the clarity of my images and the affection with which I have always preserved them. But there is probably more that I do not comprehend, some longing that I attach to those images and which endowed them with emotional force, and which may even have obstructed my emotional maturation. In some sense the images and the land *are* the mother. The earth is indeed her sweet breast, as poets and psychiatrists alike have described.

The story of my life and of my family needs those images to begin. Tracing the threads of my life not only explains much of me but also chronicles a changing America. I and my families arose from Adair County, Iowa. On a map of the United States, Adair County is a square, thirty miles on a side, about fifty miles west and slightly south of Des Moines, the Iowa capitol. It lies adjacent to and just west of Madison County, known for its covered bridges, John Wayne's birthplace in Winterset, and for a recent romance novel that is light years from the story I have to tell. Adair County is bounded today on the north by Interstate 80 going due west from Des Moines. Des Moines lies halfway between two great north-south rivers--the Mississippi and the Missouri. Adair County extends on its west end about halfway from Des Moines to the Missouri River. This entire farming region was an aperture toward the western prairies and is still today one of the least populated areas of the United States. Adair County has no cities. Its most populated town is Greenfield, its present county seat, with 2074 citizens. Greenfield lies near the center of that 30-mile square, a town once so renowned for its middle-American charm that it was named in one poll the most beautiful town in America, and was chosen as venue for Dick Van Dyke's film comedy *Cold Turkey*. To Greenfield my grandfather and grandmother Kembery[1] retired from their farm, and in Greenfield my grandfather Kembery worked his last--custodian for the Greenfield Presbyterian Church, earning spending and heating money for his final years.

Six miles further west is the smaller farm town of Fontanelle, the nexus of my families and of my boyhood imagery. The east fork of the Nodaway river bisects the line connecting Fontanelle and Greenfield, draining the land southward toward the Missouri River. Fontanelle was the first county seat of Adair County, where the first County Courthouse was built in 1856. It was here that my parents were born in farmhouses within walking distance from town. Fontanelle was named for Chief Logan Fontanelle (1825-55), son of a French fur trader named Lucien Fontanelle and an Indian woman, daughter of Big Elk, of the Omaha tribe. This homage of town name resulted from Chief Fontanelle's role as an intermediary in the establishment of the Omaha Indian reservation. This was possible because Logan Fontanelle went to St. Louis for an education. He then promoted agriculture and education within his tribe. Chief Logan Fontanelle was killed in 1855 during a skirmish with Sioux, one year before construction of the county courthouse in the Adair County town be named Fontanelle.

To me, Fontanelle was the heart of the farms, because it was to Fontanelle that farmers traveled for trade, shopping, milling of grain, and social life. The Fontanelle square seemed an exciting place on Saturdays to this boy, because everyone came here after Saturday chores were done. Its gazebo bandstand at its center hosted music and other events. The stately elm trees around the square were awe inspiring until their death by 1965 of an epidemic of Dutch Elm Disease. We would walk dozens of times around the square every Saturday, cruising I guess one would say. In 1879 Fontanelle became the terminus of the Chicago, Burlington & Quincy (CB&Q) railroad out of Creston, Iowa. In 1918 the town had a population of 318 citizens, and the depot's telegraph wire bore the code "FE". For decades this made the town a farm market for shipping. That's all gone today.

first portrait of the author, at age 18 months.

These gentle rolling hills had lain under four continental ice sheets, or glaciers, during the Pleistocene geologic age beginning about one million years ago. The last ice sheet disappeared about 10,000 years ago. Their recessions left behind a jumble of fine clay, sand and gravel, upon which rained dust layers of countless dust storms. It was then taken over by the tallgrasses. Over 150 varieties of native tallgrass grew annually and were buried annually for 10,000 years by snow. Those tallgrass prairies make mythic images of the American past, reaching then in height to the tops of the pioneer wagon wheels, and for as far as the eye could see. Annually they rotted and regrew, until after those 10,000 post-glacial years their compost provided the best sod in the world, the great Iowa prairies, reaching hundreds of miles west from the Mississippi River, beyond the Missouri River to the Great Plains of the west. Iowa was endowed with one quarter of the entire Grade I agricultural land of the United States. As I learned these things later, I found that my affection for that Iowa soil broadened to embrace its natural history, and indeed, that of the whole earth.

At least six different prehistoric Indian cultures populated these lands, yielding several descendant tribes for the advent of the first French explorers from Canada. In 1682 the land was claimed for France and named Louisiana for its king. Then in 1803 France sold Louisiana to the United States, and from it Iowa was carved and, in 1846, joined the union.

The Townships

The 30-mile square that is Adair County is today cut up into twenty-five smaller squares, each six miles on a side. These are called *Townships*, and were the center of government at levels lower than the counties. Before the transportation explosion that changed our demographics so profoundly, hierarchical governances were practical and needed. My relatives populated four of the westernmost of these townships: Summerset, Eureka, Jackson and Washington. Summerset Township extended westerly from Greenfield to about one mile west of Fontanelle. West of it, to the Cass County line, lies Jackson Township, like Summerset extending about three miles north and three miles south of the east-west line between Greenfield and Fontanelle. Jackson township is bisected north to south by the meandering middle fork of the Nodaway River, which exits Jackson township near Bridgewater, the other town of Adair County and the remaining center of my family lands. North of Jackson Township lies Eureka Township, and to its south Washington Township. The grid of two-mile roads, with country schoolhouses every two miles in both directions, was frequently bisected still further by intermediate roads at the one-mile marks. It was onto this chessboard geometry that I was born.

I arose in ways that typify the settling of the American midwest by European immigrants. This cultural root is part of me. Great grandfather Thomas Kembery arrived in 1853 from his family farm in Somersetshire in England. I possess his touching letters home[1]. Thomas married a Scot whose McKenzie and Campbell[3] parents had arrived even earlier to an almost unpopulated territory save for Indians. Their son, Thomas Franklin (Frank) Kembery, married into a line of German immigrants. Those Keisel and Westphal families had come over from Pommern. The Clayton family was English but of unknown history. My great grandfather Robert M. Clayton married Mary Elizabeth Yerkes, of a distinguished pre-Revolutionary Germantown, Pennsylvania, Dutch family that comprises my earliest American line. Robert Clayton fought with Sherman's army in the battle of Atlanta. In western Iowa, when people were scarce and precious, these immigrant streams had merged in the nineteenth century melting pot.

Bear with me for a recount a few of the sites marked on my map, for it demonstrates how tightly my ancestral families farmed near one another. All of these sites were farms, as is the entirety of Adair County save for its few small towns. The 1929 plat of Summerset Township, with Fontanelle town in its left center occupying blocks 17 and 18 (out of 36 blocks per township), shows the farm of Frank Kembery, my grandfather, adjacent to the northwest corner of town. Here my mother, Avis Kembery, grew up. Just north of that farm is the farm of Herman Westphal, my grandmother Kembery's uncle. Three miles north of Fontanelle is the farm of Arthur Gross, who married my mother's sister, Elva Kembery. It is now farmed by my cousin Larry Gross, with whom I enjoyed boyhood fun. In my childhood I spent many summers on that farm.

Five miles west of Fontanelle, in Jackson Township, lies the farm to which my grandparents Kembery subsequently moved, and where they were living when I was born, and which is the other of the two farms I knew and remember so well from my earliest years. John Gruber, its owner, had traded the Jackson Township farm to my grandfather in return for his Summerset farm because Gruber already owned the farm adjacent to grandfather's Summerset farm. My grandmother, on the other hand, wanted the Jackson farm because it was so much closer to the farm of her mother. To understand this switch recall that in March 1929, when this move occurred, country roads were unimproved, often muddy.

No money changed hands. Farmers in those times traded farms straight across, provided the acreage was equal; and March was the trading and moving time. Neighboring my grandparents' Jackson Township farm were the farms of Herman and Annie Rechtenbach, whose farm I often visited and whom I knew as uncle Herman and aunt Annie, and, on its other side, the farm of my grandmother's half sister and her second husband, Gustav and Augusta Rechtenbach. These lay two miles north of Bridgewater.

Kembery farmhouse, 160 acres on the road between Fontanelle and Bridgewater, Adair County IA, about 1940

Nearer to Fontanelle, but slightly north of it in Eureka Township, my mother was born in 1913 on what locals knew as the Jacobsen farm, since they were its subsequent tenants. That farmhouse is pictured in *The History of Adair County*. Dr. Sweet of Fontanelle attended her birth. One mile north of it lay the farm of Paul Virgil Clayton, my paternal grandfather. My father, Delbert Homer Clayton, was born in 1913 on this farm, and grew up there. At the time of my father's birth there occurred a fire at the stable in Fontanelle, and Dr. Sweet, who had attended my mother's birth one month earlier, was unable to get his team to come up to assist the birth; therefore, Mrs. Bakerink from the farm across the road assisted. One half mile further north is the country school attended by my father. And just west was the farm of John Gosnell, who married my father's sister, my Aunt Eunice Clayton.

The six-mile square of Washington Township lay immediately south of the town of Bridgewater. Here family history also abounds. About four miles south of town lay the farm of my mother's aunt Mae Williams, sister of my grandfather Kembery. She had married Ben Williams, who owned the farm just across the country road. I never knew them, but I knew and have a photograph taken by me of their son Elto Williams, who looked much like my grandfather Kembery. A couple of miles straight south were the farms of J. P. (Jim) Kembery and of William Edwin (Bid) Kembery, the two elder brothers of my grandfather. My grandfather, Frank Kembery, was born (1884) on Jim's farm, as had been Jim himself, while my great grandfather Thomas Kembery farmed it after the Civil War; and he later lived on it again during his first married year before being able to get land of his own. Adjacent lay the farm of J. W Campbell, inherited from James Campbell of Scotland[3], my great-great grandfather, whose love life also comes down through a family letter. Close to the Campbell farm is the rural Garner Cemetery, where my great grandfather, Thomas Kembery of Somersetshire[1], England, lies buried. Thomas had, according to his word, walked the entire distance from Des Moines to his future farm seeing only one house *en route*.

These, my family, worked a goodly fraction of the farms of these four townships, four adjacent squares of six miles each. The family folklore of them sits huge within my consciousness.

CHAPTER 2.   UP AND AWAY

My father had a huge impact on my life. In the eyes of this boy he cut a heroic figure indeed. My life could in no way have evolved as it did were it not for his spirited impulses and romantic vision. His example taught me even as a small boy that you could uproot your life for far-away adventures, perhaps even adventures in space. It taught me that leadership consisted of taking decisive actions based on prepared knowledge. He taught me to seek correct explanations. An oft repeated story in my family is of the man who asked me when I was four years old what work my father did; and I replied, "My Daddy doesn't work. He flies a plane."

My mother and father were unknowing accomplices in the vanishing of the all-purpose family farm. They did not see that coming, but they were frontrunners of a huge demographic change. They were to be swept up on wings to participate in a new American business. Consider the life that they were to leave, and how their leaving it came about.

Courtship and Physics

My parents had their first date on March 7, 1928. They were fifteen. They went in Greenfield "to the picture show", as they called movies, with another high school couple having an available automobile. Although vivid in her memory, my mother has[1] no idea what that movie was. But that day shines in her memory, one of those days that seem to have made the rest of life possible. She can describe and name the other couple. Dad had asked her during classes at Fontanelle High School if she could go on a date. My mother had already set eyes on my father, admitting happily that "he was the only one for me". They were both sophomores. She said maybe, and my Dad asked her to ask her Mom when she got home. Not to delay, she called home from the Principal's Office, which was equipped with a telephone. It was a turncrank phone to the farm. First crank was to the central exchange, then three shorts connected to Kembery farm.

"I called right away because your dad wanted to know right away!"

Permission was granted, but they had to be home by 10:00 pm. So after her dinner the other three picked her up in an automobile. Nor did this pattern change; throughout high school she was required to be in at 10:00 pm. She was allowed only one date per week, and this was usually on Sunday. Standard routine was for all to go to Church at 7:00 P. M. for their social program, Junior League, with the date following.

Mom's only dates were with Dad. Though others were interested, she never dated another. Photographs reveal her as a beautiful girl, arguably the handsomest in the school. With instinctive teenage social awareness, everyone knew when a pairing was made. Such pairings were often final-- the first date leading to marriage. If the first date was satisfactory, it led as on a straight country road to a wedding after graduation. So simple it was. Most of their dates were events with my father's brother Royal and his girlfriend, Louise Stark, whom Royal later married. Royal was three years older than my father, and together they could persuade my grandfather Clayton to allow the use of his automobile, a green Paige. These dates often included driving at top speed over "the whoopie hills". The rolling hills of these Adair-county dirt roads provided many an abrupt top, and the sense of levitation while going over one was a delight. They also tried repeatedly to climb the long hill north of the Kembery farm in top gear, without shifting down. To do this they got back as far as they could, to the bend in the road beside the Kembery farm, and picked up as much speed as possible. Many tried, and few succeeded. Screaming along at top speed, the passengers would coax the car along as the long uphill dirt path inexorably slowed their pace. Defeated at last, they would shift to low gear:

"Almost had it that time!"

"We'll get it next time!"

Many cars stuck on that hill following a rain.

My parents and most others that I have known of their generation retained a lifelong love of the automobile. Maybe it was with them that this American love affair started. They also shared the idea, today questioned, that only people, not machines, can do wrong. The curse of the automobile became increasingly clear in the subsequent decades, however. It changed American life with a power that could not be resisted by individuals or by communities. My parents, though avid social critics in other ways, remained incredulous at the thought that the automobile in any way undermined life's quality. We who have lived in the cities can see it. But my experience is that my parents' generation retained its initial infatuation, like young love, with the fun machine that created "whoopie hills" and freedom.

My parents never met before their ninth grade in Fontanelle High School. Because the Kembery farm of her early years lay on the northwest corner of the town, my mother, Avis Kembery, attended all twelve grades in "the town school". The Clayton farm, however, lay several miles north of town, up the so-called "Highlands road", so my father attended the Highlands Country School. Children graduated from the country schools with diplomas. Many were then admitted to High School, and my father was one of these; but many others began full-time work on their family farms. In her personal family history, my mother wrote:

> *Delbert had to ride a horse named Dot the seven miles to the High School after graduating from the eighth-grade country school. The horse was stabled at the vet's, who lived across from the High School. After school Delbert would catch up with me by the time I had reached the*

*end of the sidewalk, which was one-third of my way home.
He would ride slowly beside me until we reached my
home. If we lingered too long in the driveway, my mother
would call to me to come in, as it was chore time. I expect
Delbert had to make Dot run the final six miles (for the
same reason). Only one time did I ever get on Dot. The
snow was deep, and Delbert had me ride through the deep
drifts while he led Dot. Horses would not move with me
on them unless led. Even as we stood and talked at the
driveway, Dot would bite at me. Maybe she was just eager
to go on home.*

My father's innocent lack of chivalry was apparently not noticed
by my mother, despite the fact that she walked burdened with an
ankle crippled by polio alongside my father seated upon Dot. My
mother's description of the town school, which I later attended as
occasional guest of my cousin Dick, further captures those times:

*The school in Fontanelle had three stories. The right
hand entrance was the girls', the left side the boys'. Down
to the basement was the gym and the grade-school
restrooms--girls' on the right and boys' on the left. First
floor held grades one through eight. The top windows
(prominent in our photographs) belong to the Assembly
Room, with bathrooms at each end. The classrooms were
at the back of the Assembly Room. Behind the gym was
the boys' woodworking shop and the girls' home-economics
room.*

The correctness for my mother's time of sexist structures was
both self evident and appropriate, and she never questioned them.
Indeed, woodworking and home economics retained their roles
two decades later. During 1953-56 when I was in high school,
we boys took woodshop and metal shop and the girls took home
economics. My mother's memoir continues:

Fontanelle High School was the upper (third) floor of that schoolhouse. All four grades (9-12) gathered in one big Assembly Room, two rows per grade, and facing the stage at front. The stage was where the principal had his desk, and where he watched over us when we had assembly. Whenever he had a class, another teacher kept the assembly, but never from the principal's desk. Things were strict. You did not move about, whisper, or write notes. If you were caught giving or receiving a note, one of you had to read it aloud in front of the whole assembly. You could not chew gum or eat candy. If you wanted to ask a classmate anything about an assignment, you could do so after getting permission. The classrooms were all along the side of the Assembly Room. The two rows of freshmen in assembly were next to the classrooms, whereas the seniors' rows were next to the windows.

In ninth grade I thought I had met the cutest boy in the school. This was Royal Clayton's brother, Delbert, who would become my husband for fifty years. My mother asked who Delbert was, and when I told her she said that she did not think that Royal had a brother, just an older sister, Eunice. But I convinced her for our first date. My mother knew the Claytons because she had worked for them before she married, when Eunice was little. She had even once dated Flave, brother of Delbert's mother. In Iowa, young women worked as hired girls whenever help was needed in the home. She and I were no exceptions.

My freshman classes were English, Algebra, Latin I, and Home Economics. The boys had woodworking shop when the girls had home economics. Delbert did not take Latin; I think he took Biology. When I was a sophomore I took English, Geometry, History and Latin II. It was in History class when Delbert winked at me across the room, and 'twas a good thing he didn't get caught. Then on March 7 he asked me for a date to a movie.

Even while dating during high school, it was evident to my

mother that, if she should marry my father, theirs would not be lives of farming. Something momentous occurred. A romantic, unpredictable change altered the entire subsequent life for them and consequently for myself. My father was taken with the airplane. Before sharing how the airplane arose to shape our family history, I note a related characteristic that I shared with my father. In our respective high schools we were regarded as the strongest physics student in many a year. The following stories were sufficiently repeated down the years that they became part of our traditions. All students in the Fontanelle High School took physics as the third year of a three-year sequence in mathematics. Ninth grade was algebra, tenth grade was geometry, and twelfth grade was physics. The physics class was taught by a formidable Mr. Knudsen, who was also Principal of the school, a position of autocratic power in 1925. Working on the day's assignment in Assembly, what we later called Study Hall, where Mr. Knudsen sat on a raised stage overlooking all grades of students in the same room, my father leaned forward cautiously to my mother, who sat immediately in front of him, to whisper,

"I can't get problem number six."

My mother set to work on that physics problem, and eventually held aloft her solution at an angle where my father could read it.

"Is this right?" she scribbled above it.

"Yes! That's the right answer," my father whispered.

Later in physics class Mr. Knudsen asked,

"Could anyone work number six?"

"Tell him that you got it!" Dad whispered to Mom.

My mother shook her head, unwilling to put her solution on the board where Mr. Knudsen could attack it. When none in the class of eighteen graduating seniors replied, Mr. Knudsen asked, as he frequently did when the class was stumped,

"Delbert, did you get number six?"

"Yes I did," my father replied, "but Avis showed me how to do it."

Mr. Knudsen sniffed, "Oh, sure. I suspect that you showed her."

This disparagement so rankled my mother and father that it became a family legend. I had heard this story on various occasions, and my mother wrote it down in her memoir. That makes it possible for me to recount it here.

Another legend-making physics tale concerned a problem that no one except my father was ever able to solve. When Mr. Knudsen assigned the homework problems in one chapter, he warned,

"Don't bother trying number eight. No one has ever got it right."

Mr. Knudsen could also not work it, and suspected that it was stated in error. In Assembly that day my father immediately worked the problem, and raised his hand to ask Mr. Knudsen if he could come up to the stage desk.

"That's the right answer!" Mr. Knudsen exclaimed. "How did you get it?"

My father explained it to him. In later years, after my physics proficiency became apparent, I asked my father about this legendary problem. Alas, he could not remember it; nor could my mother. But both remembered that Dad's solution involved a common thing in physics; instead of starting at the usual point, the problem required one to first work out a secondary relationship which could then be applied to the main topic. According to my mother, my father had a considerable reputation in this physics class. I have no doubt of this, because he engaged me in physics debates up until his death in my forty-eighth year. I suspect that my success in physics courses drew directly on the relationship to my father during the years that I was growing up, although I cannot recall from those years instances of "talking physics", which my father thought of instead as merely "why things work as they do".

I compliment my mother in this context. She was the scholar of the two, the one who completed all assignments and succeeded in all of her classes. She would have been Valedictorian of her class (out of eighteen graduates) but for an unjust 10% penalty in physics, arbitrarily assigned to her by Mr. Knudsen when my mother once missed physics class, excused with severe menstrual cramps! Throughout her life my mother spoke of only two unfairnesses in her life. First that Mr. Knudsen treated her so demeaningly. Secondly, and much the more serious, that childhood polio left her with a crippled ankle, which hindered proper walking. Limping on her turned ankle was a humiliating source of emotional pain that could, ever after, bring tears to her eyes. Of course, she was lucky that it had not been much worse. For many polio victims, polio was very much worse.

A flying machine

My father began flying an airplane while a senior in high school and continued after both he and my mother graduated in 1930. I would be born five years later. We do not know exactly how it was that his fascination for the airplane began. But it is not too surprising, because the newspapers of the 1920s were full of aviation stories. One might go so far as to describe that era as the beginning of an American infatuation with flying[2]. I find it remarkable, however, that it was possible for a farm boy of this era in remote Adair County, Iowa, to actually begin to fly an airplane. To do so required substantial encouragement from his father, Paul Virgil Clayton; and I still find myself surprised that this was forthcoming.

My Grandfather Clayton was a stern but loving man, one who knew sin when he saw it. This included cards, dancing, and drinking, although, thankfully for my own existence, did not include dating. Perhaps the churches' social roles endowed courtship with the stamp of acceptability. My Grandpa Clayton even loaned out his green Paige auto for the Sunday evenings. A more astonishing event, however, was his loan of $1000 to my father, a high school senior. Even today this would be a substantial sum to loan to a high school senior. In 1930 it was huge. This loan was secured by my father's share of the Clayton farm on the Highlands Road, a debt that was ultimately repaid through the terms of my grandfather's will. With this $1000 dollars my father struck a deal with a local barnstormer, Calvin Murray, who agreed to sell to him his *Travel Air* biplane, an OX-5, as well as to provide the twenty hours of airborne instruction that were required prior to soloing for a flying license. The *Travel Air*, numbered NC 9075, was, according to my mother, "a pretty blue plane with silver wings". My family has a wonderful collection of photos of it from fall, 1930, including shots of my father looking very dashing indeed in his aviator's cap, and photos of the flying buddies that he was meeting. The first entry in my father's log books, which we retain, is July 7, 1930. With 50 hours of solo flying, he obtained his private license in Des Moines on March 16, 1931.

My mother had many rides in that *Travel Air.* The Kembery rule was that both my mother and her younger sister, Elva, were not allowed to fly at the same time. This was insurance in an arena where tragedy was not uncommon. My mother wrote a short description of this in which it takes no genius to read her emotions: "He would fly down to see me, circling the house until I came running out. Then he would land in the nearest pasture, tie down the plane with stakes and ropes, and come to the house. My mother's orders were that he not take Elva and me up at the same time. The seat in front of the pilot could hold two." The High School Principal put it more bluntly in light of my parents' roles in the senior play: "Del, you and Avis stay out of that contraption until the play is over!" People then were not accustomed to the idea of flying, and even Dad's mother, Verna Porter Clayton, refused to watch him fly.

After mother graduated from high school she spent considerable time helping her Aunt Gusty (Augusta Rechtenbach Bateham, half sister of my Grandmother Kembery). In January 1931, while mother was working there, my father flew over the farmyard and dropped a note for her. It was in a weighted pipe tobacco can, and my father had tied his handkerchief to it as a parachute of sorts. When he pulled the *Travel Air* up, my mother recalls, he barely scraped the top of a tree and "scared her stiff". The note asked her to come to the field in which he would land, to talk. He wanted my mother to take his request to her parents that he be allowed to use the Kembery farmhouse basement as a workspace to recover the wing fabric. The Clayton farmhouse did not have a full basement--just a small room for storing canned goods-- being a much more Spartan pioneer house, now destroyed, but which I remember vividly. He also wanted my mother's help with the work. His request was granted; and then the first bad luck came.

The first tragedy was, fortunately, only a financial one. The

Travel Air, with its newly covered wings, caught fire. My father had been trying to warm the engine with a blow torch on a cold day, March 22, 1931. Sparks flew from it and caught the fabric on fire. The ribbing of early planes was covered in cloth. One photograph shows my father, hat in hand, leaning jauntily against the exposed pipe ribbing of the fuselage. The *Travel Air* frame on the ground beside him, its flesh gone, looking like some ancient trilobite, a family fossil. Amazingly, my grandfather Clayton offered him another loan, with which my father purchased his second plane, a *Robin*.

My mother resumed housework for her Aunt Gusty, who had broken her wrist trying to start their car when the starter crank kicked back. She was paid $4 per week and worked very hard. This type of employment was normal for those times, because the farms had so much need for extra labor. The nature of that lifestyle may be sampled by quoting mother's diary entries during a single week in 1931:

> <u>Tuesday</u>: *made breakfasts, washed dishes, swept and dusted. Uncle Earl and I picked cucumbers, and Gusty and I canned 17 quarts in P.M. I got supper, peeled potatoes for breakfast, and did dishes.*

> <u>Friday</u>: *I got breakfast and washed dishes. Then had nine washersfull of clothes to wash and hang out on the lines. Got dinner, did dishes, and ironed. Quit ironing to get our supper. Washed dishes and was good and ready for bed. Washed Dale (infant adopted son of Gusty) and myself.*

> <u>Tuesday the 11th</u>: *We went to the neighbors, the Davidsons. We canned 93 quarts of corn. Mrs. Oaks and Mrs. Wolfe were also helping. Men brought sacks of corn from the field. We had to shuck it, cut the kernels from the ears, and pack it in jars.*

Wednesday: We went back to the Davidsons and canned 65 quarts of corn. Men were there working and we had to get dinner for 16. Then I set Mrs. Davidson's, Mrs. Oak's and Mrs. Wolfe's hair. Came back to Gusty's and picked cucumbers, and got supper. I'm so tired.

Tuesday 18th: Went to Davidsons to can more corn--50 qt. I helped her get a bushel of crab apples ready to can. Back to Gusty's, where a dog killed a chicken, so I dressed it.

My father was flying in Des Moines at this time. He went to Des Moines and joined F. C. Anderson, chief pilot for Des Moines Airways, and built up more hours. Once, when my mother was visiting him in Des Moines, a photograph crew was at the airport to make an advertising film on learning to fly, with the purpose of showing it in the movie theatre. They asked her to pretend to be a student, so she did. There were dual controls in the front seat for students. On that very day my mother did what she described as "a brave thing"; she did not return to Aunt Gusty's farm until the next night. She writes, "My folks came and got me, and I am sure they were told what I did." On September 29, 1931, my father earned his Limited Commercial License, and had logged 180 flying hours. With the Limited Commercial License, he could begin carrying passengers for hire.

My father's photo album pictures some of the risk side from Des Moines in 1931. Amid the dashing young men pictured there stands a photo of Max Diebold, a flying acquaintance of my father's. Adjacent to that photo is a second photograph of Max Diebold, a handsome young man in suit and tie pictured in a newspaper clipping. The news photo is captioned "MAX DIEBOLD. Piloted death plane". The newspaper story begins:

Max Diebold, 21, of Casey, Iowa, injured in the airplane crash Saturday afternoon which claimed the life of William Lawrence, 19, of Norwalk, Iowa, died at 3:35 P.M. Sunday at Mercy Hospital. Little hope had been held for his recovery when he was taken unconscious to the hospital on Saturday. He suffered broken legs, a dislocated

*hip, severe lacerations, and internal injuries. A punctured
lung was thought to have prevented recovery.....*"

Below that photo is another of his passenger, William Lawrence,
who died at once. It is a full story, of a size and scope given to
things of high community interest. It encapsulated the fear and
fascination of aviation to a community of farmers. My mother
carried that fear throughout their lives together. It always seemed
a little unnatural, being so high.

By end of 1931 my father needed a better airplane. On October
1, 1931, he purchased a *Stinson, Jr,* call letters NC 900W. Two
photographs of it adorn our scrapbook, one of the plane, a second
showing my handsome father with two flying cronies standing in
front of its propeller, bearing the handwritten caption "Barton,
Delbert, Andy & Stinson 1931". He then started barnstorming,
going to different Iowa towns and giving a ride for $1.00 for anyone
who wanted to pay. He would make the County Fairs, flying out
of the nearest cleared field, or pasture. He obtained his Transport
License on March 26, 1932, and in May he moved to the Iowa City
Airport. We still have that original Transport License: #19253,
United States of America, Department of Commerce. "Age 19,
Height 5'11",Weight 160, Hair brown, Eyes blue". This data is next
to a small ID photograph, about half the size of passport photos,
showing my father smiling broadly in his aviator's cap. Evidently
good identification was not a high priority. My father is wearing a
tie. His signature is next to that photo. Under it is printed:

> *This Certifies, that the pilot whose photograph
> and signature appears hereon is a Transport Pilot of
> "Aircraft of the United States". The holder may pilot all
> types of licensed aircraft, but may transport passengers
> for hire only in such classes and types specified in the
> accompanying Pilot's Rating Authority, which is made a
> part hereof.*

Many photographs and memorabilia exist from the airport in
Iowa City. One is a photograph of my father standing proudly in

front of his Stinson Jr. He has on a cap and a tie, in short shirtsleeves, with his hands folded behind his back. The Stinson Jr. appears exceedingly handsome as well, with its large enclosed fuselage, and is parked in front of a cement-block building bearing the large printed word SERVICE. That service hangar was topped by an attractive cylindrical roof. Airborne views of Iowa City taken from his plane show fascination with the view from on high. Another reminder of these times is the local pride evident from a story reported in the 1932 Iowa City newspaper:

CABIN PLANE BROUGHT HERE

Clayton to Operate Four-Place Stinson at Airport

Mr. Delbert Clayton, formerly of the Des Moines Airways, has brought his Stinson junior four-place cabin plane to Iowa City and will operate from the Iowa City Municipal Airport, it was announced Thursday.

Mr. Clayton will operate from the airport on cross country or local trips. The plane, equipped for night flying, will carry three passengers and the pilot. Mr. Clayton brought his plane here because of the facilities offered by the Iowa City port, one of the most up-to-date in the middle west.

When my mother accompanied my father to Iowa City in summer 1932, my father had found a job for her. Evidently he wanted to keep her near until they could marry. The wife of Deke Brownlee, manager of the Iowa City airport was expecting a baby, and cooking made her ill. She could eat after the food was cooked. So mother's job was to cook meals, do dishes, shop for the food, and be her prenatal companion. I do not know what she was paid for this, but it could not have been much. But it was enough to remain in Iowa City.

My mother tells a romantic story of her time there. One moonlit night my father and another couple picked my mother up from work. They drove to the airport and wheeled the Stinson out of the hangar. They flew to Cedar Rapids to see Greta Garbo in *Grand Hotel*. It was a light and beautiful night, with stars and moon shining when they flew home. Because the runways were not night-lighted, they had rotating beacons to mark them. After the Brownlee baby was born my mother was no longer needed by them, so my father flew her home to the Fontanelle farm. My mother and father were both 19 at this time.

My father kept his hand in farming because of the need for extra cash and because flying was such an uncertain livelihood. His brother Royal was at that time farming just north of the Clayton farm of my grandfather. But his wife, Louise Stark Clayton was not strong enough. Already infected with tuberculosis, she was growing weaker. So my dad rented forty acres from Royal and planted it in corn in early 1933. In return for using Royal's tractor and plow, he plowed Royal's property at night, working until it became too dark. The corn was sold, and the cash was useful; but my father did not want to farm. He wanted to fly.

My father continued working out of the Iowa City airport until April 1933. It is clear that he wanted to try all of the airplanes, because in April 1933 he traded planes again. This one was a Great Lakes open biplane, NC 847K. He did some barnstorming in it. In August he met a pilot owning a Ford trimotor, and he logged five hours on it. My mother remembers getting to go for a ride with him in that Trimotor one day in Boone, Iowa. Unquestionably one of the romantic images that colored my family romance was the remembrances by my mother, and passed on to us, of my father landing near the Kembery farm to sweep her away for good.

So on September 24, 1933, my mother and father were married. Their simple ceremony was held in the parlor of the Kembery farmhouse. A few months later they took up residence in Shenandoah, IA, in order to pursue some business plan, and perhaps I should use the word "scheme" here, related to flying. This was the heart of the depression, and work of all types was hard to find. My father was employed for ten dollars per week by the Earle May Seed Company. Because the house on Walnut Street was large enough, they took in two boarders who provided some small cash flow. And it was into this setting and place that I was born.

Nine months before my birth, at the time of my conception in summer of 1934, my father decided to recover the Great Lakes. He ordered the fabric and my mother sewed the slip covers for the wings on a borrowed sewing machine. My father and a friend sewed them onto the plane. Then the material was "doped" to shrink the fabric to fit. My mother said that the "dope" smelled like banana oil, and she did not know what it was. Then they painted the Great Lakes red and white. With it my father did a lot of barnstorming in southwestern Iowa. Our scrapbook shows him airborne in the Great Lakes open cockpit, taken by his copilot from the front seat. He wears a leather aviator cap and goggles. That photo is accompanied by this small newspaper story:

> *Mr. and Mrs. Delbert Clayton had engine trouble Monday afternoon and their airplane was forced down in the Chris Campbell field. He was able to remedy the trouble and they resumed their homeward trip. Those who gathered to see the plane were.......*

followed by a lengthy list of the names of the local curious. Those were the days of aviation romance--the days of my birth. But it should not be thought that my parents had the time and money to simply fly around. There was always work to be done on both sets of farms. My mother relates that she and my father worked hard for their keep on both farms. At the same time my father was trying to get a job in Shenandoah IA through the help of a friend, Bud Vance, whom my father was teaching to fly. I again turn to my mother's diary for a capsule of the work they did in early 1934 on those farms. These daily entries reaffirm their way of life, an old way of life inherited from European farmers, a way of life that none could see would quickly pass. Read just a few:

> *Feb. 5: Mr. Clayton butchered a cow.*

> *Feb. 6: Eunice and I cut and packed 24 jars of meat to* can.

Feb. 7: Del sold his corn for 38 cents a bushel. (He had planted his own acreage.) Eunice and I canned 50 quarts of meat. Cut up scraps for hamburger. Del ground it after supper. I rolled it into balls and packed it in jars.

Feb. 9: Eunice and I had the corn-shellers for dinner. I helped and did dishes.

Feb. 10: Del and I put over $500 in the bank. He went to Shenandoah. Royal got me to go up and help Louise.

Feb. 14: Louise felt better. I scrubbed kitchen, baked a cake.

Feb. 15: I churned butter.

Feb. 16: Got breakfast, washed the cream separator, sorted and washed clothes, dried dinner dishes, ironed, made beds, changed collar on Del's shirt, got supper.

Feb. 17: Helped do up work. Went to Greenfield in P.M. Got set of dishes for service for nine for $6.95. Got material and cut out a dress.

Feb. 19: Tore off paper in dining room. Earl and dad butchered two hogs.

Feb. 20: Full day. Earl came over and we worked on meat. Del went to timber on Mom's 80 acres for a load of wood for Earl.

It goes on daily in such vein; but I skip forward a month to events for me.

Mar. 24: We hitchhiked to Shenandoah. Had good luck.

April 1: Beautiful Easter. Borrowed Bud's car and went home for our stuff.

April 2: Unloaded stuff. I unpacked it in Steven's apartment. Cleaned and scrubbed. Baked a cake. Neil moved in.

My gratitude to my mother for this diary is twofold. Her diary of a young married couple helping on the family farms in 1934 was typical of the time of my birth. Those brief notations speak with an historical value that transcends my life and hers and speaks to the American experience. Secondly, it records the prelude to my birth in Shenandoah. My parents had hitchhiked to Shenandoah to look for some work having steady income.

Shenandoah lies in the very southwest corner of Iowa, almost at the corner of Nebraska and Missouri. The East Nishnabotna River passes Shenandoah, joins the West Nishnabotna River ten miles southwest of Shenandoah, and drains that corner of the great plain toward the Missouri River, twenty miles further west, which it joins at the tristate corner, and from where it flows south and east towards its confluence with the Mississippi River at St. Louis. This great "peninsula", as it must have seemed to the pioneers, between the narrowing thrusts of these two mighty rivers, gave birth to much in the American self image. It was there that I was born at the end of the 1934-35 winter. It was a winter of exceptionally harsh cold and mounded, drifting snows. My mother's sister, Elva Kembery Gross, wrote of that winter in which her own first son, my cousin Richard Gross, was also born:

> *We were snowed in good. The neighbors all used to get together with their scoops and clean out at least a one-way lane. By the fourth of January, we were once again snowed in and it was so bad we couldn't get to town for weeks. That is when Art(my uncle) would put the cream can, full of cream, on my little old sled and start out kitty-corner across the fields, down to Harrison Means corner, where Dad (my grandfather Kembery) would meet him in the car and they'd go on to town, where he'd get everyone's mail, a few groceries, what we could buy with a can of cream, come back across the fields pulling the load on the sled. We thought it was five miles each way.*

My mother wrote of Shenandoah and my birth there:

In Shenandoah, Iowa, we first lived in the Stevens Apartments. Most of the town and stores were within walking distance. We had no car. We were still in the depths of the depression and times were hard. Our rent was $20 a month, and included utilities. That was half of Delbert's pay. This was a furnished apartment, so we didn't have to buy furniture. It consisted of a living room, sunroom, and bathroom. The living room had double doors which closed over the small kitchenette when not in use. A drop-leaf table and chairs were in the sunroom, as was the Murphy bed. At night we let the sides down on the table to have enough room to let the bed down from the end of that room.

Delbert's friend from barnstorming days, who followed us to Shenandoah, lived with us. Neil paid $5 a week for his room and board, which was half his salary. His bed was the daybed in the living room and had to be made up each night. Both of them worked for The Earl May Seed Company, loading and unloading heavy bags of seed, which farmers bought.

Neil's $5 bought enough groceries for the three of us. A dozen eggs cost 10 cents, a loaf of bread cost 8 cents, and hamburger was 8 cents a pound. We had three grocery stores, including Safeway and A&P. I would check the ads from all three and make out a list for each store. If something was a penny less at one store, it went onto the list for that store. You see, if I saved 8 cents by walking to the three stores, that meant an extra pound of meat or a loaf of bread.

Delbert sometimes made extra money by taking passengers for a ride over town in the Great Lakes--for a dollar. Gasoline was 19 cents a gallon. Delbert was teaching Mick Vance to fly. Mick later became a pilot for

TWA. I made extra money by washing and ironing shirts for Mick for 10 cents each, usually five shirts a week.

Washing clothes was not easy. I had to go down an outside wooden stairway to the basement with the dirty clothes. There were stationary tubs down there, hot and cold water to fill them, and a washboard on which you scrubbed the clothes. After rinsing, I had to carry the basket of wet clothes up two flights, and hang them on the clothesline to dry, after first starching the shirts. Then things had to be brought in, dampened and rolled up, and put in the basket to iron later.

My parents drove the 70 miles down from the farm to see us once in awhile. They would bring us eggs, potatoes and meat. Sometimes they would take us out to eat at noon. You could get "The Blue Plate Special" for 25 cents. That was meat, potatoes, gravy, vegetable, bread and a drink.

During this time Del decided to recover the Great Lakes. We bought material for $144.50 and rented an empty garage to work in. We had to pin the material to a section, like the fuselage or wing, and cut it. Then I sewed it with flat-fell seams on a borrowed sewing machine. Neil helped him every evening and weekend.

We later moved to a furnished duplex on Walnut Street. We were now expecting a baby, so Neil and Brownie paid for sending the laundry out in exchange for my doing their ironing. Brownie operated the movie projector at the theater and many times he let us go up with him. One show we saw was Myrna Loy and William Powell in "The Thin Man".

The morning of March 18, 1935, was a lovely one in Shanadoah. My labor pains started when I got up to fix breakfast for Del, Brownie and Neil. We called the doctor to alert him. He promised to stop by the house at

*noon. I fixed lunch for the guys but told them they were
not to expect supper. I started really cleaning the house,
scrubbing floors, leaning on the mop or broom handles
during a pain. I had to clean the house, you see, because
my mother was coming! The doctor came at noon and
my pains stopped the whole hour that he waited. Wilma
Vance came to wait with me and Bud Vance borrowed
his mother-in-law's Oldsmobile to take Del the seventy
miles up to the farm for my mother. They left at 4:00 P.
M. and were back by 6:00 P. M. with her. The doctor had
come at 4:00 P. M. and made me get into bed, and the
pains got worse. He started to give me a whiff of ether, but
found that his can was empty. So he got a few drops on his
handkerchief and held it a foot from my nose.*

*I was so glad to see my mother and Del back at 6 P.M.
At 7:00 P. M. the baby was born. His head was covered
with little spit curls all over it, and he was so hungry and
sucking on his fist. So my Mama fixed him a little "sugar
tit", as we called them, which was sugar in a little piece
of cloth. Mama and Wilma then dressed the baby and
brought him to me. We named him Donald Delbert.*

My grandmother, who had arrived shortly before my birth, told
the general practitioner, who had arrived, "That's barbaric! You
can't do that" to his proposal for a modest episiotomy owing to my
good size. Mine was a dry birth. The water had broken on Saturday
and I was not delivered until 7:00 PM Monday, March 18. The
doctor had intended to offer sedation by ether but exclaimed after
only a small amount of ether came out onto a washrag, "My can
is empty". My mother raised her head right into the ether-doused
washrag. Wilma was told to hold it a few inches from her face. My
delivery was thereafter uneventful, and the placental sac emerged
spread across my face.

"He has a cowl", Wilma told my mother. "That means he will
be famous someday."

So my life began, a stone's throw from the farm, a stone's throw
from the sky. The sky was yet to win, but Shenandoah was not the

best place for a future in flying, and another twist of fate altered our flight path.

Our Davenport, Iowa years and fading farms

In the spring of 1935 Ansel Dunham of Davenport, Iowa, hired my father for some cross-country flying terminating in Davenport. When this work continued my parents decided to leave Shenandoah. On May 3, 1935, when I was 6 weeks old, my Kembery grandparents drove to Shenandoah and picked up my mother and me. For some time we stayed on their farm, with my father visiting when he was not flying. There was always farm work to be done, and my parents pitched in. During the fall my father decided to remain in Davenport, so we moved to another new family residence. My father had become acquainted with many pilots at Cram Field (Davenport) and also at the Moline, Illinois, airport, just across the Mississippi River from Davenport. He also began logging time on Taylor Cubs, which were made in Bradford, Pa. He started a flying school at Cram Field, and we have one marvelous 8 x 10" print of my father standing beside one of his Cubs. His business card says:

STUDENT INSTRUCTION   DIAL 3-2814

Clayton Aircraft Sales

CRAM FIELD

PRESENTED BY

DELBERT H. CLAYTON, DAVENPORT, IOWA

Father (age 25) with his Piper Cub (1938)

Near my first birthday my parents ventured from the temporary boarding house and sought a better home. My mother was again with child, my brother Keith, who was due in August. What they found was the first home, other than the farms, that I am able to remember. It is a poetic memory. When we went to look at 607 1/2 East Sixth Street, we found that it was an old church. The owners, Frank and Pansy Wilhelm, had renovated it and furnished it as a duplex. They lived downstairs. Pansy Wilhelm was very excited by our arrival, which she had, for reasons of her own, anticipated. It seems that she had been to a séance and it had there been predicted that a young couple with a baby would appear to rent their upper floor, and, furthermore, that they would be honest people to whom she should immediately rent it. So she did. We would live there about three years[3], and not only my brother, Keith Eugene Clayton, but also my older sister, Carolyn Ruth Clayton, were born in this converted church. I have many memories, faint disconnected images, of that home. Its pitched roofs sloped down severely and foreshortened the walls upstairs, making three-foot-high side windows near the floor. A screened in porch had been added across the front to make it more homelike. Its rooms came in pairs as one went from the front to the rear--a kitchen with a bedroom opposite it; a dining room with a bedroom opposite; and a living room with another bedroom having a bathroom opposite it. In the basement was a washing machine and two furnaces, one for each dwelling. We had to buy the coal and, of course, shovel it into our furnace. The rent was $25 per month.

It is fitting that a house of God should have a resident angel. My mother always believed in guardian angels. It may have begun that hot summer day when I stumbled, running toward the wall, and crashed against the flimsy screen on the open, small, low window. It flew out, and so too would have I but for its rebound trapping me, half in and half out, some twenty feet above the ground. My mother almost died with fright, but to me it was just another fall. I earned my first stitches in the following year from another legendary fall, this one from hanging upside down, at three years of age, from the iron railing that lined the front sidewalk. I fell onto my head, gashed open my scalp, which bled abundantly, and was stitched shut in a local doctor's office. My mother has described the relentless discomfort the dogged our first summer in this otherwise delightful place:

> We really enjoyed living there, but 1936 was a summer of record-breaking heat. Top temperature was over 100 F every day from July 4 until August 20. With the sloping roof, as in an attic, the upstairs was stifling with heat. The side windows were almost at floor level and therefore not effective. The only full-sized windows were at either end. And my baby was due in August. Finally we took Donald's baby-bed mattress to the airport and kept it stored in a shed along with some blankets and sheets. Each forenoon I would fix things for our supper, and Delbert would come in and get us about 5:00 P.M. We would sleep at the airport, or try to, until sunrise when the flies pestered us.

*Then Delbert would take us back into town, eat breakfast,
change clothes, and go back to the airport.*

*Sometime that summer Del got a student who had a
DeSoto car dealership. He traded us an air-flow DeSoto
for flying time, so we now had a car. There was a bar at the
airport, where we sometimes sat until Del finished with
the students at sundown. You couldn't fly after sundown
if your plane had no lights. Donald was able to say big
words, and the men would put him up on the bar and
give him big words to say. They thought it fun to hear this
sixteen month old say "rhinocerous" and such.*

Those may qualify my first speaking engagements. In such heat,
amid such delirious living, it was time for my mother's angel. She
continues:

*On the evening of August 19th we had visited the
Dunhams, getting home about midnight, still hot. We
opened the newspaper onto the table so we could both look
at it a little bit. Then I got my first pain. At 5:00 A. M.
Delbert went down to the Wilhelms to use their phone. He
called Dr. Shorey, who arrived about 6:00 A. M., and did
so along with a fresh breeze, and rain! Keith was born an
hour later.*

My father was often in the newspaper that fall. When heavy
rains flooded the entire course of the Mississippi, he took to the
air to photograph it:

Davenport Men Take Plane
For Views of Flood

Delbert Clayton, 24, 637 1/2 E. Sixth St, co-owner of the Clayton & Dunham Flying Service at Cram Field, and L. A. "Fuzzy" Carlton, 28, 3033 Grand Ave, warmed up a two-place Taylor Cub at the Cram Field this morning and took off on a sightseeing trip of the flood area. The tiny ship soared away from the Davenport airport at 8:15 A.M. carrying the two members of the Quad Cities Airmens Association to Louisville, Cincinnati, Evansville, and other scenes of flood disaster.

Next day's edition carried their harrowing escape, seemingly of interest to the entire Quad Cities region:

Davenporters, Returning from Air Tour of Flood, Forced Down By Fog

Blinded by fog as they were returning from a two-day air tour of flood-devastated cities on the Ohio River, two Davenport men, Delbert Clayton, 637 1/2 E. Sixth St., and his companion were forced to make an emergency landing in a farmer's field near Alton, Ill., on Saturday.

The two reported to relatives in Davenport of their safety after a narrow escape from crashing into the treetops as they sought to break through the penetrating fog. The couple made a pictorial record of the trip, using an aerial camera.

The Davenport newspaper began a regular column entitled **What's New At City's Airport**, in which my father was the most cited celebrity. This is not surprising, considering his single minded attention to flying and to his Piper Cub distributorship and flying school. That distributorship brought him acquaintance with Mr. W. T. Piper, whom he met at the 4th annual Cub Convention in Loch Haven PA, on Dec 5-6, 1939. Mr. Piper came to Davenport

IA and accepted our invitation to dinner. This was neither fancy nor abundant, but that did not bother my mother. She remembers clearly that in those years one would ask an important person to share your family dinner with none of the modern trauma over its splendor. In a span of weeks the following appeared in the Davenport newspaper under that column's banner:

> *"A busy weekend of flying rounded out a lot of activity at Davenport Airport. Delbert Clayton and Ben David had been out to Lockhaven, Pa., again, and flown back with a couple of new Cubs."*

> *"Eight new students were added to rolls at the Davenport Airport during the past week, and one was soloed. Signing up for the courses with the Clayton Flying Service were....."*

> *"Everything's up in the air out at Davenport Municipal Airport. Delbert Clayton flew more student work Sunday than on any other day in the past two years. This week one of the new Cubs will be on display at the Blackhawk Chevrolet Company. When this plane comes off the show floor it will be put into service at the field, doubling the ability of the Clayton Flying Service to take care of all comers."*

At about the same time a newspaper feature was built around a photograph of fifteen men, including my father in suit and tie, standing at attention before one of their aircraft. The headline above it reads:

15 Planes Take to Air to Advertise Air Mail Week

> *Fifteen aviators, eight from Davenport, three from Moline, two from Cedar Rapids, one from Clinton, and one from Philadelphia circled and recircled over Davenport Saturday noon in the first public demonstration in connection with National Air Mail*

Week, which began at 12:01 this morning. In the afternoon the planes were parked on Cram Field where hundreds inspected them and absorbed an endless amount of information on planes and aviation in general. The pilots were there to answer questions and they were bombarded with plenty.

Air Mail Week began just after midnight Sunday morning and clerks immediately began applying the cachet stamp in vivid red to every piece of air mail and began the count which it is hoped will total 60,000 or more before the week ends at midnight next Sunday.

Under strict injunction from Washington not to sell the new air mail stamps before May 15, the Davenport Post Office officials prepared for early buyers by placing the new "six centers" in the Neufeldt Pharmacy substation at 1430 West Third Street for sale today. This was done in order that collectors could purchase stamps for first day covers.

On those hot, muggy, Iowa summer days and nights we spent many nights at Cram Field to escape the heated upstairs. My mother and brother Keith and I were always at Cram field in the late afternoons when my father's flying work would end. Each day, he strapped us into his cub for a couple of spins around the field. If my young mother feared harm in that, she apparently decided that the experience was worth the risk. As the ground receded beneath the plane I stared in amazement at the shrinking ground. Every day brought the same miracle, fresh and seemingly unfathomable, until the lessons of perspective and distance sank in. I could see my mother, tiny as an ant, waving to us as my father waggled the wings. I saw the sky all around, clouds to the side as well as above, blue emptiness, yellow sun, amazed that we were no longer of the earth. And then down, back to that good earth. Climbing out I marveled that this thing, heavy enough to fall, could go up and stay up. Throughout my childhood that experience had me seeking reasons why it did not fall, reasons that are even today not understood by most adults. I could not solve it, of course, but it

seemed related to the wind that the propellers slung behind, and
that we especially loved on a hot muggy afternoon.

*Father has just said, "OK then, you fly it." July 1938.
Laughing, I am 3 1/3 years old*

A special center of family life was provided by the Philco
Console radio that we bought during our years on E. Sixth Street.
The big event that my parents retold throughout life was the night
of halloween 1938 when NBC broadcast the invasion of aliens from
outer space. Meant as a joke, millions heard it and believed, not
without some fear. An announcer came on at the end and said "This
is NBC's way of saying BOO!" On the way to the home of friends
later that evening my mother had false labor pains which she thinks
were induced by her agitation. My sister Carolyn Ruth was born
nine days later. Today I joke[1] with my mother, "You have believed
in flying saucers ever since!" My mother enjoyed that Philco, which
was later to go with us to Texas. Her favorite programs from that
period were *Ma Perkins, Fibber McGee and Molly,* and *Fiction and
Fact from Sam's Almanac.* One of the recurring family tales about
me was that I loved to entertain by making up my own "Fiction
and Fact".

But the favorite family story, one that may sum up these years in my eyes, was my response to the man who asked where my father worked. "Oh, my Daddy doesn't work," I replied, "he flies an airplane." So it seemed to me. My favorite photograph of myself from this period shows me seated at the controls of my father's cub, laughing hysterically. My father sits behind in the passenger seat, his hat tilted at a jaunty angle. He has just said, "OK then. You fly it."

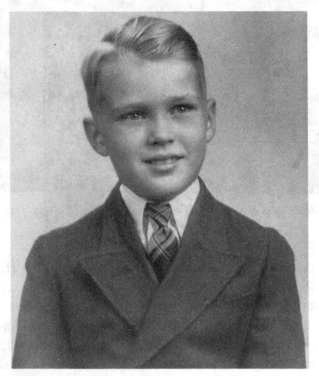

Portrait of the author at five years of age, in Davenport IA

CHAPTER 3. A STRANGE, FAR LAND CALLED TEXAS

People have left the farms for different reasons at different times. My great grandfather Kembery had left the Somerset farm to go to the new world because he was not the first son. My great-great grandparents Westphal and Grünwald had left for their reasons, perhaps to avoid military conscription as did so many Germans near 1850. During the industrial revolution many had left for the cities and the new factories that promised work of a different kind. In some ways my father's leaving of the farm was of that last type, for no matter how romantic early flying was and seems to have been, at root it was a new way of life that technology had spawned. It produced the airline industry. Today many leave the farm for yet a different reason, that the family farm is driven nearly to extinction by the apparent economic efficiency of agrabusiness--the industrial megafarm. This particular exodus has has grieved many who knew the old farm culture. It may have been a catastrophe for the stability of both American culture and of American topsoil, all for the pyrrhic economic victory of selling bushels of corn, wheat or oats for a few cents less than can the family farmer. This is described as a "victory for the American consumer" by many politicians and economists, a wisdom that is mistrusted by those who have witnessed the destruction of the rural way of life that gave such quality to those living it and such backbone to the American soul and to my own. The town of Fontanelle is today nearly moribund, lacking its reason for being that was so compelling seven decades earlier.

In February 1940 my father decided to go to Chicago in order to obtain an instrument rating in hopes of being hired by an airline. His wife and three small children could then have a more secure financial future. My father came home weekends during this effort, and when the instrument rating was achieved and airline discussions held some promise, we sold the furniture in the Davenport home that had meant much to our family, loaded our DeSoto with dishes, linens, clothing and special belongings, and retired once again to the Kembery farm. From there my father departed again and drove to Dallas, Texas, where he was hired by Braniff Airways on May 7, 1940.

During our last months in Davenport my life may have been saved by science. About March, 1940, we all contracted scarlet fever. My baby sister Carolyn got it first, but not as badly as we older ones. This was the dangerous scarlatina streptococchus, and all but my father were very sick. The bacterial struggle raged particularly intensely within me. On the second day of high fever, my temperature soared to 105 F, where it stayed for days. I became totally delirious. Dr. Shorey was called again, and after watching me suspended on the edge of possible death, he said, quoting here as well as my mother can remember, "I can cure him, but I will have to give him a shot of a medicine that has not yet been approved." Trusting in his concern and confidence, my mother agreed it would be best. She describes a syringe containing a milky white fluid[1] that Dr. Shorey injected into me. It was miraculous. After many hours my fever broke, my sense returned, and the next day I was apparently totally recovered. The red spots had vanished. I then tried to help my mother and siblings through their sickness. I cleaned up things, and brought them food and drink to their beds, where all lay, still in fever, including my mother. I was the only one who received the injection.

It must have been penicillin. Sulfa drugs had been introduced in 1935, the year of my birth. Penicillin had been discovered in 1928 by Sir Alexander Fleming, but it had hot been approved for use. It was much tested in the late 1930s, and was approved for use by our troops in World War II, where the need was great and options against infections were few. I will never know if I would have survived my bout with scarlet fever or, like some, died of this disease. Even if I would have survived, the penicillin may have saved me from the destructive ravages of rheumatic fever, which sometimes follows other strep infections. But I think it must have been a close call for me. This episode makes it very possible that science, to which I later devoted my life, saved my life. Alternatively, it may be that Dr. Shorey himself saved my life. Though he is gone, I thank him for his loving concern for a small boy in a simple good family, and for taking a risk that few would take today. The lawyers and their threats of malpractice, their active persuasion of grieved families to litigation, have made it unthinkable for a physician to use a nonapproved drug that he believes to be the best risk for a seriously ill patient.

I contracted the usual spectrum of childhood illnesses that were then so common: whooping cough, 1937; Scarlet fever, 1940; red measles, 1941, causing me to miss three weeks of the first grade; German measles, 1942, causing me to miss one week of school.

The years 1935-40 in Davenport on East Sixth Street were formative in our family. They were our first years together as a growing family, living in our own house rather than at the farms of family members. My brother and sister were born there. From there my father got the airline job that changed our geographical axis. Our home was the second floor of a renovated church, and had charm by its uniqueness. I myself can remember only a large house, in which we lived upstairs, and our neighbor, whom I called "The Lady". I remember visions of the few-block walk to the banks of the Mississippi. In my mother's memoirs[2], she wrote this more clear-eyed account:

> When we went to look at 607 1/2 E. Sixth Street
> we met Frank and Pansy Wilhelm. They had bought
> an old church, built an upstairs in it, furnished it, and
> were willing to rent it out. They lived in the downstairs.
> A church makes a long, narrow living quarters. The

rooms were side by side the length of the building. It had a kitchen, with a bedroom off it, a living room, with a bathroom and bedroom off it. All along the front was a big screened-in porch. The rent was $25 a month. In the basement was a washing machine and two furnaces, one for each. We had to buy our own coal, and also shovel it into our furnace.

We really enjoyed living there, but 1936 was a record breaking hot summer in Iowa. My parents came (from Bridgewater) to see us over July fourth. It was over 100 degrees every day from July 4th until August 20th (the day of my brother's birth). The only full sized windows were at either end. All along the sides the windows were only about 3 feet high, as the ceiling sloped downward, like an attic room. The heat became stifling there, and my baby was due in August. Finally we took Donald's baby-bed mattress and kept it stored in a shed, along with some blankets and sheets to spread for us to lie on. Each forenoon I would fix things for our supper, and Delbert would come in and get us at about 5 PM. We would sleep at the airport (Cram Field), or try to, until sunrise, and the flies pestered us. Then Delbert would take us back into town, eat breakfast, change clothes, and go back to the airport. Heat seemed to remain upstairs, with no breeze or rain. On the evening of August 19, I got my first pain. At 5 AM Delbert went down to the Wilhelms and called Dr. Shorey on their telephone. He arrived at 6 AM, along with a cool breeze and rain! Keith was born at 7 AM.

We lived here for a couple more years. In fact Carolyn was also born in this house, nine days after Orson Wells' famous "War of the Worlds" radio broadcast on Halloween night—NBC's way of saying "BOO".

A DeSoto dealer traded us an airflow DeSoto for flying time, so now we had a car. There was a bar at the airport, where we sometimes sat while Del finished with students at sundown. You could not fly after sundown if you plane had no lights. Donald could say big words, and the man would put him up on the bar and give him big words to say. They thought it fun to hear this sixteen month old say rhinoceros and such. We had taught Donald his name, address, and where his Daddy works. Once when someone asked him these questions, he answered with his name and address, but when asked where his Daddy works, he said, "Oh, my Daddy doesn't work. He files an airplane."

There was a Quad Cities Flying Club, which met once a month. At one meeting I was seated next to Amelia Earhart. Delbert said that he was always jealous about that, and I don't remember it at all.

I was rocked to sleep by my mother every night in that Davenport churchhouse. I have wondered, but cannot fathom, the effect of this nightly embrace on my life and personality. It ended in my fourth year after my sister Carolyn was born, when there was just too much to be done. I am not able to discern any trauma associated with the sudden stopping of what I had surely come to expect. But it has always been clear to me that in my core I feel profoundly loved, especially by my mother. I have long been very grateful for that, especially when seeing the sadness of many children. I later had a deep unquestioned fountain of love for my own children; but I could never bear to hear of cruelties to young children. And, perhaps significantly, I later experienced sexual conflict and difficulties with women. I link these emotions of my life without clear knowledge that they actually are linked.

Leaving Iowa farms, May 1940. Brother Keith (age 3 3/4, left) and worried author (age 5), in front of sister Carolyn (age 2) and mother (age 27)

Not only did my father go to Dallas to seek the job he eventually obtained with Braniff Airways, but with his unbounded confidence in the future he packed the DeSoto with our dishes, linens and other items to get them to Texas with him. He was aware that he did not have the required two years of college that was the industry norm in 1940. But he was confidant that he could parlay his ten years of experience with the public and his abundant hours and experience flying into a good argument for himself. We still possess the telegram, dated May 7, 1940, to Mrs. Delbert Clayton, c/o Frank Kembery, Bridgewater, Iowa: "Employed by Braniff today. Will write full details later. Love, Del". We also have the letter, on Braniff stationery, of that same date, proudly proclaiming in the logo Braniff's 12th year, which said:

Dear Mr. Clayton:

Our company regulations provide for employment of pilots on a probationary period of one year, during which time they are considered as temporary employees on probation or training and not as permanent employees.

During this one-year period, a pilot's actions and performance of duties are closely observed and graded. Also, any time during this period any pilot whose services are in any way unsatisfactory may be dismissed from service as not possessing the desired qualifications.

At the end of three months you will be called in and your record of service checked over, with a full discussion of your virtues and faults, with suggestions for necessary improvement and correction.

At the end of six months service your record will again be reviewed and you will either be dismissed from service as unsatisfactory or considered as eligible to continue with the Company on a further probationary period.

If still in our employ at the end of one year's service, your entire record will be carefully checked and decision made as to whether you will be kept in service as a regular employee.

The first week or ten days of employment will be given to training purposes and during this time you will be paid at the rate of $125.00 per month. When you are assigned to regular co-pilot duty, you will be paid at the regular co-pilot scale.

*We want this matter fully understood before you
accept employment.*

Yours very truly,

BRANIFF AIRWAYS, INC.
(signed)
R. C. Schrader,
System Chief Pilot

We cherish my father's first Braniff business card, with its red-white-and-blue logo in the upper left corner, otherwise stating simply DELBERT H. CLAYTON   FIRST OFFICER. My mother retained his flying log books. It is of interest to report an early entry in his Book 3, Instrument Rating, 24 April, 1940. The next entry in that log book from Love Field, Dallas, May 22, 1940. He piloted for Braniff a DC-2, which held fourteen passengers. Its call numbers were DC2 NC13724. He flew his first DC 3 on September 11, 1940. It held 21 passengers, and was the workhorse upon which the successful airline industry was built. A beautiful example hangs today in a place of honor in *The Air and Space Museum* in Washington D.C. Hanging from the ceiling on cables, it cuts a tremendous figure, especially in my eyes. If I walk under it alone, as I have several times, and look up, the words "Hi, Daddy!" form on my lips, even as tears cloud my eyes.

Our family travelled to Texas on the first of a long history of Braniff passes. Grandpa and Grandma Kembery drove us to the airport in Kansas City, which was one of the earliest destinations for Braniff Airways. My mother cannot remember the airplane that flew us to Dallas. It could have been either the Douglas DC-2 or the new DC-3. My mother was very discouraged with the sight of Love Field, Dallas, in 1940. We taxied up to a tacky little terminal barn with dirt floors! Fortunately the new terminal, a wonderful art-deco marble-walled beauty, was already in the works.

The first summer in Dallas was hard for my mother owing to the Texas heat. This was before air conditioning, and my mother has said that our bodies never seemed to dry. She was frequently bathing us, but it was impossible to put the same clothes back on. She had only four changes of clothing for us, and she changed our clothing three times a day. Every morning she washed the previous day's clothes by hand, hoping the three outfits would be dry by noon. My mother said that the natives called the summer of 1940 a cool one, because it seldom if ever reached 100 degrees. That was the first trouble with Texas.

We established contact with a doctor. His name was Dr. Tittle, and he was to be our pediatrician throughout childhood. His office was near Lee Park, between Oak Lawn Avenue and Turtle creek Boulevard. It was my sister who was ill this time. She had to have a cast removed from a leg that she had broken in a tumble out an upstairs window at Grandpa Clayton's farm. It had been a very bad fall, and it was a miracle that she had not been killed. Later, playing in the yard something bit her leg, and it swelled enormously and turned purple. It may have been a scorpion, or it may have been a black widow. We had both in abundance. When we went out to play barefoot, we had to exercise some caution to avoid stepping on the scorpions. Right away we had learned the second trouble with Texas. But in later years when I visited my mother in Dallas, scorpions and black widows are no longer in evidence. I feel certain, without scientific support, that they could not survive civilization. Not only did everyone kill them, but the chemical lawn treatments are probably antithetical to their life, which was so abundant in 1940 in Dallas.

After a temporary stay in a nice house at 6129 Worth Street, which we rented from a teacher and his family who were away for the summer to their Colorado cottage, we found a house to live in. My parents answered an ad for a furnished four room house for $25 per month. It had two bedrooms, a living room, a kitchen and a bath. This was on Dyer Street, just off Greenville Avenue, just north of Mockingbird Lane. It was an old house on a dirt road with a rural route number, Route 5, Box 318A. Across the field behind the house lived Mr. and Mrs. Rowland, who kept goats and chickens, and from whom we bought eggs.

Riding tricycles on Dyer St., Dallas. Author left at age 6

Today it is all replaced by robust commercial buildup; and justly so, because it was a very meager setting. Fortunately for us children, two other families on that dirt street also had children. The Kiowski family, who were great friends to us when we were new, had two children, Jeannie and Johnny. My most memorable moment at the Kiowski's was setting the carpet in front of their fireplace on fire by playing with the flames, and with their parents out too. Fortunately they returned and got it out without much damage. This is a clear visual memory to me from age five. The Mills family at the end of the street had several children. My lasting memory from their house is of tumbling into a tall and thorny rosebush at the rear corner of their driveway while riding one of their bicycles. To this day the angry scars along the inner side of my right bicep attest to the bloodletting of that fall. That required Dr. Tittle again.

My profound sense of loss

One day my father announced that he was going to get rid of our DeSoto. I remember my intense grief. My mother described it as a trauma for me. I did not want to give it up. I wished almost desperately to hold onto whatever we had. My father said he would trade it for a new one, but my grief was inconsolable. I ran after the departing car, tears streaming down my face, hollering "No! No!" But he did it anyhow. That story had a happy ending when my father returned with a new DeSoto, a 1941 model. My eyes shone with joy when I saw how wonderful it was. Could anything be more beautiful! However, I have found it almost impossible in my adult life to part with my cars.

During this transition to Texas something important had happened to me, and was still happening on Dyer Street. To this day I do not know exactly how to describe it accurately. But because it was to follow me throughout my life I describe it as best I can. I discovered *sorrow*. I discovered *loss*. Share these recollections of a five year old, recalled at age sixty. Some thing, or things, that I clung to were lost. The two great forces of my young life were my mother and her mother's farm. That 160 acre farm, with its great white farmhouse, its magnificent red barn, its corn cribs and oat bins, its water tank, its animals, its chickens, its apple trees, its rolling hills and its agriculture, had been my playground since birth. It, not Dallas, was *home*. Despite living in many houses for short times, and especially because they were short times, we returned repeatedly to the simple familiarities of that farm and of my Kembery grandparents. There we had lived while my father went to Texas seeking work, and it was from there that we later departed.

It was to be forever.

Even from the outset I sensed it might be forever. During all of the memory-filled repeat visits, replete with their fun, their boyish poetry, their familial warmth, I knew it was to be forever. From that day forward I have clung to the past, just as I clung to that old DeSoto. I have saved the past, turned it over in my mind to view it from every perspective. I have stared at it in photographs, photographs dear to me, as if they could save it all for me. But I knew, I knew. It would be forever.

Health care in rural Iowa in 1940 was so sparse that even my parents did not know as we drove away that my grandmother Clayton was dying, and dying quickly. We knew of her goiter, because it was so noticeable, but we had no idea it was part of a malignant metastasis. We did not even know those words. She would not live through the year; but we were unaware. I have kept the pencil-scripted letter that she wrote to me and to my four-year-old brother in late August 1940:

> *My dear little boys,*
>
> *How do you like to live in Dallas? Is it a nice town?*
>
> *Do you remember when you were here at my house and we used to walk down to the orchard and look at the tiny little apples that were just starting to grow on the trees? Some of them are good to eat now, so I am sending you a box full and I hope you will all like them. The big ones grew on the first tree and the little ones grew on the next tree. There are lots of grapes too, but they are still green and sour.*
>
> *The white kitties have grown until they are almost as big as their mama. They are not little soft fuzzy things now but they are still pretty. They are so white and their ears are pink.*
>
> *We never see that snake out in the garden anymore. I guess it went away.*
>
> *Don't forget that I love you as much as any grandma could love little boys. And of course I love Carolyn too. Write me a letter when you get big enough. Goodbye for this time.*
>
> *From Grandma Clayton.*

She had included a note to my parents to say that she was feeling "somewhat better" and was "going to the doctor in Des Moines on Monday and will write you from there." That note did come, and it told them that she had a malignancy. My parents did not know what malignancy was, and wrote her to say so rather than asking someone medically informed. Her answer came back: "It is cancer." All knew that dreaded word, though none knew what it really was.

So, in our first winter on Dyer Street in Dallas, Grandmother Clayton died. I remember so well my mother coming to my bed and telling me of this unfathomable event. *My Grandma Clayton was gone? Gone where?* We drove to Iowa in our new DeSoto during that bleak November. Though but five years old, I picture vividly walking into that old Clayton farmhouse, only a few miles from the farms of my uncle and of my Kembery grandparents, to view her dead body. She lay exposed in a simple coffin in the corner. She looked sweet, silent, mysterious; but I could tell that she was gone. The small parlor of that primitive farmhouse where my father had been born was cold, and she was cold rather than radiating the warmth that I had known. We all walked past. We stood. We told Grandpa and Aunt Eunice we were sorry. I cried. We all cried. I can see that room to this day, see her inanimate.

My mother has handed down one of her treasured stories from that visitation. Standing solemnly by the parlor door as a neighbor dropped in, hat in hand, I said, "My grandma died." To which my brother Keith added, "Mine too."

Searching my meager five-year-old store of memories and feelings always leads me back to a third chapter of this trauma. At winter's end my Kembery grandparents drove to Dallas to visit us. The farm could now wait, in its own repose, blanketed with snow, with Merle feeding the animals. So they visited to see our new home. Because I wanted to do so and because my mother somehow thought it for the good, I afterwards returned to their farm with my grandparents. My sixth birthday, March 18, 1941, was spent in their farmhouse east of Bridgewater that I knew and loved so well. The weather warmed, and the call of the farm's duties was strong. I went out with Grandpa Kembery often. He regaled me with his subtle humor that I later discovered to be recognizably English. Sometimes I went out with Merle, their adopted son, who was then perhaps eighteen years old and a good farm worker. I gathered eggs with Grandma. I collected corncobs for the great stove. Most significantly, however, my grandmother helped me to write tiny notes to be included in letters to my mother.

Sadness came over me, unseen by others I believe. I missed my mother. She had rocked me to sleep every night until I was two. Even after my brother Keith was born, she rocked me to sleep after he went down. I needed her and she was not there. My happiness on the farm was challenged by periodic doubts over my first separation from my mother, a separation that would last until my father's summer vacation. My grandmother certainly recognized something of my blues, so she gave me stationery for my little notes to her. Grandmother Kembery asked me what I wanted to say, and helped me to write correctly. I could already read simple things, but this was my first try at writing. My awkward letters quickly filled the page, which was inserted with Grandma's letter. Writing to my mother surely helped. But my heart was aching, and I was not sure why. I was sure that she could not have deserted me? Did she not want me with her? Surely that could not be so. But still I ached. I often sat at windows and stared out across the farm scene. It was a terrible pain.

Being invited by grandpa Kembery to help with the chores in the spring warmth helped me more than it helped him. I became a very introspective boy during this period of loss. I watched many bird nests, robins of orange vest and bluebirds and sparrows, lots of sparrows. Of course, I had incomparable fun too, especially when my cousin Dick, would also stay with us so that we could play together. Sometimes we stayed at Aunt Elva's farm, where I helped Dick with his chores. We strolled its rolling hills north of Fontanelle. But I was a worried boy slowly pulling myself together.

I cannot know that these recollections correctly state my condition. Psychiatrists will say that if I remember it as so, if my feelings about it are strong, then it was so. Until age 50 I dreamed of Grandma's farm. Tears would regularly come to my eyes at the slightest recall of it and of some emotional thought that surrounded it. These came suddenly and unexpectedly, and then were temporarily gone, allowing my return to normalcy. Throughout graduate school years these episodes occurred. I was not embarrassed to visit a psychiatrist, and I did, and it helped. Still, I could never really shake these tears until my love with Nancy, my wife from age 48 onward, softened their blows. Visiting that farm in 1964 at age 29 with my first wife, Mary Lou, I stared with tear-filled eyes at grandma's farmhouse, where someone else now lived. I could hardly speak with Mary Lou. I could see that my departure had been forever. There was no way to go back and to relive the joy and *at the same time make some pain turn out right.* It was indeed forever, just as I had feared at age five. Visiting Uncle Herman and Aunt Annie at the neighboring farm, where they still lived, on that same 1964 occasion, I rejoiced to see them again; but the tears would not leave my eyes, and I was embarrassed at my condition. Mercifully this was a partial catharsis; thereafter the tears never again came so easily. I am kin to William Butler Yeats, who wrote in his *Reveries over Childhood and Youth* of speaking openly for the first time with his sister of his longing for Sligo: *I know we were both very close to tears and I remember with wonder, for I had never known anyone that cared for such momentos, that I longed for a sod of earth from some field I knew, something of Sligo to hold in my hand.*

Bear with me for one final 1941 moment, a tiny one forgotten by all but me. Finally driving back to Texas, reunited with my mother who had flown up to drive back with us, my grandmother Kembery said in the car, "Donny, now that you are six you will go to school in September." I said nothing, but I thought "I don't want to go to school. I want everything back like it was."

I actually am one of the lucky ones, fortunate to be able to recall what happened and even get to the bottom of it. It is preferable to suffering sad feelings and emotional inheritances that can never be identified, much less analyzed and dealt with. And I also suppose that I am not such an exceptional case, that each of us guards a tender core left from the time when we became a person, and that that core carries unspoken fears and sad, sad losses as well as strengths and joy.

CHAPTER 4. SCHOOL DAYS

I cried on my first day in school. How could it have been otherwise, with my one sad inner thought festering since my Grandma Kembery's innocent question on the drive down from Iowa? Going to school was another evidence that one cannot go back. I was, of course, too young for that big thought. Unresolved doubt left me only half willing to go forward. There was no tantrum on that fateful day when I had to leave home; no sense of panic. But tears filled my eyes and made the bright daylight of September 1941 a painful squint. My mother took me to University Park School in our new DeSoto, now beloved by me. And there she left me; unsure, backward looking, regretful as she pulled away. An understanding teacher took charge of me and spoke gently to me and led me to my class. How could I know that I would love school as much as I loved the farm?

Although I was six years and five months of age, unlike children today I had never been to Kindergarten; never been to play school; never experienced the good fun and the good adults to which school would introduce me. I had no training in social graces save those from home; I had undergone no social normalizations; I had no experience with groups. My own children would later shoot through this first-day trauma with nary a blink. They would have averaged about two years school each by the time they reached six years of age. But I was a family animal, half fox and half hedgehog.

I had been captured on a farm where foxes and hedgehogs knew simple rules and harbored simple hopes. I had been transported in a two-ton steel box on wheels to a far away blazing prairie with new houses next to one another. And now I had been deposited into school. Where would it all end? I was a Mama's boy, content to be with my mother, where I had always been.

Owing to my good genetic luck and to the incredibly competent level of high school education of my parents, I was more than ready intellectually. My mother always enjoyed telling in family gatherings one of the several validating and oft-recounted tales of my preschool and school years. In some of these stories I anticipated school: "Mama, what is that?" I asked, pointing to a windmill. "That's a windmill, Donny", she replied. Three miles further along the farm road, "Mama, what is that?" I asked again, pointing to another windmill. "That's a windmill, Donny", she replied again. This repeated. She concludes with how this exchange repeated half a dozen times until I at last gave up, and how she later suffered owing to her guilt at finally realizing that I had wanted to know what a windmill *does*, rather than its name. Retelling this oft told story touches me deeply, and I admit that it shows me as I was, and as I am today. I sympathize with the sense of regret that only a conscientious mother can know.

Another family folklore arose from my second grade: "Mama, teach me how to do long division". "How do you know about that?" my mother asked; "You do not have that yet in school". "I know," I replied, "but when I get there I want to know how to do it". Mother said she may have then shown me an example. She probably did, because she well knew how to do long division. Like any bright and loved child, I heard frequent tales of this sort, testimony to either the depth of my curiosity or to simple motherly pride. People always laughed, and admired me. But with life's hindsight it is not difficult to see an element of my chronic fear of failing.

University Park Elementary School stands on the north side of Lovers Lane, an east-west artery that divides the Park Cities, University Park from Highland Park. To the south of Lovers Lane were the gracious and substantial estates of the older section of

University Park and of Highland Park, a settlement of affluence within the city of Dallas. To the north side of Lovers Lane lay the solid but smaller housing of most of University Park, of what was then middle-class America, but an upwardly mobile middle class, families far removed from the ravages of the depression and perhaps aspiring to move across to the south side of Lovers Lane. It was the northern side that my family joined and with whom we belonged, though many of my later school chums attended older Armstrong Elementary, built in 1914 and destroyed in a 1951 fire. It was located on the more affluent side south of Lovers Lane. University Park Elementary School sat toward the eastern end of Lovers Lane, almost to Hillcrest. A few blocks to the south lay Snider Plaza and Southern Methodist University. A second major east-west artery, University Boulevard, descended back to the west from SMU in 1915 as the first paved street in University Park. This school district had already established itself at the time of my entrance in 1941 as one of the great school districts of the state of Texas, an accolade that has continuously grown in the half century since. Books and nostalgia now exist on *The Park Cities*[1].

The University Park Elementary School was constructed and opened for use in 1928, at a time when the Park Cities represented a northward thrust of the growth of Dallas. The building itself was in the eyes of this boy a very grand one, sand colored stucco atop masonry, classical in architecture, solid in its images. Its two floors must have totaled almost 30 feet in height, with huge panels of wooden double-hung windows that were themselves at least seven feet tall. Those double-hung windows were routinely raised for fresh air during my school years. Eye catching even to the six-year old was the carved stonework friezes towering four feet above its entrance and their beautiful counterparts mounted between the first and second floor windows. If I had to go to school, this was the place for me! This classical solidity and appealing form expressed the optimism of early 20th century America and the importance which its communities assigned to education. When I entered in 1941 the enrollment was rapidly increasing, with 894 pupils in 1941, in contrast to the 165 when it had opened in 1928. Double sessions had to be held throughout the 1940s owing to the pressure of numbers. I remember every street in University Park being full of young school-age children. Students in morning classes attended from 8:45 until 11:45, and those in the afternoon attended from 12:15 until 3:15 with a different group of teachers. This school pressure in 1941 was not uncommon at that time, as WWII broke upon us at a time of residential expansion and growing young families. Indeed, WWII halted the construction work on the newly planned Hyer Elementary School, designed to alleviate that pressure and whose foundation had been laid in 1941 prior to a wartime halt of the construction.

A Home on Sherry Lane

Transplants as we were from the farmlands of Iowa, my family barely made it into the University Park school district. My mother's memoirs[2] recount those details with more facts than I knew at the time:

> *During the summer of 1941 we went house hunting. Donald would be starting school in September. We found a builder, Sam Lobello, who had bought a block of land to build houses on. He named it Vincent Street for his son. It was just outside the city limits of University Park, outside Dallas proper, in the town of Preston Hollow; but it had been accepted into the University Park school district. The total price was $5,500. It was three bedroom, one bath, with $500 down payment, and payments of $39 a month. We had a new home in Texas. We moved into this house August 25, 1941, and lived there for seventeen years.*

What I recall most vividly of those early school years was living essentially in the country. We were surrounded by undeveloped fields of tall grasses, fields that provided many an adventure to us. The northern edge of the City of University Park lay just one street south from ours, at Colgate Street[3], which also ran east-west. The next east-west street to the north of Colgate was then Vincent Street, where we lived. Our street extended for but two blocks east-west between Preston Road and Douglas Avenue (2), the two major NS arteries in University Park. There existed no other houses between the northern edge of University Park and the Northwest Highway (Loop 12). We were virtually in the country north of Dallas. Legally we were in the town of Preston Hollow, which had a little town hall and a police station on Preston Road. The important thing for me, and for which I thank my parents deeply, was that it established me in the Highland Park School District. To the west of Douglas Avenue, itself just three houses from ours, there stood not another house between Douglas Avenue and the Cotton Belt Railroad, which was located where the North Dallas Tollway now lies.

Throughout my elementary school years, my brother and I and other kids on our street roamed the fields north and west. When we crossed to the western side of the Cotton Belt Railroad tracks, we encountered more abundant trees among the fields, some hilliness and rocky outcrops, and, wonder of all wonders, an abandoned rock quarry on the creek. This rock quarry made a swimming hole for us that was all our very own. We never recognized the potential dangers of this fun. Slightly further west were rather high and steep chalky cliffs, which we approached from the top and climbed down and up often, but not without a sense of extreme caution that fear of steep heights can induce. These cliffs were near the intersection of Loop 12 and Inwood Road. These places were very happy ones for us, providing the same sense of free adventure that I had known on the farms. And several neighborhood children were friends to Keith and myself in these ventures. In the *Dallas Morning News* story[4] written about the Clayton family, Diane Galloway reports my mother's words: "This little two-block street out in the country was a great place to raise children." And indeed it was. I recall more than twenty children living in that two-block stretch of new homes, obviously peopled by young families carving a new life on what was then the edge of the city.

When my father's Stinson *Voyager* needed repairs, he cavalierly landed it in the open fields to the west side of Douglas Avenue and taxied it across Douglas Avenue to the repair shop of our family friend, Morris Jacobs. This shop was near Preston Road on the beginnings of what became Luther Lane. Later he took off again along Douglas Avenue. He simply waited until no cars were coming! Having been born to it I found these landings and takeoffs to be unremarkable; but they were the talk of the neighborhood for some time.

Our house had four different mailing addresses, and these caused us much confusion as kids, but reflect the growth of suburban life. First it was 4225 Vincent Street, enumerated to parallel the numbers on Colgate Street just two blocks south. It took its name from the son, Vincent, of the builder, Sam Lobello. The intervening blocks[2] were then one great field of wild grasses with a large water pond that had probably been a cattle tank for earlier ranchers. But in 1945 the city of Dallas annexed our street (from Preston Hollow) along with an area still undeveloped to its north. This annexation was approved by Preston Hollow residents in 1945, so that the town and its Town Hall on Preston Road disappeared. Dallas changed our address to 6012 Vincent Street to parallel the numbering in Dallas rather than in University Park. At the same time the residents of University Park and Highland Park voted against consolidation with Dallas, a turning point of considerable historical interest for the subsequent life of The Park Cities. After some months the postman discovered that Dallas already had a Vincent Street in the Oak Cliff area. Before ZIP codes this ambiguity caused numerous problems, so my parents were allowed to meet with neighbors to choose a new street name. We became 6012 Sherry Lane. But friends and delivery men could still not find this strange Dallas number, so it reverted back again to the University Park numbering system and we settled in as 4225 Sherry Lane. I discovered that $4225=(65)^2$, so I enjoyed claiming to reside at $(65)^2$, the only square address on Sherry Lane.

At about this same time the Episcopal Church bought a nine-acre tract of land at the corner of Colgate Street and Douglas Avenue. On that cleared space the Sherry Lane gang played baseball endlessly. From our back-yard fence on Sherry Lane we looked right at this vacant lot. The field was rough, and bad bounces were common, but it seemed to be ours. Well-known architect Marion

Foshee designed and built from honey-colored Austin stone *St. Michael and All Angels Church*. It is a masterpiece, and even then, it possessed characteristics out of the ordinary, design elements that captured my six-year old fancy. I loved architecture from an early age. The firm Foshee and Cheek had already achieved architectural fame for their design of Highland Park Village. Fortunately for us, the baseball lot on the east side of the church remained, and was even improved when the church planted grass.

Otherwise, the physical surroundings of our little street did not change for about nine years. Then the Preston Center Shopping Center was first laid out and paved. My mother remembers this as being in 1951. One of the first buildings built was a post office. Diane Galloway, an unofficial historian of The Park Cities, quotes my mother's recollections of this in her newspaper story[4] about our family: "They cut a north-south street called Westchester (extending its University Park counterpart) through our two-block street. That was done so that the mail trucks would have easier access to the post office and also aided the traffic that began to flow into Preston Center. But it also caused much new traffic on Sherry Lane. After a few years of this, our street was zoned commercial, as this was the only direction that Preston Center could grow. By 1958 almost all of the homes had been sold." Today Sherry Lane consists of shops where the homes had stood, and east of Douglas Avenue are high-rise buildings. I quote from my mother's memoirs[2] her summary of our ten-year family life there prior to this change:

> *This little two block street was a great place to raise children. The only traffic was from those living there. At one time there were about forty children. I want to list a few names, as sometime you may wish to recall who lived there. On the north side, starting at Douglas Avenue, were Siegal, McElreath, Charles, Schaeffer, Manglesdorf, Hitt, and Birenbaum in the first block. The second block housed Halperin, Levy, Williams, Hambleton, and perhaps another before the lumber yard at the east end of our street. On the south side from the same Douglas Avenue end were Jones (a cranky man if a ball went*

*into his yard), Rogers (who had a beautiful Collie dog),
Clayton, Rosenbloom (later Lambert), Goldstein, Hildreth,
and Curry. In the second block were Newman, Viets,
another Goldstein with two older boys, and a couple of
more houses. At the Preston Road end was a Sinclair
filling station and a liquor store. And don't forget the
big old gray unpainted house beside the liquor store on
Preston Road, which was kind of spooky. There was also
a round building on Preston Road called The Big Top, a
drive-in for hamburgers. The boys and Carolyn liked to
roller skate around its circular sidewalk.*

Most of these families remained there for the seventeen year
existence of Sherry Lane as a residential street. Remembering back
on these names, all of whom I knew not only from playing up and
down the street but also because in junior high school I became
their morning-paper delivery boy for the *Dallas Morning News*, I
am struck by the number of Jewish names. I count eleven as likely
in retrospect; however, we did not speak of any of them as being
Jewish. I was unaware of this distinction as a boy. It cannot be
coincidental that this area of north Dallas became a stronghold
of Texas Jewry. It would not be long before a beautiful synagogue
would be built nearby. And while I speak ethnically, I would repeat
that of course there were no black families for miles in any direction.
This area was totally white, and the only black people ever seen by
me were the yard men who came to mow some of the lawns, and
Alice Thompson, a dear old woman who helped my mother with
the cleaning.

University Park Elementary School

Vivid in my memory are the swing sets at University Park School.
I tried them right away, and played on them almost every day.
Because I was a good sized boy, keen and already able to read,
they immediately moved me forward half a grade. For that reason
I began in the High First, and was ultimately a January graduate
of each year. Later, after 3rd grade, they suggested moving me

up again. The teachers said that I was finishing the assignments too quickly. "I was a nuisance to the others", they told my mother. But we all resisted. Especially I resisted. I was always a little bit fearful. Mother did not want me to be the youngest boy by such a large amount. And I did not want to leave the friends that I now found so secure. So I probably became less of "a nuisance". One famous incident involved my moving the clock hands forward by 5 minutes. This caused class to end 5 minutes early. This may not sound a crisis, but in WWII years, parents and teachers expected, and got, good discipline. Mother told the teachers that she would take care of my punishment. She didn't, of course. She was too soft to really punish. So I got off free. Mother cautioned me that I was very lucky, because "when she was a girl, one had to hold one's hand out flat on the desk and have it whacked with a ruler for punishment".

My elementary school scrapbook stores all of my report cards. Why did we keep all this stuff? Was it my fear of losing what I had? The scrapbook also contains my school photographs, letters to my mother written by me in 1944 from my grandmother Kembery's farm, family snapshots from those years, and summer photographs from the farms. Official portraits of my class are there, each face looking friendly and known to me as it can be only in a small stable community. Or was this my mother's doing, these abundant details of my life? Was it her wish that this time would stand still? The report cards, from Miss Pierson in the High 1st to cards for each

University Park 1943 school

subject and teacher through the High 6th, confirm that I was an excellent pupil. It was in Mrs. Jones' High 4th arithmetic class that

I first found myself realizing that the teacher could be mistaken. The challenge was called "stated problems", which were verbal statements of situations that required formulating as arithmetic. In class I innocently raised my hand and corrected Mrs. Jones' logic on more than one of these. My classmates were horrified. I laugh now that this class yielded my worst grades, B's, ever obtained in arithmetic. But it started my reputation as a math whiz, and later teachers called on me quite openly for help on tough problems.

At about this time I was administered an IQ test. The school asked mother to not tell me of my result, which they told her was quite high. This reinvigorated the idea to move me forward a full grade. But the school guarded against sharing results with the pupils. When my mother turned 80 years old I asked her about it, and she said that I had scored 140 as best she could recall. But I was very clear that I did not want to be advanced a grade. I wanted to remain with my classmates.

During the 1940s we practiced penmanship daily in elementary school. This involved correct utilization of the entire forearm rather than the fingers, lengthy drills of repeating ovals done while maintaining that form, and drills at the recommended technique for forming and connecting letters. This teaching style focused on repetitions. I developed a lovely flowing penmanship similar to that that had been learned by my parents, and my scrapbook contains my certificates during 1944-46 for membership in *The Good Writer's Club*. As an adult I later regretted modifying that strict training in order to utilize more individualistic writing traits. I also noticed as a university professor that later generations of students came without such training, and it showed in their written compositions.

My mother did a lot of driving, not in distance but in numbers of trips. First grade let out at 12:30, so when my sister Carolyn entered first grade in 1944, she needed a pickup at 12:30. 2nd grade pickup was at 1:00, 3rd at 1:30, 4th at 2:30, 5th at 3:00, 6th at 3:30. Mother usually had to take us, just like a modern mom; but our drive required only seven minutes. Carpooling was made unworkable by having staggered letout times. Mother therefore picked Carolyn up at 12:30, and returned at 3:00 for me. Keith would just play in the playground the extra half hour until I got out. So it was constant effort for my mother, but she never complained. Caring for her family was her life.

The school was also a collection point for waste metal. We saved tin cans and foil and all metal as part of the war effort. We had to remove the paper from the cans, insert the lids, and flatten them for deposit in the bins at the school.

I have two handwritten poems from unknown dates in upper-elementary years. Not good poetry and gawkishly sentimental, they reflect my childish personality during WWII. One of these, marked A+ in red ink, reads:

I'm Thankful

I'm thankful for my home so sweet,
I'm thankful for the things I eat,
I'm thankful for our great big schools,
I'm thankful for all fish in the pools.

I'm thankful for my mother, father, brother and sisters so dear.
I'm thankful for freedom of worship, want, speech and fear.
I'm thankful for Columbus that discovered our land.
I'm thankful for our teachers that are just simply grand.
I'm thankful for the birds that sing.
I guess I'm thankful for everything.

Donald

My sister Carolyn was able to transfer to Hyer Elementary after its construction in 1949, by which time I was riding my bicycle daily to Highland Park Junior High School. These changes ameliorated mother's driving requirements. Hyer Elementary stood on Greenbrier behind the University Park Methodist Church and within easy walking distance. Many children walked to school. This new school was named for the first president of Southern Methodist University. I had begun riding a bicycle in late elementary school. I remember riding it to school fairly often, about 2 miles. Keith started at almost the same time. Almost all of my friends lived between the school and our house, so the bicycle allowed me to become familiar with the homes of many of my friends. To this day among my fondest childhood memories is that of riding my bicycle all over University Park. These were practical machines, totally different from the cycling fashions of today. Mine had but one speed forward and it had coaster brakes. The tires were large, with inner tubes that we had to learn to repair with patch kits. Punctures were frequent. I became quite good a riding without hands, controlling the bicycle by subtle leanings. I could ride many blocks, with many turns, without once touching the handlebars.

One incident taught me the difference between my family history and that of many Highland Park residents. Some of my bicycle rides led me to the home of my friend George Lee. George lived in the Volk Estates, a Highland Park enclave near Turtle Creek in which all lots were at least an acre in size. George lived in a Colonial Revival mansion at 6801 Baltimore. I was dumbstruck entering their foyer for the first time, with its marble finish and inspiring staircase. As I exclaimed later at home, "Mom. They even have an elevator!" There was much of old blood and old money in these estates, and I was made conscious of being a farmboy. George's family did not make me uncomfortable; rather, it was the evident disparity between our wealth and traditions that did so. This made me sufficiently uncertain that I subsequently preferred to visit the homes of friends in the younger housing north of Lovers Lane.

World War II on Sherry Lane

For many reasons the second world war looms vividly in my memories even today. It caught the imagination of schoolboys. I say "schoolboys" intentionally, because the boys liked WWII much more than girls did. Our family dinners suffered enforced silence during the war news on the AM radio. My father hushed us to listen to Edward Murrow's *London Report* or Robert Trout's

WWII reports from Europe. Every day we followed the drama. This led me to take up construction of model airplanes as a hobby. Each model that I built was of WWII planes that were then in the news. These were not the easy assembly kits of today, but rather balsa wood blocks of the correct size for shaping into the well proportioned model. The shaping was done with a special sharp knife, using enclosed cardboard templates that could be matched to the evolving forms to get their shape right. The balsa forms once shaped were then glued together to produce the model, which was then sanded to smoothness and painted with the proper paints as designated in the instructions. Finally the decals could be soaked and slipped from their paper onto the plane to dry. I constructed many of these, but I most remember my P-51 Mustang, the B-25 bomber and the P-38. I staged imaginary bombing raids and fighter dogfights with these models. Even today I will turn out to watch in awe as remaining specimens take part in air shows.

It became necessary to ration supplies of many products that were important to the war effort. This was achieved by government issue of booklets of stamps, fixed numbers of which had to be exchanged at the time of purchases. One booklet of stamps was for shoes, one for dairy products and meat, one for canned vegetables, and one for sugar. The household was allowed one book of each per year for each family member. Three kids grew fast, so our greatest demand was, surprisingly, for shoes. Mother traded sugar coupons with grandmother Kembery for her shoe coupons. Each page had 20-50 coupons, depending on the size of that stamp. Books were small, about 4 x 6 inches. We still own many, kept as souvenirs. We qualified for 5 books of each, one for each person in the family (my younger sister Barbara was not born until 1944).

We were always aware of the gas rationing, probably because it had large impact on our driving. According to my mother, rationing was strict. As Braniff co-pilot my father needed to make many trips weekly to and from the airport, but he was given no extra gasoline stamps. Love Field was about five miles from our home on Sherry Lane, so Dad bought a motorcycle to get to work. The purpose was to conserve gasoline stamps rather than money. Not only did the motorcycle use less gas, but delivering him in the car required round trips before and after each flight, whereas the motorcycle remained parked for his return. Having two cars was out of the question. But we cannot remember what the gas stamps looked like, and do not have any in our collection because we had to use every one. An additional advantage of the motorcycle was that my mother no longer had to take my father and pick him up from his flights, which was a strain on her in our busy family. For fun my father would take us for rides on the back of the cycle. Neither he nor any of us wore helmets. It was an innocent time.

In the family we were all very aware of the scarcity of things and the need to use stamps wisely. Once at the dinner table my brother Keith objected to my sister's liberal sprinkling of sugar; "You can't use that much sugar because they are gas rationing it now!" Our plight was actually not as bad as in small families because we received five books for ourselves and a sixth for Dale Bateham, an Iowa cousin of my mother who lived with us for two years. He had to fork over his book to my mother.

Mealtime at the Claytons' House

This WWII conservation instinct with things was second nature to my parents. On their farms they had never had much. Things broken had to be fixed. I had been born in the height of the great depression, so of course they always stretched things as far as they would go. So war stamps were to my parents the latest example of the need to conserve everything. This made a big impression on me. I grew with an innate conservatism over things. We repaired broken appliances. We handed down clothes. We cooked dinners because that was the cheapest way to feed the family. The result of this in my later life was that I continued in those attitudes even when their need had passed. Even today I find it hard to throw away broken things to be replaced by new. My wife Nancy later came to appreciate an honored family story, how my grandfather Clayton when asked if he wanted gravy for his potatoes would invariably reply, "It's good enough without."

Grocery shopping during the 1940s was very different from today. Just a few blocks north from the east end of Sherry Lane at

Preston Road stood the Safeway, where I often accompanied my mother on shopping. It stood at the Northwest Highway, what today is called Loop 12. Being on a budget, my mother bought food with cash set aside for that purpose. It was the inside that differed so from today's supermarkets. Packaged and prepared foods did not exist, except for canned goods. Snack foods were rare. The items that fuel today's supermarkets were just not there. Imagine no frozen foods. There was a butcher department for meat, and a greengrocer section where she bought our vegetables and fruit. There were ample canned vegetables and fruits, and frequently she bought them. But the entire dinner was prepared daily by her from basic goods. Keith, my brother, did not like the peas, which were served frequently, so my mother prepared a cream sauce to obscure their flavor and to keep the peas from rolling from our plates onto table or floor.

Some traveling stores drove along Sherry Lane. Manor Bakery had a bread truck, which "the Manor man" always stopped to see if we needed bread or rolls. The milk man brought glass quarts of milk and left them in the early morning. Less regularly came the greengrocer truck and the fish truck. In November and December farmers drove by in trucks from the Rio Grande valley, selling bags of grapefruit for $1. Our entire family developed a love for grapefruit that remains in me to this day. Home service was handy because it reduced the numbers of items that had to be brought home from the Safeway.

Mother baked deserts almost daily, perhaps because she was herself so fond of them. We all loved apple pies. So my mother often bought the apples, sliced and peeled them, and cooked them, rolled a pie crust made from flour, milk and butter, and baked the worshipped object in the gas oven abundantly sprinkled with cinnamon and sugar. We got fresh peaches in summer, when she did the same with them; but she used canned peaches in winter. Another favorite was puddings baked in a pie shell. She made a custard from eggs and cream and mixed in pudding mixes, chocolate, butterscotch or lemon, topped with meringue made from the egg whites, and sugar of course, and baked it in the same pie crusts. My parents' bachelor friend Morris Jacobs dropped in about 6:30 daily to "see how we all were"; but we all knew it was to be offered a piece on Mom's desert of the day. All six of us sat daily, eating together, at our dinner table until we graduated from high school. Alternative dinners or other things to do were nonexistent. As in a Norman Rockwell painting, we all sat every day in our places. It was a highpoint of every day, perhaps especially for my parents who relished having their four children at the dinner table. There we talked and joked, without any video entertainment to divert us. This was a special time of family laughter and family stories.

In later years I asked my mother to help reconstruct the basic menus that she prepared for our dinners. This was no challenge for her, and she quickly wrote down the pairings, which were repeatedly weekly. I list the top five:

(1) Meat loaf, scalloped baked potato slices, baked beans, jello salad, bread and desert. She mixed eggs and crumbs into the ground beef and baked it. This was Keith's favorite menu throughout childhood.

(2) Pot roast with small red potatoes, small boiling onions, and carrots. Accompanied by cole slaw, fresh baked bread or rolls and desert.

(3) Fried chicken, mashed potatoes with chicken gravy, green beans, sliced tomatos, bread and desert.

(4) Lamb chops with mint jelly, boiled potatoes, creamed peas, Waldorf apple salad, bread and desert.

(5) Baked chicken with dressing and gravy, cranberry sauce, corn canned or on the cob, and often chocolate-chip cookies, baked "because the oven was hot". This was my favorite menu.

These basic menus were complemented from time to time by baked casseroles, either ham and rice or tuna and noodles. Her standard recipe for the former was this: 1 cup of ham, diced; 2 cups cooked rice; 3/4 cup grated cheddar cheese. Over this was poured 1/4 cup sauteed onions, 1 can cream-of-mushroom soup diluted with 1/2 cup milk. Buttered bread crumbs were scattered on top of the whole which was baked 30 minutes in the oven. As a tease we boys called the casserole made with canned tuna and boiled noodles "tuna noodlefish."

My brother Keith and I devoted our childhoods to play. Sandlot sports were our thing. We played baseball with University Park friends at the vacant lot beside the Episcopal Church on Colgate or at Caruth Park, where there was a diamond but to which we all had to ride our bicycles. Given insufficient numbers for an organized game, Keith and I played our own game, in which he stood glove in hand along the Sherry Lane front lawns and to whom I hit fungos, some being base hits and some outs according to our rules, with me all the time announcing the batter and the game situation like a radio sportscaster. A variant that we often played alone was pitching and hitting with a small plastic ball having holes in it. We could pitch it fast, because it rapidly slowed in flight, and we could hit it with full swings even next to the houses. We developed rules for outs, hits, doubles, home runs, *etc*. What fascinated me most in this game was the excellent curve ball that I could throw. I threw the same pitch with a baseball, but its curve was so gentle that one could not be sure of the actual amount of curvature. This practice ball really responded to the spin that made it curve undeniably. I spent a lot of time thinking about how spin could cause the ball to curve. Countless hours after school were spent this way. We played touch football, at school, at Caruth Park, and at University Park YMCA. And we played basketball, one-on-one, or two-on-two, or three-on-three at the goal above our garage doors or at Bob Viets' similar setup in the next block of Sherry Lane. I was extremely good at all of these, being top dog. At each sport I was invariably one of the two expected to choose the sides. But my skill did not translate to playing with my Highland Park High School teams, to dismay from some friends. The truth is that some emotional risk held me back from those commitments, from exposing myself to talented competition. I was, for whatever reason, the guy that could have been great.

Foggy Junior High School Years

After the simple tranquility of grade school, the Highland Park Junior High School (grades 7-9) provided significant new stress for me. Today I would speculate that this had to do with being a teenager, but it was a distressed time. I was forced to leave the elementary school that I loved. The leaving of a loved place was difficult throughout my life. I was confronted by many unfamiliar students who had come through the other elementary schools of the system and merged at grade seven. Most of these became friends, but I was concerned with how I would stack up against strangers. And the school, whose building was the original Highland Park High School, with similar architecture to the newer high school that I would attend later, was older and seemed darkly cheerless to me. It had become the junior high in 1936 when the new high school was constructed on Emerson, and it was demolished in 1997.

Curiously, I can hardly recall my junior high years, save for two spankings in the woodshop administered by its teacher. I was a prankster, and physical spankings were neither unheard of nor forbidden in those years. Much more frequent was detention hall, in which I was forced to remain after school and study quietly. These frequent detentions were educators reactions to my childish defending of myself against hormones. During English class, reading poetry, I would devilishly burst out with a replacement stanza that would set the class into hysterics. In retrospect I think puberty was upsetting me, with my preoccupation and fear of sex. I felt it coming so blanked it out. My Junior High experience is mostly blank. My mind never strolls the halls of that school in my reveries. I can not picture my teachers, nor remember activities in my classes. Something very different churned inside me.

A couple of my report cards from my seventh grade even surprise me when I review them today. Seventh grade brought our first final exams, and although I cannot remember them, my report cards reveal an uncharacteristic problem. In Miss Burger's class in Texas History I made A's on the term grades but a C on the exam. At the same time in her Geography class, my exam is marked as D. These were the worst grades of my life, although they were averaged with the good term grades to yield B's. There has to be a good story to that C and D on those exams; but I do not even remember it happening except on one occasion before Geography class when I arrived early to draw a map of England on the blackboard and drew a swastika within it. I have no idea why I did that. I was hastening to my seat as the teacher arrived. My entire performance was evidently more erratic than in elementary school. Perhaps I could pretend that Miss Burger simply disliked me. I was becoming a nuisance, a jokester in my classes. But I know another thing in my heart, and can state now what lay hidden many years and which I would have been too terrified of to then admit. There festered within my character an element of panic at having to perform. To surmount that tendency became a task for my life. It showed up in my reluctance to compete in organized school sports, even in sports that I was very good at. It troubled me later when I entered the sexual world.

Funny stunts brought more disapproval than humor. One hilarious stunt at home was directed at my younger sister, Barbara, then age five. Going out after dinner to play with Keith, Barbara following along, I said to Keith with a worried face that we had better keep an eye out for the gooch. He caught my drift at once and added a sense of caution. For weeks, going out to play, we would look first for the danger of the gooch, being very quiet, standing at attention like two meerkats until all seemed safe. But we stressed the horror of ever being caught by the gooch. Then one day I put Keith up to the denouement. I had him

Practicing my golf swing at age 13, on Sherry Lane, Dallas.

hide in our garage dressed in a black plastic raincoat and wearing a mask. Leading Barbara out lightly to play, stopping for one cautious glance for the gooch, we played bounce passes with the basketball. Suddenly the gooch appeared at the garage door, and I shouted, "Run, Barbara, it's the gooch!" She ran screaming, terrified, until, doubling up in laughter, Keith removed his mask and shouted to her.

I began playing golf with my father in junior high. I remember those numerous days very well. Golf was the sport of choice among airline pilots, and my father encountered many of his colleagues on the courses. Our most frequent golfing venue was Tenison Park on East Grand Avenue in old inner East Dallas. It sounds extreme to those who have not been motivated repeatedly by such experiences to know that even to this day, five decades later, I can in my mind's eye walk over the entire 18 holes at Tenison Park, and remember how I tried to play each hole. And yet, I can not remember a single thing about classes or teachers in junior high. Are these years a black hole for all teenage boys? I developed a beautiful swing by practicing it in the yard at home, paying close attention to Ben Hogan's book *Power Golf.* I became good enough to captain my high school golf team; but I remained an anxious competitor. I wanted to be better than I was rather than simply do the best I could. Outside of competitions I subtly improved poor positions of my ball in order to show a better score at the end. This was one of the many faces of my fundamental weakness.

It probably was very good for me that during junior high I was allowed to become a newspaper carrier for the *Dallas Morning News.* I remember this very well, much more clearly than school. I think the clarity reflected its being my goal and my idea. My parents were persuaded to let me do it, and I think I liked its responsibilities and regimen. It was not a job for sissies. For five years, until my senior year in high school began, I arose at 4:00 AM every morning—no exceptions except for planned times away when substitutes could be arranged. I had acquaintances in University Park, and we generally subbed for each other at those times. I began this in eighth grade as a bicycle service. My newspapers were dropped by a distribution truck at the Sinclair gasoline station at the corner of Preston Road and Sherry Lane, where I cut the binding cords of my precisely counted number of customers, packed counted numbers into my canvas shoulder bag, and rode to the areas, dismounted, and walked the routes delivering each paper by hand to the respective porches. On bad weather days, especially in winter, my mother arose to drive me around in the car, but I still walked each route to deliver the morning paper. My father had bought a Crosley station wagon for economical transportation to Love Field for his flights, and when I turned fifteen I was able to get a special beginner's permit to drive it myself for these morning deliveries.

I was good in this occupation, my first job, and took pride in recruiting more customers by visiting their homes and in my monthly collections door to door of the monthly bills. This diligence won recognition from the *Dallas Morning News*. My scrapbook contains two photographs published in that newspaper in a special section called *Carrier News*. One undated photo shows a selected group of carriers collecting their prizes, frozen turkeys, at the newspaper offices; and the second from November 23, 1951, shows that group of awardees getting our free tickets to the SMU-Texas football game played to be played in the Cotton Bowl. Also in that November, when I was 16 1/2 years old, I was awarded a nice gold-embossed certificate *Award for Carrier Achievement* stating that I had "distinguished himself in faithful and effectual performance of his contract for the purchase and distribution of the *Dallas Morning News. The News* extends congratulations to this young man for his business acumen, integrity and meritorious service."

It is the Highland Park tradition that in the ninth grade small groups of students would host large formal parties, an official sign of our maturation. These were mostly dances at Arlington Hall in Lee Park. The invitations' being to a "Formal Dance" indicated that we would wear coats and ties. I wish I could have enjoyed these more, but in truth they caused me constant anxiety. I did not know how to dance, although my mother staunchly ordered dancing lessons for me. Instead of suggesting that the main thing was to have a good time, and that how well I danced would be little noticed, these lessons increased my anxiety by instructing me in the precise sequence of steps and the counting of their routines. I was instructed in differing rhythms—fox trot, waltz, rumba, samba, and perhaps others, even their names totally new to me. In short, I would have to perform on the dance floor, and with a girl on my arm, one whom I would politely ask for that dance. This seemed a horrendous challenge, and it happened unceasingly throughout our final year in the junior high school. My scrapbook displays sixteen such invitations between November and March. All but one were hosted by girls in my class; one was hosted by a group of five boys who knew more what they were doing than I. Their invitation stamped them as self-confident and sophisticated, whereas I felt inadequate. It escaped me how one would even know what to do to organize such a party, reserve the Hall, book the music, print the invitations. In no other circumstance did I so feel myself to be a simple farm boy of farming parents; at no other time did I so endow my friends with imagined graces and sophistications. Nonetheless, I did attend; I did dance despite anxiety at requesting the dance from a girl; I did have fun—if only I were not required to perform!

During Junior High School I was allowed to catch the bus to downtown, and there I discovered the Cokesbury Book Store. I bought my first chess books, by champions Capablanca and by Fine, and I studied them much more than school books. It was in my character to read about something before trying it. But an even greater wonder was my purchase of a small blue book entitled *Tables of Logarithms.* By studying it I discovered that it was possible to multiply numbers by simply adding their logarithms. In math class I demonstrated that I could calculate the number of seconds in a year by the multiplication (365days)x(24 hrs)x(3600 seconds/hr). By adding the corresponding three numbers looked up from my tables I obtained 31,540,000 instead of 31,536,000. This amazed a few, but many more were amazed that I would even bother with such stuff! It disturbed me more, however, that I could not explain why it worked. But when I bought my first slide rule four years later, I again amazed a minority by multiplying two numbers by measuring two successive distances along a stick!

My Father seeks God and Technology

My father was a big influence in my life. He was gentle, understanding and wise in human relations. No harsh words ever passed between us. He forgave my every flaw. Just one problem marred his beneficial effect. During my school years his fascination with religion blossomed and in the end went too far. It began simply enough. He persuaded my mother that we should start attending church or Sunday school. This began at the Lutheran Church on Luther Lane, just 200 yards from our house. Then in search of something he took us to a couple of tent revivals. Later he switched our Sunday school to the University Park Methodist Church on Greenbrier. It was there in the adult Bible class that my father became acquainted with its teacher, Henry Hinsch, who introduced him into a path of belief in which USA and Great Britain were described as the long lost but reassembled northern tribes of Israel. Led by a tip from Henry Hinsch my father sought out a fringe Dallas church near downtown led by the Reverend John A. Lovell. It met in a ballroom of the Maple Hotel. This church was named "First Covenant Church", the name reflecting the group's conviction that God was speaking to them when in Genesis 12 it was written:

> *The Lord had said unto Abram, "Leave your country, your people*
> *and your father's household and go to the land I will show you.*
> *I will make you into a great nation, and I will bless you;*
> *I will make your name great, and you will be a blessing.*
> *I will bless those that bless you, and curse those that curse you;*
> *And all peoples on earth will be blessed through you. "*

The *first covenant* referred to that covenant between God and Abram. The modern progeny of Abram, or at least the dispersed northern ten tribes, were supposed by this church to have reassembled as England and the United States. Potentially a harmless notion, this conviction resonated so strongly with my father that he could not let it go. He could not maintain the healthy skepticism central to common sense and to science. It grew to the level of obsession and dampened my relationship to my father in later years. I tried to steer clear of international affairs in our family discussions, because all current events tended to be reinterpreted as prophesy revealed. All of politics now lurched firmly to the radical right. For we children the beginning church experiences had been comforting in encouraging us to embrace the Christian underpinnings of western civilization. I did that then and still do today. But when a teenaged boy perceives his father's irrationality, his feelings about his father must change. It became an embarrassment to tell friends where we went to church, because their families attended one of the several grand traditional churches of the Park Cities. Both of my sisters in later adult years openly regretted our participation in that church. It was near this time, after his first visit to England, when my father speculated that he would like to retire to Glastonbury, where Jesus had visited as a child (so he believed) and where Joseph had planted a legendary thorn tree. When J. A. Lovell, minister of First Covenant Church, died of a sudden heart attack, my father assumed the ministry of this church at the request of its members. But during my school years, he was but a member of its congregation. One of my most persistent pleasant images is standing next to him during the hymns and hearing his very good singing voice.

A different side of my father was fascinated by technological gadgets, putting him half a century ahead of his time. Perhaps this had begun with airplanes. Always the first to try new technology, during the 1940s we acquired a 78 rpm record player, from which I fell in love with music. We acquired also a television with a 7" or 9" screen, where we saw some of the early efforts in news, sports and politics, including live scenes from the 1948 presidential nominating conventions. We also encountered wrestling, which we gathered to watch with mounting incredulity. A blessed outcome arrived after some months of this during a fiercely staged brutality and retaliatory justice when my father, suddenly sensing that this was rigged entertainment, stood up and abruptly turned it off saying, "This is a bunch of damned nonsense!" Then looking at the cigarette in his hand, he practically shouted "And this is a bunch of damned nonsense too!", and stubbed out the cigarette and threw the ash tray and his pack into the garbage. That he never smoked again says something about my father.

He also purchased a wire recorder where our voices were saved for the future, and later that evolved to a tape recorder, which lasted longer in the new waves of technology. Then came a larger TV. Then came the most wondrous magic of all, an air conditioner. Then a camera, first a box camera, then a professsional Speed-Graphic, then a SLR. Then a darkroom printer, chemicals and a darkroom made by black paper and tape over our kitchen windows. My father was an excellent photographer, and I learned much from him. Then a woodworking shop in the garage, where he built from scratch a boat to house an outboard Evinrude engine. Assembled lovingly from varied woods of quality, the final product was by a very wide margin the most beautiful small outboard boat on Lake Dallas. I joined him in making some pieces of furniture. His pieces survived sixty years, until today.

Of all technological wonders I was most fascinated by the air conditioner. The window console that my father bought in 1948 stood in our living room front window. It puzzled me that a motor could remove heat from the cooler air inside and deposit that heat in the hotter outside air. This seemed to me like water flowing uphill; and in this sense my instinct was sound. We would run in from play, hot and perspiring, to stand before its cool air outlet. And outside I had placed my hand in the expelled air, definitely hotter than the outside air. Exposure to motors and engines had shown me that one running always generates heat. I argued to my father, who agreed, that running an indoor fan only made the room warmer, not cooler. It only *felt cooler* if it was blowing on bare skin. But the air conditioner cooled the room with the aid of a motor. So the flow of heat uphill toward higher temperature was a constant mystery to me. My father appreciated the puzzle and helped somewhat by saying, "I think it is similar to a refrigerator."

This analogy did not help me as much as it might others, because I had puzzled over the refrigerator as well! It may have presaged my career in physics to wonder how things worked, from the windmill as a small boy, to the flight of an airplane, to the curve ball, to the air conditioner. I see in retrospect that these were actually very good puzzles, but ones that did not seem to bother others as they did me.

As far as my father is concerned, when not flying I picture him reading, his nose, like mine, always in a book, usually a religion book, seated for hours in his reading chair. For the remaining thirty years of his life we had many conversations about the physics of how things worked. He loved physics. I remain grateful for that experience. I realize that few boys discuss physics with their fathers, and I am certain to have profited from that unusual experience. My father devoted much of his non-flying time to my brother and me. He was always ready to do things with us, and was loving and wise in daily matters. He explained to my mother that his plan was to spend spare time with us so that we could remain close as we would quickly grow out of childhood. Most of all in this regard, he took my brother and myself to play golf on days when he did not fly.

Still seeking Iowa

Our emotional involvement with the Iowa farms persisted undiminished throughout my school years. My mother spoke with my grandmother Kembery every Saturday night. To share the cost they alternated placing the call. The main goal was to share news of our family with them; but the calls also kept my mother close to the people from whom she had come. She learned the gossip of Fontanelle and Bridgewater and passed some news on to us. My mother's occasional homesickness was evident, although she never became mired in that and was always forward looking with great cheer toward our own family life. But at Christmas she would cry at Bing Crosby's rendition of Irving Berlin's "White Christmas", remembering the white winters of Iowa snow and family love. I was moved by that annual occurrence because I identified with it. The mail truck would arrive each December with a cardboard box of presents. It was very exciting. The second example of my saved elementary school poems reads:

> *I woke up Christmas morning and what do you think,*
> *Bunches of presents all blue, red, and pink.*
> *Of course I was happy all of you know,*
> *I got a gun, a book, and a play midget show.*
> *I got a ball, a kite, and a model airplane,*
> *I got a bicycle, a desk, and a pre-war train.*
> *If I was up in Iowa where there was ice and snow,*
> *I would want a sled that would really go.*
> *Of course my Christmas was lots of fun,*
> *And I wasn't glad when it was over and done.*

> *Donald*

During some winters we were able to travel to Iowa for Christmas. My father's flight pattern was the main determinant. I recall arriving one winter in Fontanelle in snowfall so strong that the main highway was kept open only with difficulty. Our car could not drive to the farm. Arthur Gross, husband of my mother's sister, brought team and sleigh over the farm roads to Fontanelle to take us safely to the Gross farm north of Fontanelle. During the long three miles to the farm we huddled together under blankets, and the black sky showed itself full of bright stars when the snow stopped along the way. This image has remained bright in my memory, larger than life.

But more often we traveled in summer when school was out. Those trips were usually by car. But on one occasion we flew on Braniff passes to Kansas City, a common route for my father, and were picked up by relatives. On another memorable 1944 journey my father flew my brother Keith and myself from Dallas to Fontanelle in his Stinson *Voyager*. It had one large gasoline engine with propeller and a single overhead wing connected by struts to the fuselage. Its windows would slide open, and my brother and I had a terrific time watching the scenes below and putting our hands out into the headwind. Our dog, named *Potatoes*, rode along for his summer at the farm, and in one hilarious moment he put his head out the window, as he loved to do in our car. When he saw the open vastness beneath him he immediately jumped down whimpering to hide beneath the rear seats. Keith and I doubled up in laughter. On this flight I became sick twice, and to this day I am easily airsick in private planes or in boats. I did not inherit my father's flying stomach. He landed the *Voyager* in a field on the Kembery farm. But to return to Dallas and his work on the next day it was necessary, in order to take off, to taxi the plane through the barnyard onto the paved two-lane county highway in front of the farmhouse and take off from it. Each car that happened along (perhaps one every five minutes) pulled to the side and stayed to watch the great spectacle of my father becoming airborne. Keith and *Potatoes* and I stayed all summer, alternating between the Kembery farm and the Gross farm. *Potatoes* had the time of his life after submitting dominance to *Laddie,* my cousin's collie.

During this summer of my ninth year, I saw many familiar things about the farm in new light. Farm machinery assumed new interest. The corn shelling machine could clean a cob and spit out the kernels in nothing flat, and I loved to watch a bushel of corn fed through to prepare shelled corn. The belt cylinder on the John Deere tractor could transmit torque from the tractor engine to the threshing machine, and to other machinery needing engine power. I watched with fascination as the long belt, pulled tense by the tractor position, would whip around the cylinders to transmit the power from the tractor to the engineless threshing machine. It was my nature to watch something interesting for a very long time. The windmill's pumping from the ground into my uncle Art's water tank perpetually drew water from its well unless it was locked. It fascinated me that water, clean and cool, could be drawn from the ground and that the supply never exhausted. It was drinking pure, despite coming from the ground. We did not treat drinking water at all. *What is the ground that it yields so much good water?*

The most vivid of these new interests occurred when Grandpa Kembery teased me with a new pail of milk saying, "Donny, I can turn this pail upside down above my head without the milk running out." When I laughed at this preposterous claim, he assured me that he really could. So I bit by demanding to be shown. Slowly rocking the half filled pail back and forth like a swing, gathering energy and speed, he suddenly circled his head with it. Repeatedly the pail orbited above his head, with the pail bottom always directed away from him. When above his head, the milk did not fall from the inverted pail. I tried it myself, with a bit less milk, and found that I could do it too. I could easily look into the pail and see the milk lying flat on its bottom in all positions, even upside down when straight up. This became my greatest fascination from that summer. I performed this repeatedly with small amounts of water, and found that if the pail passed slowly above my head, almost stalling there, the water would gush out. I demonstrated this trick to others repeatedly during the summer. They were also amused, but did not dwell on it as I did. Uncle Art would joke during the summer, "Donny, have you been out to twirl your bucket today?" I was amazed, frankly. I tied a rock to a string and circled it above my head, finding that the string remained taught unless the overhead speed was too slow, in which case it would fall toward me. I knew there had to be a significant reason here, but it was not until high school that I discovered myself to be in the footsteps of Isaac Newton.

Our summer leisure in Dallas was dominated by swimming at the University Park Pool. The years 1943-52, when we did most of this swimming, lay in a shadow of fear of polio. We all feared it. My mother had contracted polio in 1916, the year of a polio epidemic. The viral disease was then called infantile paralysis. In 1916 in New York about 9000 people, mostly children under five years of age, contracted the disease. Western Iowa was not safe from the virus, and one morning my mother could not walk. We heard that story several times during childhood. The number of national cases annually reached a maximum of 58,000 in 1952, when I was a junior in high school. Throughout those long hot summers after WWII, while swimming at University Park Pool we all secretly feared the disease. It gave no warning. More than once the pool was closed for a couple of weeks when outbreaks were occurring. A turning point occurred in 1949, in my 14th year, with the Nobel Prize winning announcement of the successful growth of the polio virus in the laboratory, which enabled vaccine development. Jonas Salk's first vaccine was released in 1955, three years later during my junior year at SMU.

These years 1941-56 on Sherry Lane were idyllic. Whatsoever dark anxieties plagued me were faults of my character, probably somewhat obsessive, not the fault of our family life. Ours was a joyful family having abundant time and love for one another. Nor is this a romanticized vision of my childhood. My mother expressed this many times and found that the love and excitement of our family rendered her homesickness for Iowa a feeble thing in comparison. She has said and repeated that she and my father wished that time could just stand still, so that we could live always in the sweet embrace of those years together. My nostalgia for those years, as for the farm, reinforced my anxious sense of heaven passing away. My emotional life was spent in that shadow and in an unrecognized mild obsession, the focus of which is unclear to me. Growing to manhood would not be easy.

Highland Park High School 1950-53

My years at Highland Park High School restored my consciousness and are, unlike those of junior high, remembered well. They were certainly a success academically. I was the recognized leader in my mathematics classes, often called upon by teachers to clarify problems that arose in class. This occurred because frequently more than one approach exists to a mathematics problem, and although the teachers were very competent I was sometimes faster than they at seeing the connection between one idea and another and these teachers had no problem with recognizing that. I was therefore chosen to represent Highland Park in the Hockaday Mathematics Tournament, competing on May 10, 1952 in a one-day event against representatives of other Dallas high schools and of those in Denton, Ft. Worth and McKinney. I remember clearly that the problems were extremely hard for me, so that I thought for two uncertain days that I had not done well whereas in fact I had won. The decisive, hardest and final problem stays with me as if it were yesterday. A plane figure consists of three sides of a rectangle capped by a semicircle; the total perimeter is L; find the lengths of the two sides of the rectangle for which the enclosed area is a maximum. Instead of despairing on a hard problem, I had reasoned from scratch to seek a way to solve it and had concluded that the length of the base of the rectangle must be twice its height, which in turn must equal $L/(4+\pi)$. It was a problem in calculus, which we did not study in 1952 in high school. I came very close to reasoning it out[5]. Speaking of this fifty years later at the banquet for Distinguished Alumni of Highland Park High School, my words to them were:

> *What I learned from this day applies to all of life. We face many hard problems for which there are no answers, or at least no easy answers. What we must do, all that we can do, is call upon all of our knowledge and training in the attempt to arrive at the best solution we can find. This is the credo that I carried through my career after discovering it during this math tournament.*

Chemistry and Physics were two subjects of my final year. Physics was my favorite course. It was taught by the same Mr. Marshall that managed the University Park swimming pool in summers. His laconic style and Socratic questioning drew us all into attempted explanations. This suited my fundamental nature, the seeking of correct explanations. So I became the class leader by simply being true to my nature. I described Grandpa's milk-pail experiment to the class, and what I thought it meant. This reinforcing experience may be the main reason that I would later become a physics major in college. My near perfect grades in these math and science courses elevated me to rank third among my 92 classmates at graduation in January 1953. I was in passing elected an officer of the Highland Park chapter of the National Honor Society.

*Author demonstrating a chemical reaction
to Highland Park High School classmates*

My lack of worldliness, my anxiety over aspects of the world that I did not understand, caused me to decline an outstanding opportunity whose magnitude I did not understand. To broaden their student base Harvard University was offering full scholarships to outstanding students of more varied backgrounds than those of the typical Harvard profile. Their representative came to Highland Park High School for student interviews owing to the high reputation of our school district. I was one of four students selected for oral interviews based on their perusal of my record and on the recommendations from our academic counselors. The interview lasted about an hour, and the next day the counselor called me happily to her office to inform me that I would be offered a full four-year scholarship. Because I did not understand how exceptional the Harvard education might be in opening future doors, I was suspicious of the offer when I sat with my parents that evening to ask their opinion. My attitude did not show a desire to accept this offer. Neither of my parents had attended college, much less obtained a degree. How could they advise? My mother was anxious about the possibility of my moving so far away when I had never even been away from home for more than an overnight at a friend's house. My father did not understand why Harvard offered more opportunity than did any college. Were not they all pretty much the same? Their doubts reinforced my own. Amazingly in retrospect, my school counselor did not say clearly that Harvard was a greater opportunity than Southern Methodist University, our local college and destination for many of my class. So I declined this opportunity with hardly a second thought. I am most surprised that no one stated clearly what a grand chance this was. It is evident how immense was my gulf to understanding of the world, no matter how strong I seemed to others.

I engaged in normal activities of high school life. I was elected captain of our golf team and played number two in Highland Park matches against other Texas high schools. My biggest day was a narrow 2-up loss to superstar Hal McCommas, whose dad was the pro at Bob-O-Links golf course. In September 1952 I won the fourth flight of match-play competition for the championship of the Dallas Public Links Golf Association at Cedar Crest golf course. As I recall, fourth flight was determined by the medal-play scores averaging 79 reported by my golf coach. I also competed in several other tournaments. As captain I became one of the student speakers at the Spring 1952 Sports Banquet, and was there awarded my athletic **H** letter, which enrolled me in that special society of those that had contributed significantly to athletic programs and which

High school commencement photo (Jan. 1953), age 17

took the physical form of the letter H affixed to a sweater. I took not only pride in this but also relief, inasmuch as I had suffered from a self-imposed guilt, seeing myself as being outside the mainstream because I had not competed in football, basketball or baseball, all sports for which I carried significant reputation.

I volunteered to be stage manager and handled sound effects for our senior play, a mystery farce entitled *Sight Unseen*, staged in the week in January 1953 prior to our commence-ment. I even orchestrated one final grand prank when I secretly slipped a note into the day's announcements that were routinely read to all at the office microphone by an unwitting student reader and transmitted through speakers on the wall of each classroom: "School recesses today at 2:00 pm owing to a staff meeting." Throughout the school worried teachers wondered that they had not been informed of this meeting while students vented their usual glee at any reprieve from school hours. Some later suspected my involvement owing to my past reputation, but I simply shook my head along with the others. In the past I had always given myself away by laughing aloud, but with new maturity I remained somber. When suspicious friends continued to ask if it had been my doing I would reply, "Not that I can recall."

Of great weight among life's impulses, I fell in love with a beautiful loving classmate named Mary. We were inseparable for two years and discovered and yielded to sexuality. It was a grand and tender adventure, sweet as spring flowers, but one that exposed my anxiousness. As I slowly developed an even worse fear, the fear that I was behaving abnormally, a pressure mounted within me. I was afraid to describe it to anyone, even to her. Seeing no other escape from an inner accusation of inadequacy I made no attempt to suggest that we should go to college together in order to remain together. That ended it; but at least I could hope that it would all be better next time. I had met my enemy within me. I had exposed a very personal manifestation of inner anxiety that would torment me until I reached a different plateau of maturity.

CHAPTER 5.   SOUTHERN METHODIST UNIVERSITY

*At that moment of four in the afternoon, the
Pendulum was slowing at one end of its swing, then falling
back lazily toward the center, regaining speed along the
way, slashing confidently through the parallelogram of
forces that were its destiny.*

-Umberto Eco, *Foucault's Pendulum*[2],
Ballantine Books, 1988, p. 4

I entered Southern Methodist University in Dallas in January
1953. I had graduated from Highland Park High School in January
because I was still on the mid-year schedule that was initiated
in 1942 when I was moved ahead half a year at University Park
Elementary School. These three education institutions lie but a
long stone's throw from one another. Strong bonds existed with
SMU, which lies at the northern border of the towns of Highland
Park and University Park, where many SMU faculty members lived.
Large numbers of Highland Park graduates entered SMU almost
as an extension of the life of that northern Dallas incorporated city.
At SMU our most famous Highland Park footballer, Doak Walker,
had become a state and a national hero, filling the Cotton Bowl

on Saturday afternoons. The Cotton Bowl on the Fairgrounds of the State Fair of Texas was commonly called "the house that Doak built". I attended every Saturday home game in the Cotton Bowl during 1953-56.

It is almost impossible to overestimate the influence and impact that Doak Walker had on Highland Park, SMU and Dallas. What is true today was also true in the 1940s—society's heroes are for the most part military heroes, business heroes, war heroes, or sports heroes. Doak Walker was the preeminent sports hero. Even his name carries a heroic ring—*Doak*. In 1943-44, while I was in University Park Elementary School, Doak led Highland part High School to contend for the Texas state football championship, which was a far bigger event in the 1940s than today when professional athletes hold center stage. After serving in WWII Doak Walker led SMU to national prominence in football for the first time. Operating as tailback in coach Matty Bell's single-and-double-wing offense, Doak did it all. He was Time Magazine cover boy in 1948. He was three times an All American selection at SMU, winning tight football games repeatedly with inexhaustible heroics, and winning the Heismann Trophy as a junior, and later chosen all-Pro four times while he was scoring leader for professional football. A then current popular song may have been written with Doak in mind:

> *They always call him Mr. Touchdown,*
> *They always call him Mr. T,*
> *He could run and kick and throw,*
> *Give him the ball and just look at him go.*
> *Hip-hip-hooray for Mr. Touchdown,*
> *Hip-hip-hooray for Mr. T,*
> *Let's have a great big cheer*
> *For the hero of the year,*
> *Mr. Touchdown U.S.A.*

But Doak was also a typical Highland Park boy, loved and respected for his humility and spiritual character. Parents loved him. Friends worshipped him. Younger schoolboys such as myself held his name in awe. In addition to football heroics, he cemented the Highand Park-SMU axis as no one had before or since.

Highland Park High School and the entire school district of the Park Cities has become a Texas legend of its own. People compete strenuously to become residents, because only in that way can children attend its celebrated school district. There is an interesting point for today, while 21st century society debates our schools and their relationship to economic wealth and productivity. Namely, there is no strong reason other than its schools to explain why people pay far more for a home there than for comparable homes outside the Park Cities (University Park and Highland Park). Houses from 1940 are today purchased for the lot on which they sit; the house is sometimes razed and replaced by a theme mansion. In the year 2005 the *Dallas Morning News* reported on January 27, 2006 that the school district raised 2.5 million dollars of private support for its schools in the annual drive called "Mad for Plaid"--*plaid* in reference to the Scot Highlander school theme. This amount equals the money raised annually by many universities from their alumni and community. These drives are to counter so-called "Robin-Hood" laws that redistribute school-tax money based on property values to other less affluent districts. That practice may or may not be socially justifiable, but there is little question that it is education and community devotion to superior schools that makes Highland Park a destination for many seeking the best environment for their children.

All of this I inherited, a simple itinerant farm boy from Iowa whose family was absorbed into the margins of that school district in 1941. Spurred on by that advantage, I entered SMU in January 1953. Had my high school been ordinary, I might not have attended college at all. This realization, as with so many others of like kind, seems almost frightening as I look back today on a charmed life.

SMU was but forty-two years old in 1953. Its Georgian red-brick buildings sat on a campus that had the look of freshly cleared flat lawns at the broad top of a large hill. It had been founded in 1911 by the United Methodist Church, which established what is today one of that church's leading seminaries, Perkins School of Theology, on the southern edge of its campus. The founding establishment was cooperatively assisted by leaders of Dallas and Highland Park, whose citizens pledged $300,000 to secure its beginnings. In appreciation, SMU named its first building Dallas Hall. With its grand central dome and majestic columned entrance steps and lovely interior rotunda, Dallas Hall has been from the beginning the very symbol of SMU. This building is the first that one will see by checking their web site, www.smu.edu. The original building was constructed in 1915 and was designed by the architectural firm Shepley, Rutan & Coolidge. The building's architectural style was particularly influenced by Thomas Jefferson's interpretation of the Roman Pantheon. The 3-story rotunda and flanking wings are typical of Georgian Revival architecture. The rotunda is topped with brilliant stained glass, and the dome is copper with a rich green patina. In 1978 Dallas Hall received the first plaque in the Park Cities from the National Register of Historic Places. Like any other student in 1953, I took for granted having my English classes being held in Dallas Hall. There I was charmed, even shocked, by David Ruffin's chanting of Elizabethan ballads in class. It was all so grand and solid and new to a socially backward boy of eighteen years. Little did I dream when I entered SMU that I would one day be designated *Distinguished Alumnus*; nor could I possibly have foreseen the reasons for that.

Conscious of wanting to limit costs for my parents, I lived at home on Sherry Lane during my first two years of college. Although practical, this was in a sense unfortunate in that it probably delayed my growing up emotionally. I also declined, more out of social anxiety than lack of interest, to join a fraternity, although most of my friends did so. I continued to be a boy at home. I was able to use our Nash Rambler as my transportation from home, as it became, *de facto*, my car. I also spoke with Registrar and Director of Scholarships Leonard Nystrom about scholarships and received one as the third-ranked member of my Highland Park graduating class. Highland Park grads were welcomed in many ways. I was, at the same time, also offered the first of my two athletic scholarships, neither of which I accepted. Because I had been captain of the Highland Park golf team, I was offered a golf scholarship. I declined this on the grounds that I no longer wanted to spend long hours practicing at a time when I wanted to study; but in truth a darker reason entered. I was always made anxious by the expectation that I perform, and I knew my golf game was not good enough, and would require retooling of my swing, and would cause me to face losses. I did not want to face that anxiety over failure. This type of anxiety, which came in many forms, was a constant of my life. It reappeared six months later when I was offered a swimming scholarship by famed SMU swimming coach "Red" Barr. He had watched me swimming competitive laps in my required P.E. course during my first year. All those endless boyhood summers of playing racing tag in the University Park Pool showed in my instinctive ability to get from here to there fast. I had in those tag games seldom been "It", save when I wanted to be. But I declined this scholarship as well, citing the same reason and suppressing similar doubts.

Another aspect of this fear is difficult to disclose. With my Highland Park High School sweetheart I had already discovered that performance anxiety was going to be an issue in intimate matters also. As our sexual explorations had advanced I discovered that this anxiety affected my performance. But I did not consciously recognize the fear, preferring to think that sex is innately difficult. Of psychological issues I possessed no understanding at all. I chose to dismiss the awareness from my mind as too scary to contemplate. I acted brave but was not. Meanwhile I was admired by all. I was a golden boy from whom much was expected and who was frightened by expectations.

Because I entered SMU as an engineering major I was enrolled in freshman physics. It was my only class that I was sure that I really wanted to take, because it had been my favorite subject in high school. And it quickly became my favorite class at SMU. The teacher for that class, Frank C. McDonald, was chairman of SMU's small department of three PhD faculty members. His academic pedigree, PhD from the University of Chicago, was very good for a modest southern university. A tallish slender man with wire-rim glasses and a wry sense of humor, I found him very enjoyable and very no-nonsense in his presentation of classical mechanics, which is where physics always begins. Just as the human race had to digest Galileo and Newton, so it seems that each of us must individually make peace with their brilliant insights. It was those insights of Newton's to which most of the first semester is devoted in physics classes everywhere. It was the only class that I always looked forward to, which should have told me something.

Not everyone agreed with my appraisal of Dr. McDonald. Dee Norton, my high school friend with whom I had shared all twelve grades as classmates and who had been quarterback for the Highland Park football team, developed a growing dislike of Dr. McDonald. This conflict between students who wanted explanations of what they need to know and professors who want each student to work it out for himself is one that I would later see repeated over and over during my own academic career. Dee could not digest the ideas on first hearing and found Dr. McDonald's explanations therefore to be weak. Dee was a pre-med major and wanted to understand what he had to know to get a good grade. Dr. McDonald, on the other hand, typified physics professors in wanting students to mull over the concepts and arrive at deeper understanding. Physics is not a subject that one can "just learn", or memorize. One must work it over, questioningly, in the mind. When Dee sought Dr. McDonald in his office for questions, Dr. McDonald would eventually suggest to Dee that he had to study harder, had to think about the concepts, think about the problems, whereas Dee felt that hints on those topics were the very reason he was in Dr. McDonald's office. It is a common impasse. Both sides are justifiable. By contrast, my experience with Dr. McDonald was very different. I was sympathetic with his approach to the subject matter. I responded to the go-think-it-through approach. So Dee got a mediocre grade, bad news for pre-med majors inasmuch as medical schools always tended to look at grade-point averages. With his considerable character, Dee surmounted this setback, was accepted to a top medical school and had a fine career in medicine. But four decades later, at the fortieth reunion of our Highland Park class, I still heard Dee muttering to others what a jerk Dr. McDonald had been. It had obviously hurt. This lesson would remain true during my own four decades as a physics professor: namely, that physics majors with aptitude for physics always loved my classes whereas those students who are required to take and pass physics, perhaps engineers or pre-meds, often gave me, in their frustration, scathing evaluations. Although there is much more that could be said about this dichotomy, this is not the place.

A chance to become a physicist

I was elated to see that I scored 100 on the first physics test. I noted Dr. McDonald's careful eyeing of me as he passed back those tests. On the second physics test I also made a 100. The following week I received a request from Dr. McDonald to see him in his office. This worried me a great deal, because as an eighteen-year old, my past experience had been that such requests were never good news. Of course I had to go. We talked briefly about the course and then he said "Don, I have never had anyone get 100 on my tests. There is always some mistake. Scoring 100 on both tests is so surprising to me that I would like to ask how you did this." I stammered that I liked the concepts, and that physics had been my favorite course in high school, and did he know Mr. Marshall (he did)." To the last question he said that he had telephoned Mr. Marshall to discuss me. Wow! I feared that he might suspect me of cheating. But then he said, "Don, I think you may have a gift for physics. I would like to help you become a physics major if you want to do that." I was stunned and flattered by this, but I explained that I was in engineering because of its co-op program that would allow me to work some semesters to earn a salary with a sponsoring engineering firms. He replied that he had been thinking about that, and that he was willing to give me a job as a teaching assistant in the physics laboratory. If I would instruct and grade those labs, he could get me a tuition waver and pay me for my work. "You have a gift for physics," he repeated, "and you should follow that gift if you want to."

I was genuinely excited by that offer. I sensed that rare feeling that comes when someone we respect owing to his position speaks to us in a way that recognizes our abilities. I said that I would like to think it over and discuss it with my parents. He said there was no rush, and we parted. I realized at once that I had been stalling, that my parents would give no advice, and they did not. But we did discuss it and my father exclaimed that physics had been his favorite high-school course in Fontanelle, Iowa. But they said the decision should be what I wanted to do. So the next day I told Dr. McDonald that I wanted to accept his offer, and he replied that he would set it up for the fall semester. I remained excited. I was now majoring in my favorite subject!

Fondren Hall of Science was one of the newest and largest of the Georgian-architecture buildings. Above its central portico a massive bell tower rose, gold domed, from the center of the roof. The physics lab had a number of interesting experiments for teaching physics (as in most universities, primarily to engineering students). My favorite was the instrument for measuring the acceleration of gravity, and for confirming kinematically the relationships between time and distance traveled that were required by the constant acceleration. A weight falling between two electric wires was electronically wired to trigger a spark each 1/60th of a second (from the standard alternating current power). Each spark burned a hole in a vertical paper strip, so that the spark holes in that paper strip located the falling weight every 60th of a second. Each student got such a strip as his data, and from it he was to measure the acceleration of gravity by the increasing distance between spark holes, distances taken to measure the velocity of fall at successive moments. It was fun to do, and the students enjoyed confirming the inferences that could be made from the classroom formulae. Another favorite was a simple pendulum, a mass hanging from a string whose length could be varied to confirm that the period was proportional to the square root of the length. While explaining and repeating this experiment and the many others, I discovered that my own understanding was deepened by the repetitive experiments, improving my explanations to the assembled students. They regarded me as some young Einstein, whereas I was in reality but a lad who liked physics and who was simply understanding it better each time through. But I was also learning something important that I would use throughout my later career. Namely, one does not "learn" a subject; one is introduced to a subject. And the learning of it comes from revisiting it, the more times the better.

Chess and Contract Bridge

I spent a lot of spare time in the Student Center. That building is now long gone. It was an old wooden building that looked more like the reception lounge in a modest but sprawling western lodge. Basically one story, with a large room punctuated by sofas and chairs and tables distributed about. It sat on pier and beam and had wooden floors. It was there that I continued my chess playing, as there was always a game or two underway. I learned of the Chess Club, joined it, and soon became Chess Champion of SMU. I had developed an instinct for positional moves by playing so many games with my father during Junior High and High School. But I quickly dropped the game after that, and played only on rare occasions during the next forty years. A photograph of me playing chess appears on p. 154 of the 1956 *Rotunda* yearbook. In this rustic old student center I also learned the game of Contract Bridge. Tables of bridge ran constantly and I could always watch or join in as a replacement for someone leaving. But two years later this charming but impractical building was gone, replaced by the new Umphrey Lee Student Center, about a block west, near the entrance to the campus. Umphrey Lee was Chancellor of the university. The President, Willis Tate, always seemed familiar to me as the father of my Highland Park classmate, Willis Tate. When the new Student Center was completed, I often spent time there among the bridge tables. I liked bridge so much that it became a lifetime hobby.

I discovered much later that my love of both Chess and Bridge was more than a love of goofing off. These two games have always appealed to people inclined toward physics and mathematics. Both games possess seemingly infinite combinations and variations. In chess I believe that the intense positional play and countless evolutions forward appeals to the conceptualist. We see the positions in our minds. Most good players begin reasonably well playing without the board and pieces, picturing it in their minds. After about twenty moves I would become lost. Bridge, on the other hand, fascinates as a study in probabilities. The correct bid or play of a card is the one offering the best odds for the good result. Good players learn the odds. The appeal of each game to me continues until today, although I ceased playing them for reasons that will be apparent later.

One of the bridge players knew of the Dallas Duplicate Bridge Club, which met in the downtown YMCA near Ross Avenue. I can not remember my first visits in 1955, but I remember vividly a year of exciting play there. In duplicate all pairs play the same hands in the competition to see which among them gets best results on those hands (usually 24, 26 or 27 deals). The deals are placed back in the duplicate board after each play, the result of each play recorded, and passed on to the next pair for their attempt at the same hands. What this appealed to was its fairness, inasmuch as all pairs had the same good or bad cards and good or bad luck in the distribution of opponents' cards during the play. I liked that. Luck was filtered out. Only the smartest did considerably better than the average, and that was our goal. Ralph Randall ran an exciting game at the downtown YMCA. Dallas was a Mecca of duplicate bridge in the 1950s, so that I became acquainted with some of the great names in bridge, although at the time I was unaware of their greatness. John Fisher[1] ("Dr. John", as we called him) ruled the club as its best regular participant, and later went on to national fame. Dr. John took an interest in me and often gave me tips after the hand was played. As a result, I frequently asked him how to analyze a bridge hand. I also met Sidney Lazard and Oswald Jacoby and Oswald's son, Jim. Then an amazing thing occurred. On two occasions, I was invited by Oswald Jacoby to play as his partner at his home in a money match against a gambling pair who wanted to challenge the great Ozzie. Oswald Jacoby did love to gamble, and in several different games. But it would be only later, when I returned to the world of bridge following my academic career that I became aware of the singular fame[1] of Oswald Jacoby. At the time he was just a gambling bridge player to me, one that others held in awe; but I was too naïve to know.

This particular episode illustrates an aspect of luck that occurred to me very frequently during the subsequent decades. Namely, I was placed by good fortune in close contact with some of the greatest and most celebrated stars of the world of physics. All too often I was only barely aware of my good fortune, comprehending only later that I had just had a unique opportunity. I give myself little credit for meeting great men; rather it was good luck that came too often to be accidental and therefore is attributed by me to my circumstances placing me in the right place at the right time, for whatever set of underlying reasons.

During this time, one of my bridge partners at SMU, Johnny Torbett, and I drove together to Midland Texas to play in our first Sectional Tournament. Johnny played lead trumpet in a famed SMU jazz band. I can not even remember how well we did in that bridge tournament. And my masterpoints from that era, which were then handed out on small slips of paper that had to be mailed to the American Contract Bridge League for recording, were never mailed and were eventually lost. But for the next four decades I thought often of bridge and grappled with the daily newspaper column, even though I had to give up playing, cold turkey, for graduate school and a career.

Early Marriage

I had begun dating a new girl that I had barely known from Highland Park High School. Our relationship rapidly became physical, perhaps impelled by my inward pressures, and just sixteen months into my SMU career, Mary Lou Keesee and I were married. We were nineteen years old. Our parents did not question the wisdom of marrying so early, for in fact they had done the same thing. At the time of their marriage my own parents had been just recently graduated from Fontanelle High School in Adair County, Iowa. They had made a lifetime commitment of happiness together, and so, I assumed, would we. In December 1954 we were married in a well attended ceremony in Park Cities Baptist Church near Preston Road and the Northwest Highway. Just a few blocks from our home on Sherry Lane, this was the church of Mary Lou's family, the church in whose choir she sang. My father loaned me our family's 1952 Hudson Hornet, which we drove across south Texas on our honeymoon trip to Monterrey, Mexico, where we had a nice hotel room in a somewhat grand old hotel in central

Monterrey. This was a great deal of fun, very adventurous for 1954. But our sex life resembled blind leading blind, both of us tentative and fearful, and me suffering performance anxieties and lodging them in my own private fear center within my brain, not to be discussed and improved and never to be mentioned but kept a secret, even between us. We would keep those secrets for ten years.

My wedding, December 1954, at age 19 1/2.
My brother Keith (left) was best man.

With help from both sets of parents, we rented for the next two years a cute small apartment at 3109 Fondren, just one-half block east of Fondren Library at SMU. It was a one-room efficiency, with bath and kitchen, and a Murphy bed that folded down each night from a shallow closet in which it was contained, with built in supports for its raising and lowering. I added a second academic major, mathematics, to my bachelor degree plans. I also really liked mathematics.

The department was chaired by a scholarly Edwin Mouzon and included Ray Hassell, the father of Joan Hassell, one of my

Highland Park classmates. Dr. Mouzon taught courses taken by me over the next two years. This mathematics was also not advanced as those things go. I studied courses in differential and integral calculus, matrix algebra, differential equations, and introduction to complex variables. But these were exactly what I needed to slowly work up some sophistication in mathematics, invaluable for my future in physics. By accelerating our degree plans, I graduated in 3 1/2 total years and Mary Lou in three. My transcript contained all grades of A save one, and I was elected to Phi Beta Kappa in my junior year. My Phi Beta Kappa photograph can be seen on p. 86 of the 1956 SMU *Rotunda* yearbook. I was a faculty favorite, with much admiration, but with an undisclosed sickness gnawing at my heart, and I could not admit why.

Physics at SMU

The SMU physics department had but three tenured faculty members. There was little research to speak of. Frank McDonald had much earlier given up his research in optics, which was already out of fashion in physics. But we had some excellent optics experiments in the junior laboratory: the tandem lenses of microscopes and telescopes; Newton's fringes of interference; the Michelson interferometer which had been used by Michelson and Morley to demonstrate that light from a moving source moved no faster than light from a stationary source, unlike bullets on trains; the diffraction grating to measure the wavelength of various colors of light. Wayne Rudmose had constructed a very nice anechoic chamber for his research in acoustics, which, like optics, is also out of fashion as a fundamental study although always confronted with interesting technical problems. I was stunned at the quietness within that chamber, produced through effective sound absorption by its baffles and absorbing walls. That stillness transported me for a moment to a clear and calm winter day on the Iowa farm, fresh powder snow lying heavy all around, no wind, no birds singing in the freezing air, just overwhelming quiet. I took most of my courses under Robert B. Lindsay, the only Assistant Professor. He was the son of a well known R. Bruce

Lindsay of Brown University, who wrote a textbook on classical mechanics. In his small classes (typically 3-5 students), I studied mechanics, electricity and magnetism, and thermodynamics. In mechanics we studied the motion of a pendulum on the rotating earth, the pendulum[2] of Jean Bernard Leon Foucault (1819-68) that demonstrated the rotation of the earth. I was thrilled that something so huge, so slow and so ponderous as the daily rotation of our earth could be demonstrated by a simple mechanical experiment. Foucault's pendulum remained forever in my thoughts, my prototype of the idea that an experiment on earth could expose cosmic events[2]. That theme gripped me throughout my career. This SMU faculty of physics was, in fairness, an old-fashioned and unambitious one, but I could not discern its lack of modernity because I had as yet no comparison of myself with physics majors at Harvard, or Princeton, or Chicago. These SMU courses, whatever their shortcomings, allowed me time to mature and learn well the elements of classical physics. I was a cautious advancer. I think, for me, SMU was just about right.

The decisive nudge toward life's plan

I was jolted from my small-world contentment in the fall of my senior year. Again it was by Frank McDonald. Called into his office, I presumed we were to discuss the teaching of the labs. At four in the afternoon he asked me to take a seat in his office, facing him and the window looking southerly outward toward the quadrangle beyond Dallas Hall. He asked if I had given any thought to graduate school, offering that he thought that I should do so if I enjoyed physics as much as it appeared, and that I had an evident gift for physics. I had not thought a single thought about graduate school. He explained that if I would like to conduct research in physics for a career I must expect to be able to do that only from the vantage point of a PhD. I had never considered that either, and if I had I would not have seen its truth. He said that he would be pleased to write letters of recommendation for me. He opined that Caltech (California Institute of Technology) seemed to be the best place just now. "They have Feynman, and Millikan. Millikan was at Chicago when I first went there but moved to Caltech for the latter half of his career. He died just two years ago."

Dr. McDonald had mentioned Millikan pointedly to me because he knew that I had so enjoyed performing Millikan's oildrop experiment to measure the charge on the electron, repeating in our junior lab Millikan's 1923 Nobel-Prize-winning experiment. In that small darkened physics lab room I peered through the microscope and found the tiny oil droplets between two voltage-fed parallel plates that caused a vertical electric field. Each drop would fall or rise depending on which force was the greater, its weight owing to gravity or its lift owing to the electric field acting on whatever net charge resided on the oil drop. By adjusting the voltage, I could suspend a specific liquid drop in space, balancing weight and electric force. The balancing voltage yielded the total charge on the oil drop. Suddenly, however, the balanced oil drop would dart off upward or downward because the violent events of the atomic world had led to a sudden change in the charge contained in it, so that the forces no longer balanced. The experiment measured *the change* of the total charge on the droplet, and I was amazed to see the celebrated pattern emerge. The change in charge was in every case expressible as an integer number of constant charges-- say, 3e, or 5e, or -2e, etc. The change was never 3/2 e, or -3.6e, or some other noninteger multiple. By carefully examining a set of such charge changes, I confirmed, as Millikan had shown thirty-two years earlier, that the oil droplets gained or lost identical amounts of charge, called *e*, known as the charge of the electron. But the electrons themselves were indivisible—no fractions of an electron. Charge appeared to be *quantized*, appearing only as whole numbers of electrons, each having the same charge *e*. My first real insight into the quantum nature of our world, into its divisibility only by integers, or fundamental units of a quantity, sank in on me as I repeatedly recalled this experience. That integer structure applies not only to electric charge, but also to momentum, to energy, to angular momentum of particles.

It would be four decades later when I would realize that the very existence of life depends on the charge *e* of the electron having just that tiny value that Millikan had demonstrated. If that tiny number were only twice as large, chemical energies would be increased sixteen fold, and the chemical reactions necessary for life would be impossible to surmount. If, on the other hand, the value of *e* were only half so great, life forms would be violently disrupted by even the gentlest sunlight, similarly to ultraviolet being lethal to bacteria. Forty years down the road I would be perplexed that the value of *e* appearing in the laws of physics was exactly the value needed for life to appear. No explanation of this coincidence exists even today.

On that afternoon in Dr. McDonald's office I experienced intense *déjà vu*. We had sat exactly so three years earlier when he had offered me the chance to major in physics. I felt similar excitement, and again said that I would talk it over with Mary Lou and my parents. But the excitement again told me that I wanted to do this. I said so. On that spot, at that moment, I realized that Frank McDonald was my *angel*, that person known to each inner heart who makes the rest of our lives possible. What he was now suggesting would never have happened were it not for his personal concern for me. He warned that there is one thing I must do immediately; namely, take the Graduate Record Exam, coming soon. I would have to do well on it to be admitted to Caltech. Mary Lou definitely wanted to go. Pasadena! Los Angeles! The Big Time! She was a *bon vivant* where seeing the world was concerned. My mother and father expressed no opinion, confessing that they could not understand the role of a PhD in a job in science. But they were supportive, and they were almost as excited as I was.

I could never blame my parents for not knowing what a PhD in physics was for. They had never attended a university, and had been unclear about what universities were for. I was not able to explain it to them. I had learned just enough physics to know that I was in the dark. Certainly there would be more courses to study, and at a much more advanced level. But that could not be the main point. I could do that myself, from books. And surely I will qualify for higher paying jobs, but that seemed to me to not be the point either. I thought I would get the chance to perform research into aspects of physics that are not yet known, but how could I say when my SMU degree had not really introduced me into what had been learned in the preceding decade. The physics I understood so well was all *old physics*, and modern physics was just something I would have to learn from its start. I fumbled trying to express this to them, not quite having the nerve to say that I did not know--that I was moving to California for a long and demanding period of study whose goals I could not clearly describe! My father had known what he wanted to do—fly an airplane—and why he had uprooted us from Iowa to Texas—for a more secure job flying an airplane. So he expected the same clarity from me. Nor could I explain to Mary Lou; but she seemed not to need any explanation. She *wanted* to move to California. And anyhow, how could she believe that I truly knew what I wanted to do when I was unsure in our intimacy. I knew no more about sex than I did about modern physics. Going forward on instinct seemed right to her.

And so the application to Caltech was mailed; Frank McDonald's letter was mailed; and I sat through a puzzling three-hour graduate record exam in physics, which was held on the SMU campus. The exam was puzzling because I knew exactly how to answer three-fourths of the questions and had little idea about the others. I kept digging back to our single semester on modern physics (not nearly enough), because I did recognize that these puzzling questions were about quantum mechanics. Not having studied quantum mechanics at all, these led me to conclude that I had failed. But, seemingly miraculously, I had not failed. Far from it, I scored in the 83% bracket nationally for physics majors, and I had done as I thought; namely, I had answered correctly all questions that I understood. And I had inferred the correct answer on about one-third of the questions that stumped me.

The news arrived in March that I was accepted by Caltech. I had hardly dared to hope for this, because I heard that they had many applicants from the nation's most distinguished universities. In almost the same mail came more good news, that my application had been selected by the National Science Foundation for a two-year Predoctoral Fellowship. What followed was a time full of elation and anticipation. It was a time of congratulations on the SMU campus, which had never provided a Caltech graduate student before. We had just four months to finish classes, go to

Commencement, and gather things for travel to California. At Commencement I grasped my papers tightly, sensing that I was the first in our entire extended family to have attended the university. My life seemed to have taken on the aura of the miraculous. It was an exciting four months, and for the duration the anxiety about my intimacy with Mary Lou was overlooked. I graduated *Summa cum Laude*, I think third in the class, dragged down by my single B in Engineering Drawing. Mary Lou also graduated with honors and a schoolteacher's certificate and a plan to teach in the Pasadena School District. They had written to confirm that she was hired to teach English Literature in John Muir High School. She was excited. Fortunately we did not take in all we had to do at one glance. Like all young people, we got things done with dispatch and were musing about how we would get to California with the many belongings we hoped to have with us when my father arranged to meet with us. He would not say why. He showed up with a light green, two-year old 1954 Pontiac Firechief four-door sedan. Its great roadability, mass and V-8 power made it ideal for pulling a small rented trailer. And gasoline was very very cheap. It was our parting gift from parents as dear as one could ever know.

In August 1956 Mary Lou and I departed Dallas for California. It occurred to me once more that my father had done the same thing when he left the Iowa farms for Dallas. It also occurred to me that my goal was as purposeful as his, even if I was not able say clearly what I was after. Ours was an exciting drive, knowing that our goal was so special and seeing lands that we had never seen--New Mexico, Arizona, and the southern California deserts. Our Pontiac Firechief carried us in fine style along old US Highway 66, a fabled route. The 3 1/2 years at SMU faded further into the distance with each passing hour. After many days, we drove into Pasadena, California, a new goal and a new home.

CHAPTER 6.   APPRENTICESHIP AT CALTECH

The Pontiac *Firechief* burned leaded gasoline, best for its large V8 engine and compression ratio. One did not want engine knock. Along US 66 we had purchased gasoline with cash from Mom-and-Pop owned roadside stations. Today's large corporate gasoline emporiums did not exist, nor did credit cards. The August, 1956 deserts were very hot in the afternoons. My father had obtained a water-evaporator that we had installed in the right front window, where it took in hot dry air and blew cooler wet air into the car. The 540 calories of heat absorbed from the air per gram of water evaporated provided the cooling. I explained this physics to Mary Lou. The motels were also Mom-and-Pop relics of early auto travel, some of which can be found today on the old highways huddled anachronistically among today's Holiday Inn, Best Western, and others. These motels came in fantastic shapes—teepees, haciendas, log cabins, horse-stable designs, and Cape Cod beach-rental architecture. It was cash-as-you-go, $5-10 per night, with a greasy grill next door for breakfast or for coffee and a cigarette. When the Interstate Highways came along, the economic fallout closed down most of these old motels, and indeed even some of the small towns that contained them and that had once relied on travelers.

The night sky amazed us. As the deserts cooled they became

almost chilly, and the stars shone in the dark sky with a brilliance that few still see. We gazed at the panorama nightly, the best thing to do after driving all day. Mary Lou asked me what the stars are. I did not know, except that they were supposed to be like the sun but further away. That sufficed for Mary Lou. But their brilliance made one wonder. On August 11 we witnessed the Perseid shower, falling stars; but I said that I did not believe they really were falling stars. My naiveté about the known world was a character trait that, in retrospect, I see characterized my life. I accepted every fairy tale from my childhood until I saw the evidence that they were simply fairy tales. Santa Claus was a real person until I was six years old. I was grateful to Noah for herding every type of animal on earth into his ark to save them during the flood. I had been slow to learn how children were conceived, and reluctant to believe the truth when told by my farming cousin. My sexual naiveté was a burden on my young marriage, but it was too scary to recognize that fact. I had lived with the naïve views of my parents, of marriage, of knowledge, of the world, and of life and death. For better or worse I had always been happy in my naiveté, until this ripe moment when I embarked on the adventure of seeking truth through physics. If naiveté was my weakness, it was also my defense against a daunting and occasionally frightening world. Even today I feel a rush of conflicted feelings over this. On the one hand it is embarrassing to have been naïve over many things. But on balance I am thankful. Many today learn too much too soon. For me it was probably best to have been a bit of a corncob.

Following the road maps to Pasadena, US 66 brought us to our last road motel in San Bernadino. Next morning Mary Lou and I were both as excited as two 21-year olds could be to drive our new Pontiac westward into Pasadena. Cross-country travel was less common then, and we were adventurous spirits. Our rented enclosed trailer contained my books, my wooden desk and coffee table, both made in my father's woodworking shop, our record player (33 rpm), and our clothes. Our early view of the San Gabriel Mountains just to our right slowly disappeared in a gray haze that settled over the area. It was our first view of smog, which was much worse in August 1956 than it is today. By noon our eyes burned slightly with its irritation. We found our way to Colorado Boulevard, the main east-west boulevard through downtown Pasadena, and found a motel. There we were able to unhook the trailer and head back out to view Pasadena. Driving west half a mile we saw an attractive old wooden-framed and shiplap-sided diner at the corner of Colorado and Lake Avenue, and ordered hamburgers from their grill and coffee served up at our booth by busy waitresses. I became attached to this diner for my coffee breaks, as it was but a few blocks from the Caltech campus. It would survive about ten years before its destruction on this prime intersection.

After lunch we parked by Caltech to stroll the campus and allow me to report to Throop Hall. There we went to the Off-Campus Housing Office and obtained a mimeographed list of possible rentals. It was, they assured me, never up-to-date, because in the pre-xerox, pre-computer age, the list was typed on a stencil and reproduced. They let me borrow briefly their list of locations that they were sure were no longer available, so that we could cross them off the mimeographed list. They gave me a sheet showing the average monthly rental rates for furnished houses or apartments in the Caltech area: 1 Bdrm--$50-100; 2 Bdrm--$90-150; 3 Bdrm--$150-250. Since I was to be a graduate student in physics, they gave me a sheet of instruction for new physics students. Its first items were mandatory written placement exams to be taken in the following week. This was solely to determine how ready each student was, so that they could be placed in the appropriate course. How daunting, I thought, to be examined even before attending class; but it was a sound idea. Leaving the campus we decided to seek John Muir High School, where Mary Lou would soon begin teaching. We found it in northwest Pasadena, on Lincoln Avenue almost into Altadena. This was a fine looking school, in what was then a strong community. Later, after the Foothills Freeway divided Pasadena, I believe the student body became largely black. For Mary Lou, teaching English literature there was a good experience, and her salary plus my NSF Fellowship rendered us well off among graduate students. In retrospect I wondered what her students thought of a cheery bright-eyed twenty year old teaching them English literature.

For two days we chased down rentals from the Caltech list; but we soon saw a rental sign at a very nice apartment building at 397 S. Sierra Madre Boulevard, just north of San Pasqual. That would be our home for our first year, and for the second year after summer back home in Dallas we would reside just across the boulevard at 518 S. Sierra Madre. These were to be reasonably happy years, if only because we were so busy and because our lives seemed so full of purpose. We were in every sense very lucky. The tall palm trees that lined the streets contributed a very California look, as did the Spanish architecture that is so common. One thing took our breath away; namely, the many orange trees, mostly residuals from the famous Pasadena orange groves, just eastward from Caltech. Here was a grove fragment filling a lot hosting no home, and there were a few remaining orange trees in front of one of the many marvelous 1920s bungalows for which Pasadena is famous. The Civic Center was a Spanish architectural jewel. But signs of change were also evident. Some areas of town were beginning to look financially neglected. We were witness in 1956 to a Pasadena evolving by the relentless pressure of Los Angeles from its history as a gracious independent city, though it retained its reputation as home to the wealthy.

Beginning Graduate Student

I was amazed at the number of beginning graduate students taking the placement exams. There must have been twenty-five of us in the room. The first exam covered classical mechanics. This was my strength. I passed easily and was placed in the graduate course, using Herbert Goldstein's classic text. Next was the electricity and magnetism exam. On this it was judged that I had passed; but in truth they gave me the benefit of the doubt, not wanting to hold me back. Most of the short problems were very basic ones, involving knowing things like the boundary conditions at the surface of a metallic conductor—the direction of the electric field at its surface, the potential inside, the charge density on the surface, and such ideas. Although I was proud to have passed, I faced what would become a personal nightmare in being assigned to William R. Smythe's graduate course and classic book, *Static and Dynamic Electricity*. It was very hard for me because of techniques and geometries that I had no introduction to. The third placement exam confirmed what I knew from the Graduate Record Exam; namely, that I was not well prepared in modern physics and in an introduction to quantum mechanics. My lack of exposure to quantum mechanics material was strange considering that the emergence of quantum mechanics in the 1920s was the greatest intellectual revolution of the twentieth century; but we had not even scratched its surface at SMU. Despite being the most profound change of philosophical thought about the physical world since Newton, I had barely heard of it. My experience was not uncommon, because in the 1950s the teaching of physics in average universities favored the classical developments of Newtonian mechanics that were admittedly beautiful and that were comfortable to most university professors who had themselves emphasized classical physics in their own educations.

So I was grateful to be placed in Tom Lauritsen's course for Caltech physics seniors, *Introduction to Modern Physics*. Even the term *modern physics* admitted the tenor of the times, for it referred to the rich world of atomic phenomena that depended on quantum mechanics for its understanding. This course was the overview of modern physics that a greenhorn in quantum mechanics really needed. My lack of knowledge is illustrated by this somewhat embarrassing story. When I had discussed my placement test results with physics faculty advisors for new graduate students, they thought that I had done quite well considering that my university background did not equal that of most of the other students. It was confided to me that my gaudy record at SMU had won my entrance and that my 83rd percentile on the Graduate Record Exam had also been seen as a testament to my good ability. SMU's rather average reputation as a physics school led them to expect my GRE score to be at least 10% smaller than graduates of Princeton, Harvard, Chicago, Cornell, Berkeley and the like. I had wrongly assumed that they had taken me despite my 83rd percentile, presuming that others would have done much better, as some did, whereas in fact they had viewed 83% in a positive light. For final course placement I met with the head physics advisor, Carl Anderson. He seemed a bushy-eyebrowed teddy-bear of a man, slow talking, gentle and friendly, relaxed with his cigarette. During our conversation about my placement exams I was not aware that he had won a Nobel Prize for the discovery of the positron, the positive antiparticle to the electron. My ignorance embarrassed me when I found out the truth from other graduate students the next day. Surreptitiously I looked up "positron" in my new *Introduction to Modern Physics* textbook; and there he was! Anderson had been but twenty-seven years old when he performed the discovery experiment in 1932, and his 1936 Nobel Prize made him the youngest Nobel prizewinner in history. Right away I had learned that things were going to be different here.

Glimpsing Science through its Faculty

Tom Lauritsen created a vivid impression during my first-year course in modern physics. He was the son of C. C. Lauritsen, founder of Caltech's Kellogg Radiation Laboratory and at that time professor emeritus, but who still came to his office daily. *Tommy*, as he urged us to address him, taught modern physics to Caltech seniors and to inexperienced grad students like myself, using *Introduction to Modern Physics* by Richtmeyer, Kennard and Lauritsen, which was Tommy's update of an earlier book by Richtmeyer and Kennard, which had in turn been Kennard's update of the earlier Richtmeyer book. This classic textbook had lived many incarnations. This was a very good course for me, helping me learn what contemporary physics was about. Tommy explained the miracles of the atom with a jaunty classroom manner that seemed to always share our own amazement at the quantum world. He threw in colorful remarks that made sure we had gotten the point. I learned for the first time that the electrons in the atoms did not have precisely located orbits like those of the planets around the sun, but that they moved with only precisely calculable probabilities of being found in this place or that, moving this way or that. I learned that light existed in indivisible bundles of energy whose magnitude depended on frequency, and that the light from a hot body could not initially be understood without those quanta. I puzzled a lot over such ideas and loved every minute of it.

After class, Tommy quickly became enveloped in a cloud of smoke, drawing fiercely on his pipe as we followed him back to Kellogg Lab, pressing questions along the way. Tommy answered our questions humorously but precisely and imaginatively. Among ourselves we joked about him as the guy from the Peanuts cartoon strip that went around with a cloud over his head--except that Tommy's smoke cloud was one of fun rather than of gloom. Each of us liked him a lot. Life-directing remarks fall on many an eager ear at such times. On one of these smoky strolls Tommy revealed that a paper on the origin of the elements in stars was being written in Kellogg Laboratory. I fastened onto his clarification that this was being written not be him but by W. A. Fowler and Fred Hoyle and two Research Associates from England. Walking with him into Kellogg Laboratory one day, he showed me the small windowless room in which Geoffrey and Margaret Burbidge had been writing a draft of their work with Fowler and Hoyle a few months earlier. Could it really be true that our atoms were forged within the interiors of stars? It was the first occasion on which I heard of Fowler and Hoyle, the two scientists that would inspire my scientific life, and of Geoffrey and Margaret Burbidge.

Outside Tommy's office in the upper ground floor of Kellogg Radiation Lab, just next to his father's office, I would pause to look at the 2ft x 3ft master drawings of the energy levels of each light nucleus, the locations in energy, as steps on a ladder, of the nuclear states having more internal energy than the ground state. Chatting about these I heard that these drawings were altered often as Kellogg faculty took into account new nuclear data on their energy levels. It had to sink in on me that there were no books in which to look up this current information, that the series of drawings lining the walls was a living and evolving display of a frontier of human knowledge. It came to me that I could aspire to causing a drawing to be modified by my own research. And so it would turn out to be. Visiting in his office, his pipe now exhausted, Tommy would open his box of cigars from his top desk drawer, Roi Tans, not Cuban, light up and offer one to me. Tommy sure did smoke! At that time I saw nothing not to admire in his cloud of smoke.

Tommy Lauritsen introduced me to his dad, Charles Christian Lauritsen, in the office next door. We learned to call him "Charlie", although that took getting used to for one so young meeting the founder of the Kellogg Radiation Lab. C. C. had emigrated from Denmark in 1917 and he and Tommy maintained strong ties to nuclear physics at the Neils Bohr Institute in Copenhagen. Because of this I would have the chance to hear one of the famous Neils Bohr's last physics lectures in just two years time; and his son Aage Bohr, also a famed theoretical nuclear physicist, was a not infrequent visitor to the Kellogg Lab. Both father and son Bohr visited Caltech because they knew it to have the same questioning attitude abut nuclear physics and quantum mechanics that motivated them. The intensity, the fame, and the dedication of all of these men impressed me deeply.

Tommy Lauritsen alone would have demonstrated the difference between a Caltech professor and an SMU professor. But the others at Caltech reinforced that same mixture of passion and achievement, so that we students moved as if among a priesthood of brilliant scientists in search of truth and, incidentally, teaching us physics. Suddenly I saw that the study of physics at this level was not the learning of physics but rather the pursuit of new knowledge. No one had explained that idea explicitly to me, perhaps assuming that everyone would know the basic thrust of Ph. D. work. My Classical Mechanics class was another example, taught by Howard Percy Robertson, who had applied Einstein's general theory of relativity to an expanding universe to discover an appropriate geometry for the space-time of such a universe. It is today universally referred to as the Robertson-Walker metric. I was not sure then what that meant, but I was sure that what I was seeing in Caltech professors was the reason that I had come.

The other dynamo of Kellogg Radiation Lab was Professor W.A. (Willy) Fowler, who had been away on sabbatical leave in Cambridge, England in 1955 and who was little in evidence during my first year there. It was he that Tommy had revealed to be working furiously with Fred Hoyle, who I learned was a famous English cosmologist rather than another Caltech Professor, on a research paper systematizing the nuclear reasons for the atoms of our world to have the abundances they are found to have. Following the lead from Hoyle's pioneering 1946 and 1954 papers, they, with Geoffrey and Margaret Burbidge, two transplanted Englishmen that Fowler had begun work with in Cambridge during his sabbatical, were assessing nuclear physics processes for the chemical elements to have been created within stars. They published a short sneak preview of their paper in *Science* in 1956, during my first year at Caltech while they were busy writing their full paper. The thought that the chemical elements of our world may have had a natural origin siezed me. In my science education concerning the chemical elements the question why they existed had never arisen. I had grown up in a deeply religious family that took it for granted that God had created everything just as it is. My father actively ridiculed infidels who maintained that man descended from monkeys.

Created in the stars! My thoughts drifted to the brilliant field of stars that Mary Lou and I had gazed upon nightly during the drive west. Tom Lauritsen's personality and this new idea attracted me enough that I resolved to take the graduate nuclear physics course from Fowler in 1957, when it would be next offered. And although we were encouraged to begin research with one of the professors during our first summer break from classes, I decided to stall by accepting a student research appointment for the summer at Mobil Research labs in Dallas. I would wait for a chance to talk with Fowler, probably during his course, to see if I could enter this research. Those fortunate choices launched my career.

Bestriding Caltech like two 1950s colossi were Richard Feynman and Murray Gell-Mann. In 1957 I was ready for the graduate quantum mechanics course, taught by Gell-Mann, an already celebrated 28-year-old full professor. Gell-Mann had been a childhood prodigy, and had learned immense amounts of material from all fields of knowledge, including many languages. By age twenty-one he had earned his PhD in physics from MIT. His legendary achievements in the substructure and classification of the elementary particles were burnished by his bestowing brilliant names on the properties of particles—"strangeness", "quarks", "the eight-fold way". These concepts became immortal, and I learned their names before I understood what they meant. Gell-Mann's uncannily brilliant naming of particles and their symmetries may be likened to a Nobel-Prize-winning poet choosing just the perfect word. Thereafter the names bound the physical concepts with the name of their creator, a type of immortality. Gell-Mann's youth was testimony to us all that great achievement could come in our next few years. It is therefore easy to imagine the excitement felt by this 22-year old in taking my seat in his class. I had never known or even imagined professors like these. The room, by the way, was a small classroom, about four rows of six desk chairs each, almost filled by twenty or so other students. There was no huge audience, as would occur later in the Feynman lectures to freshmen, no sense of a cultural happening. The course was difficult to understand, and I regret admitting that I did not master it. One problem was lack of a textbook. Such lack is a teaching mistake. We learned from Gell-Mann's lectures. Insufficient time exists in during classtime to lay its cultural backgrounds, to systematically define the mathematical concepts, to build from the ground up to prepare the student to absorb the advanced concepts. It was sink or swim. I was not sufficiently mature to realize that I should obtain a good book on quantum mechanics and read it in parallel to the lectures. As a result, the intellectual structure of quantum mechanics was not fully grasped by me. Gell-Mann's lectures came to be called by me "Murray's bag of tricks". His display of these techniques was awesome, but I was not ready to grasp the overarching structure. Another problem was the brilliance of some of the other students in the class, so that it was tempting to Gell-Mann to lecture to their level. One of these was Sidney Coleman, a first-year student of great reputation, and I still remember one exchange while Gell-Mann was describing the scattering of a particle from the potential energy provided by a target particle:

Gell-Mann: *So what is the spherically symmetric part?*
Coleman: *sinkr/r.*

Gell-Mann: *Right, and at low energy that is all we need to consider.*

My glance around the room showed that I was not the only one simply trying to comprehend the question, much less the reason for its answer. I struggled to keep up, to fill in the conceptual underpinnings of the topic, to divine its logical structure rather than its techniques. I passed the course well only because its two tests were basic enough that they could be prepared for by study of the notes; but I did not truly grasp *quantum mechanics*. Still, I would not have traded this experience for the world. I had sat through perhaps forty one-hour lectures about twelve feet from one of mankind's geniuses; and I did understand that.

To understand Gell-Mann's enormity, the reader must compare listening to J. S. Bach describe the art of the fugue, or Einstein lecturing on the physical principles of general relativity. I can not exaggerate the immensity of such experiences. Above all, Gell-Mann and other Caltech professors helped me to see clearly that physics was the act of struggling with nature's puzzles rather than of learning what was known.

Graduate Student Life

During my first two years at Caltech I shared a very attractive student office on the top floor of West Bridge Laboratory of physics. This historic building was designed by New York architect Bertram Goodhue around 1917 as part of the initial campus plan. The 1922 aerial photograph on p. 96 of Judith Goodstein's engaging history, *Millikan's School* (Norton 1991), shows the Norman Bridge Laboratory as one of three main buildings-- Throop Hall, Gates Laboratory and Bridge Laboratory. My office, which I shared with Melvin Daybell, Donald Groom, and Charles Peck, other beginning graduate students, had an east office window on the fourth floor, with an excellent view of the dome of Throop Hall (demolished 1973). With strong Spanish architectural influence that characterizes the Caltech campus, we walked up a spiral staircase having wrought iron railings over tiled floors. In this historic building in which the weekly Thursday

Physics Colloquium was held following tea, I studied my courses. I loved those first two years there prior to moving my office into the Kellogg Radiation Laboratory for my research apprenticeship

I discontinued hobbies that I had cherished heretofore. I played golf twice, then decided that it must be given up as too time consuming. I swam twice in the really nice Caltech pool south of California Boulevard. I gave that up because of the reaction of smog in the lungs. Fast swimming requires great gulps of air, and the smog was so bad that deep breaths were impossible. My lungs would catch up. I played bridge once, at an ACBL Sectional tournament in Pasadena. It was too time consuming, and in fact too thought consuming to accompany a PhD effort. Physics at this level requires extreme devotion. I played a couple of chess games with friends, won those handily, then quit that too. Quitting a skill was hard for me, because I cherished each skill. I hung onto them as part of myself. To give up a skill seemed to waste the preparation that had honed it. It seemed to almost give up on who I was. But it had to be done, and I chose to give them up. Choosing that dedication was one of the first great actions of my life.

I was now feeling a higher calling. I loved my physics challenges in class, loved being at Caltech, and devoutly wished to succeed. I realized that this opportunity came to only a fortunate few. So I worked constantly, and even when not working I thought about the physics. I thought about it during dinner, or while watching TV with Mary Lou. It never left me. I talked about the physics with Mel, Don and Charlie, my office mates. We would meet in the evening to have a beer and talk about physics. It was at this time that I learned the skill of intense concentration, day in and day out holding a complex topic in my consciousness. It taught me that I could sometimes grasp a new concept just by thinking about it while doing other things. I learned that if I wrestled with a problem before bedtime, it would sometimes be solved during sleep! This last experience came to me repeatedly during my later career as well.

Despite my somewhat monastic dedication, Mary Lou and I had special days and nights of fun. We became fast friends with a young faculty theoretical physicist, Fred Zachariasen and his wife,

Nancy, especially after we learned that our two little boys were contemporary with their daughter. On frequent Sundays we took a thoughtfully selected picnic lunch and bottles of wine up the mountain into the Angeles Crest Forest, where we walked, talked, and enjoyed our gourmet spreads. We even broached the subject of the difficulty of marriage, but were too insecure to expose our fears. That may have been an opportunity lost. We befriended a graduate student in engineering and his wife and had many social occasions with them, including some fun weekends away from Pasadena. With other friends we established a special routine for New Year's day, when we would rise at 4:00 AM, usually hung over from the wilder than average party, and drive the several blocks to South Orange Grove Boulevard, where we parked and walked unimpeded among the floats lining up for the Rose Parade to follow along Colorado Boulevard. The floats were so beautiful in the awakening sky, flowered constructions seemingly from a different world. These floats were acts of love. Then, day breaking, we gathered at someone's house for lots of coffee, breakfast, and all day lounging to watch the bowl games on TV, ending of course with the Rose Bowl. Caltech played its own small football games in a cavernous and deserted Rose Bowl. The first Rose Bowl had actually been located as a simple small stadium on the Caltech grounds, just south of California Boulevard. We performed this ritual for five straight years. And, above all we partied. Almost every weekend was a zinger. Music, dancing, alcohol, indiscretions, and readily forgiven embarrassments. Very little room exists for conventional preoccupations when striving day and night at the top intellectual levels. The farm boy never left me, but he was receding fast, a blurring recollection of my origins. I did not visit Iowa for seven years.

*At Base Camp for climb of Mt Humphries
in Sierra Nevada, Oct. 1958*

As Christmas 1957 approached my father telephoned me to suggest that we meet him in Diego, where he was flying to pick up a new airplane. Excited at the thought of spending Christmas with my family via such a romantic transport, we agreed. He informed us that we could take the train back from Dallas to Pasadena—he had checked into it. We took a Greyhound to San Diego and left dad's San Diego hotel in late afternoon to taxi to the airport. There Lockheed pilots had left awaiting him a shining new Lockheed Electra, a four-engine turboprop airplane having higher cruising speed than the Douglas DC-6. Following a mandatory checkout of its systems with a Lockheed pilot, we took off into the San Diego dusk. I was invited to ride jump seat, the small fold-down seat between pilot and copilot. I realized that I had not ridden with my father in an airline cockpit since high school, when, in order to play golf together in Denver, I had accompanied him riding jump seat in a Braniff Airlines Douglas DC-3. And in a quiet moment I remembered a childhood of numerous airplane rides with my father.

How wondrously swiftly the Electra rose into the sky, the familiar diminution of sights on the ground receding with unexpected swiftness until the dimming and setting sun brightened at altitude into the golden ball of afternoons. After banking over the Pacific Ocean, with me looking along the dipped left wing at ocean waves beneath, we leveled and headed east while the sun set at our tail. Both Dad and his copilot had time to explain to me the banks of instruments on the panel before them, a multifaceted array of gauges, dials and lights, unlike anything I had ever seen. Three hours later, as we approached Love Field in Dallas I could not see the ground. Darkness and fog and light rain obscured any groundlights as we descended gradually while my father checked with ground control via radio. My eyes alternately on the altimeter and looking for ground out the window, I read its needle, 4000 feet, 3000 feet, 2000 feet—still only blackness out, compounded by bright reflections of the airplane's landing lights off the seeming wall of fog. I was a bit frightened as my father calmly set our flight path crosshairs to align with the radar landing beam. Now 1000 feet, then 500 feet, and suddenly, almost magically, the runway lights appeared directly in front of our path. Their rows of yellowish lights edged the runway, now speeding by us as all became clear and safe once again.

I could see why my father loved it.

Master and Apprentice

In fall 1957 William A. Fowler reappeared in full force to teach his awaited course in nuclear physics. Mastering its material was the prelude to doing research in the Kellogg Radiation Lab. I enrolled and enjoyed the course more than any other. On the first day he told us to call him Willy. I was not alone in having trouble making his 8:00 AM lectures, however. The blackboards were already filled since 7:30 with his detailed notes so that he could go over them in class. Fowler later told me that he preferred an 8:00 AM class time in order that the remainder of his workday not be interrupted by teaching. Frequently I sneaked in 5 minutes late with the remainder of my morning coffee in hand. I was not alone in that. I tried while Willy explained his notes to write down physical insights that were not contained in my set of his printed notes, which he had mimeographed and distributed to enrollees. But simultaneously concentrating on what was being discussed and writing clear explanations on the notes proved a stiff challenge. To this day I prize my dog-eared copy of these course notes, 300 pages of a practical man's guide to the nucleus. Imagine today a nuclear physics graduate class with 25 enrolled students! It just does not happen except at a few special places. One is lucky to have five today in the nuclear physics programs of most universities. So much did we joke about this breathtaking start to each day that someone had a brilliant idea; namely, to spend the night in sleeping bags in his West Bridge classroom and awake stretching and yawning, hopefully to a clanging alarm clock, just as he walked in the door. This might get our message across to this tornado. Being married, I did not join the overnight group, but I came early to share the fun. Willy just roared (he loved a joke), then continued with new determination to lay it on thicker than ever.

Fowler displayed an infectious sense of fun. Everyone loved him for that attribute. He was intensely motivated to work tirelessly on the nuclear physics of the stars; but this never prevented him from having an uproariously good time in the process. Fowler's gift to all about him in Kellogg was his sense that every research task was both exciting and fun. And his attitude and leadership did in fact make Kellogg Laboratory's reputation that of a fun-loving place for research. His style would later be adopted by me in my own academic career. At this time his paper[1] with Geoffrey and Margaret Burbidge and Fred Hoyle appeared in print, and I was thrilled when Willy gave me a reprint of it while I was talking to him about my excitement over the idea of nucleosynthesis within the stars. "Read about the *s* process", he suggested. His enthusiasm showed, but he sobered me by saying that he never offered a research fellowship under his guidance to a student who had not done very well in his nuclear physics course. Relief flooded me at the realization that I was now enrolled in that course. Work hard, then come back to see him, Willy said.

Fortunately I had studied Fowler's material fervently and was able to write a very good final exam, which earned a coveted A; so when I went to his office to ask about a commitment to research with him, he showed enthusiasm and agreed to take me on as a student apprentice. My stipend would hereafter be paid by the Office of Naval Research, which in those years supported nuclear physics research in Kellogg Radiation Lab. Further course work would be optional as my attention was turned to helping Fowler formulate mathematical ideas on the origin of the elements. This commitment between student and research advisor is the decisive crossroad in each PhD student's life. In my case an opportunity of extraordinary magnitude had just been handed to me by Fowler. Almost instantly I could feel my transfiguration from student to apprentice. Perhaps I pulled my shoulders back and stood taller—I can't say—but that is how I felt. I had read portions of his paper with Fred Hoyle and the Burbidges, especially about the *s* process, dumfounded by the notion that the atoms of our chemical elements here on earth had been assembled by nuclear reactions occurring within the hot gases of the interiors of stars. What Fowler first sought from me was the putting of flesh onto the bones of ideas outlined in that review paper, which he already called *B²FH*.[1]. Nothing tickled Fowler more than a cute acronym. Many of the *B²FH* ideas were in turn older, having been described by Hoyle in a 1954 article in *Astrophysical Journal*; but I had never seen that work. We were, working together, to take these concepts and run with them, flesh them out, give them quantitative power so that useful astrophysical consequences could be calculated. And when I say *calculated*, I mean hand calculated, or by slide rule, or cranked on the mechanical calculator in the basement of Kellogg. We had no operating computer in 1957. My first digital computation would not occur until 1960.

Leaving Earth

On October 4, 1957, *Sputnik 1* was launched into earth orbit by the Soviet Union. It was the first earth satellite. Newton had described an orbit as a state of continuously falling around the earth. *Sputnik*'s launch changed my life profoundly, although I did not know it at the time. I was simply amazed by this human object's continuous fall toward earth without ever hitting it! We all knew of the Moon's orbit around the earth, far from the earth, but this was the first man-made object placed in orbit just barely above the earth. I can describe its impact at Caltech in October 1957 no better than by quoting my own description, written[2] in 1974:

> *During the weekly Thursday physics research colloquium it was remarked excitedly that Sputnik 1 would pass overhead visibly before sunset. As a body, 200 physicists—eminent professors, post-doctoral researchers and students—filed to the roof of Bridge Laboratory to watch for it. Many sudden shouts and pointing fingers later, it came like speeding Venus across the twilight. The chatter died quickly away, and in five minutes of silence we watched its pass. Just as suddenly it was gone, and we all stared dumbly at each other, just barely comprehending that mankind would never be the same again.*

I was as stunned as anyone; then slowly a smile touched my face as I recalled Grandpa's orbiting milk pail.

The space age had begun, and within four months a frantic US effort had successfully placed a satellite in orbit[3]. We followed this in the newspaper and on television news. Called *Explorer 1*, this satellite obtained the first scientific discovery from space when its radiation detectors recorded repeatedly passing through belts of high-energy protons and electrons trapped on the earth's magnetic field lines. These "radiation belts" had been predicted by no one. This event also galvanized me, so much so that for the next five decades my thoughts would never stray far from the origin of the chemical elements or from scientific discovery from space.

The state of cosmological knowledge in 1957 was primitive by today's standards. Little was known. The *Big Bang* had not yet captured the scientific stage and was not even seriously discussed.

That had to await the discovery of the three-degree microwave radiation that filled the universe. There was no consensus on why it was that the stars seemed to be composed almost entirely of hydrogen and helium. That puzzling fact was just accepted as somehow reasonable. What was clear was that stars born more recently in the universe had been born from gas that contained more of the elements heavier than helium than did the oldest stars. New astronomical observations of stars were showing that the oldest stars, those earliest born, were at least 99.9% hydrogen and helium (hereafter H and He). It seemed that the universe had begun with only H and He. It had been Hoyle's thesis that the stars themselves were the ovens in which heavier elements were assembled from H and He by nuclear reactions.

A major hurdle to the plausibility of stellar nucleosynthesis was the lack[4] of any stable isotopes at mass 5 and mass 8. The hot gases of H and He seemed impotent to create new elements. This obstacle had been leaped about five years before I arrived, when Edwin Salpeter, visiting Kellogg Lab, calculated that the Be^8 nuclei created transiently when one He scatters from another actually were held together long enough that a third He nucleus might be captured by them[4]. By radiating a photon, the triple-alpha complex could become the stable ^{12}C nucleus. Shortly thereafter Fred Hoyle pointed out in Fowler's office that this would not work unless the ^{12}C nuclear structure had a spinless excited state at precisely the right energy. Hoyle had brazenly suggested[5] that such a state must exist or we would not be here! This prediction gave Kellogg new nuclear reactions to measure. And when Fowler's team[5] discovered that prediction to be correct, it hooked Fowler forever on nuclear astrophysics by his own admission.

The picture for stellar nucleosynthesis that I inherited was that as stars came to the ends of their lives, they return most of their gaseous matter to the interstellar gas, into which it is mixed. But the stellar ejecta contained the heavier elements than H and He, such as carbon, oxygen and silicon, that had been fused from H and He in that star. The return of the enriched material thereby enriched the interstellar gas in carbon, nitrogen, oxygen and all the other heavy elements. Hoyle's theory of the entire universe was that as the universe expanded, new H and He atoms were created to replenish the lowering density at just such a rate that the amount of matter per unit volume remained constant even while the universe expanded. This was *the steady-state theory.* The physics of the creation of the new H and He was not known, and therefore the reason for their being observed specifically in the ratio 10:1 was also unknown. The steady state was a beautiful theory and had strong intellectual support in 1957. Then in 1954, Hoyle described his picture that the elements from carbon to nickel in atomic weight were synthesized in stars much more massive than the sun[5]. It was in this froth of excitement that Fowler hoped that I could help evaluate the processes by which the stars, using nuclear physics, had made the remainder of the chemical elements from the H and He. What an opportunity for me!

Nuclear Reactions within the Stars

On that first day in his office, Willy was animated and forward looking. He wanted to discuss not one but two problems in nuclear astrophysics with me, and he wanted me to take them both on! Fowler was so taken with the science of nucleosynthesis that he wanted to have a research student doing its theory. The main goal would be a quantitative theory of the *s* process. This is a theory that holds that about half of all the nuclei heavier than iron were created in the interiors of red giant stars when nuclear reactions liberate neutrons from nuclei and those free neutrons are in turn captured by the nuclei heavier than iron. The main process for liberating a neutron would occur when common ^{13}C nuclei are imbedded in a hot helium gas, as occurs in stars after hydrogen has been fused into helium. One writes that nuclear reaction as $^{13}C + {}^4He \rightarrow {}^{16}O + {}^1n$, where the ^1n is the symbol for the mass-one free neutron. That free neutron may in turn be captured by a nucleus of iron, $^{56}Fe + {}^1n \rightarrow {}^{57}Fe$, or by a different heavy nucleus. The net effect is the increase of the atomic weight by one unit on every occurrence. Given enough free neutrons, one could envision iron being transmuted to barium! *B²FH* had called it the *s* process, with the *s* standing for "slow", referring to the slow increase of atomic weight as each heavy nucleus captures a free neutron every century or so. The same thing happens in a nuclear reactor on earth. This could be my main task. I had no doubt of wanting to work on that problem. His offer allowed me the chance to evaluate nothing less than the physical origin of heavy atomic nuclei. It offered to me the chance to create a mathematical theory for a sublime question. I was very up. We held weekly discussions of my progress for the next three years.

In that same conversation Fowler also described to me his worry about one of the key nuclear reactions that that occurred in hot hydrogen in the stars. It was one reaction of the carbon-nitrogen cycle in stars. This series of nuclear reactions occurred between the carbon and nitrogen nuclei in the stars with free hydrogen nuclei (called protons), and the net effect of that sequence of proton reactions was the conversion by nuclear fusion of hydrogen into helium. In his nuclear physics class Fowler had explained how a carbon nucleus captures the proton, namely by forming a new compound nucleus and then emitting an energetic gamma ray to expel the excess energy. It became thereby a radioactive isotope of the element nitrogen. Fowler himself had discovered that radiative capture reaction in Kellogg Lab. At the time it offered proof that protons could be captured by a nucleus if the new nucleus could get rid of its excess energy by radiating it. This nuclear alchemy, Fowler had explained, was occurring at the center of the sun. The sequence of four such reactions that ensues was discovered and named the *CN Cycle* by Hans Bethe, who had shown that its net effect was the conversion of hydrogen into helium.

Nuclear fusion liberates a huge quantity of heat energy because the helium nucleus has slightly less mass, 0.7% less mass to be precise, than does the sum of the four hydrogen nuclei that are fused to yield it; and the disappearing mass appears as heat energy according to Einstein's celebrated equation $E=mc^2$. This was, Fowler said, what they believed kept the brightest stars hot and shining, for otherwise they would cool like any hot ball in cold space. Fowler's course had also introduced me to the basics of thermonuclear reactions between charged ions, so he launched quickly into his current worry about one of those reactions. He was concerned about the destruction rate of mass-13 isotope of the element carbon, written ^{13}C. This is the heavier of carbon's two naturally occurring isotopes, the one having seven neutrons in its nucleus rather than the more common six neutrons. It is also the repository of an excess neutron that could be liberated for the action of the *s* process, as described above. What worried Fowler was a recent experimental measurement of the energies of the excited states of the ^{14}N nucleus. One of those suggested the existence of a troublesome nuclear state, namely one possessing 7.60 MeV more energy than the natural ground state of ^{14}N. At that 7.60 MeV excitation energy the associated increase the stellar rate of the destruction of ^{13}C would be large. It could not survive the hydrogen burning that creates it. Willy could then not understand how ^{13}C could be as abundant as it is observed to be. One earth natural carbon consists of one ^{13}C atom for every 89 atoms of ^{12}C, the most common isotope of carbon.

I was speechless. Fowler was noticeably worrying about what seems at first glance to be only an obscure nuclear detail, a detail concerning those states of ^{14}N having more energy than its normal ground state. His experience led him to leap to a gigantic conclusion, that if that reported experiment were correct, we would face a major problem in nucleosynthesis theory, namely too much natural abundance of an altogether different nucleus, namely ^{13}C. Willy saw that I had not followed the entire argument, so he repeated it. As I heard it again I thought, *Wow! What an interplay between nuclear details and nature.* Willy then expressed the hunch that the reported nuclear state in ^{14}N was an experimental error in the experiment reporting it, and he hoped that I could find out. I agreed at once, captivated by the stunning circuitous argument that the nuclear state must be in error because ^{13}C was so abundant! That amounted to arguing about some detail of a nucleus on the grounds of the natural abundances observed in the world. I had never before heard such an argument. Imagine my surprise when I learned that Fowler himself had first become hooked on nucleosynthesis when, a few years earlier, Fred Hoyle had made the first such argument in Fowler's office about a different nuclear detail of the carbon nucleus. That manner of thinking about natural abundances and nuclear physics details captured me heart and soul.

Fowler then offered a specific suggestion that could settle the matter. My measuring the energy spectrum of ejected alpha-particles from the nuclear reaction that occurs when ^{15}N nuclei are bombarded with energetic ^{3}He nuclei would settle it. Alpha particles are nuclei of helium. That nuclear reaction, written in nuclear shorthand as $^{15}N+^{3}He \Rightarrow ^{4}He+^{14}N$, would expose all nuclear states of ^{14}N in this energy region. He asked me to calculate the situation and check his worries about the CN cycle. Feverishly I did this calculation that afternoon; and Willy had been right. I could not wait to perform this alpha spectroscopy experiment on the three-million-volt Van de Graaff[6] accelerator. But first I would have a lot of work to do.

I was on cloud nine as I hurried home to explain to Mary Lou the amazing good fortune of the research topics that Fowler was willing to entrust to me. I was to become his *apprentice*. This is the route that young scientists must follow to become producers of new knowledge. None of us have the savvy to perceive the ripe problems on the frontier of knowledge. Each needs the guidance of a world leader to break into world leading research. That simple but profound realization, evident in retrospect, is largely unrecognized by college graduates. Even today most regard graduate work as a chance to learn more, and perhaps increase one's marketability. As an aside that is very amusing in the light of later developments, Willy confessed personally that he was not confident that studies of the *s* process in stars would constitute sufficiently "solid physics" for a PhD, so he wanted me to do a good experiment as well. At least I would then have training in something useful! That confidential remark correctly conveys the *Zeitgeist* of 1957, namely that musings on the origin of the chemical elements were the domain of cranks and prophets. By contrast, Willy and the Kellogg Lab already had won a big reputation for measuring the rates of nuclear reactions in stars, which had become a part of laboratory nuclear physics. But Kellogg Lab had as yet no reputation in nucleosynthesis theory, only the current publication of that milestone review paper with Hoyle that extended the early ideas of Hoyle and others. My formulation of the *s* process was to be the beginnings of Caltech research into nucleosynthesis, the theory of the origin of the elements. I would be Fowler's first Ph. D student in that theoretical arena.

Nuclear reactions within the laboratory

For the experimental part I had to learn everything. I rejoiced to find that the Van de Graaff accelerator[6] had already been fitted out with a ^{3}He ion source capable of producing charged ^{3}He+ ions within its charged dome. The high dome voltage would then propel those ions toward the target. At least I would not have to build or buy an ion source. I studied how it worked and soon became proficient in that aspect. I was clueless about how one might obtain a nitrogen target of the mass-15 isotope, which would be required for the experiment that Willy had suggested. Normal nitrogen, such as in air, contains less than 1% of its mass-15 isotope, which is the relatively rare nitrogen isotope. Willy was prepared for that question. He pulled from his desk a vial containing ^{15}N-enriched nitrogen, which he had obtained as ammonium nitrate from Eastman Kodak Company--67% ^{15}N! He suggested that I talk with Dale Hebbard, visiting scientist from Australia, about preparing a nitrogen target from this enriched ammonium nitrate. Much to my relief, this help was forthcoming, as Hebbard was at that very time preparing titanium-nitride (TiN) targets by evaporating Ti onto a nickel backing and then exposing it to nitrogen. Hot nitrogen gas reacts with titanium metal to make a tough titanium nitride solid. Dale volunteered to help me do that procedure with my sample of ^{15}N-enriched nitrogen. Hebbard's skill yielded a very good TiN target. But I realized that I would have had to work a very long time to have done this myself. Armed with that TiN target I was ready to bombard it with 2.763 MeV (million electron-volt) ^{3}He ions and perform the magnetic analyzer spectroscopy of the alpha-particle groups that were liberated by the nuclear reaction. The ^{3}He ions after their acceleration by 2.76 million Volts struck the ^{15}N target at 5% the speed of light. This enabled them to penetrate the electric repulsion of a ^{15}N nucleus, pierce that nucleus, and cause new nuclear fragments to be formed and ejected. By measuring the momentum groups of these fragments I would address Fowler's worry.

But before running that Van de Graaff accelerator, I needed on-the-job training as a Van de Graaff maintenance man. In those days there was no staff of engineers and operators to do what needed doing. I practiced with some measurements by scattering protons from a metallic nickel foil. That allowed me to become proficient not only with the accelerator but also with the magnetic momentum analyzer that determined the momentum of the scattered protons. It is necessary in experimental physics to practice, to get the routines right for the apparatus before trying the important measurements. That was especially true in my case owing to the uniqueness of the ^{15}N target. To have prematurely ruined it would have been a disaster. While I was engaged in these things Tommy Lauritsen came to me with a brainstorm for improving the performance of the accelerator. He thought that the beam focus might be improved by installing an electrostatic quadrupole lens below the exit ground plate from the tank but above the electrostatic energy analyzer. That meant designing and installing a device in the beam tube. And although my first inclination was to avoid the delay of my experiment, I was lured into the task by the hope of better data. Another Kellogg student, John Domingo, worked with me on the project. I brought out my copy of Smythe's *Static and Dynamic Electricity* to help me think about the electric fields between four parallel cylinders.

My apprenticeship in the machine shop came next, where machinist George Fassel showed me what I needed to know about the metal lathes. For related maintenance John and I crawled several times into the huge emptied steel tank, and, after a three-month diversion, we eventually installed our electrostatic lens into the beam tube. We were warned sternly by Tommy to enter the tank only in pairs, as only recently another student, John Pearson, had collapsed in the tank before its nitrogen had been replaced with breathable air. The tank was kept at high pressure of gaseous N_2, the most abundant molecule of air, to discourage sparking by the high-voltage dome. The sparking was more frequent in natural air owing to its oxygen content. So whenever the high-pressure van de Graff tank was brought down to atmospheric pressure one had to exhaust the pure nitrogen with fans before entering. Fortunately an overcome John fell by the open entrance plate and was able to laugh about it later, although we made frequent remarks to him about evidence suggesting he had suffered brain damage.

Kellogg Lab maintained a lively *esprit de corps.* Central to this was our Friday night party, which was *mandatory!* We were not really free to do as we pleased. The pressure to display "the right stuff" was palpable. At 7:30 every Friday evening in the ground-floor seminar room of Kellogg Lab someone presented the nuclear physics discussion of the week. We were all expected to be there, so we were there. The speakers were either visiting scientists, including very famous ones such as Aage Bohr, Ben Mottleson, Hans Bethe, Jean Humblet, Ed Salpeter, or Kellogg faculty and research students such as myself. My own first assignment as presenter was embarrassing. Pretending to better understand what I was talking about, I twice overstepped my knowledge and was gently corrected. Bluffing represented an insecurity that I suffered from for some years. As my competence grew my know-it-all posture became needless, and was in fact antithetical to the scientific impulse.

Perhaps as reward a Friday night party occurred after the seminar, at which wives were also expected to attend. The wives and girlfriends were Willy's requirement. He loved them all. This was not a hard sell because it was the best party going in Pasadena on Friday night. Bob Christy, Alvin Tollestrup and other Caltech faculty might join in. Each party was hosted by a faculty member in their home, with Tom and Marge Lauritsen the most frequent hosts, but Willy and Ardy Fowler and Charlie and Phyllis Barnes close behind. The format was lots of beer, lots of joking, a wonderful spread of food featuring baked ham and good bread and mustard, lots of flirting, lots of music, singing and dancing. Charlie Barnes could always be counted upon to play the piano to accompany the singing, which he often did with his pipe in his mouth, with Willy inevitably insisting on the Kellogg Fight Song, "Ching-gung goolie-goolie-goolie-goolie-watch-na-ching-ga". One might wonder that

Kellogg Lab needed a fight song. This was not just to motivate the laboratory troops but reflected Kellogg's team spirit in battles on the softball field against other labs and departments. Willy was determined to win if at all possible. We were the only department having its own fight song. I happened to be good at softball, as in sports generally, and helped occasional Kellogg victories, further cementing my relationship to Willy.

Willy could be counted on for comic shenanigans at these parties, usually directed toward one of the wives or girlfriends. An example illustrates. Willy would appear in Trilby Hat with a large mirror. After much beer and after persuading a victim to see if she could blow the hat from his head, he would sit on a chair in the corner of a room and hold the mirror at a very acute angle dividing his face into a mirror image of the direct half, then say to her "Now blow!" She would giggle, and he would repeat, "Blow" When she did so, the free hand behind the mirror would lift the hat from his head, just a little, and he would say "Blow harder!" Giggling more, she would blow harder, and he would lift the hat higher, etc. All other action stopped to watch this comical and brilliant stunt. Everyone roared as the girl blushed. But along about 1958 Willy was supplanted as chief stuntman by Phil Seeger, a licensed magician, who would amaze us all with his confounding magical displays. Phil and I were to become collaborators on the formulation of the r process, a much more "rapid" process of heavy-element nucleosynthesis achieved by the capture of free neutrons in stellar explosions.

Such goings on helped make my apprenticeship both fun and memorable. After a couple of years of very hard work a subtle transformation happened with me and within each of my fellow students. As our knowledge and abilities grew, so did our confidence and ultimately our creativity. Such is the magic of the process that a credible scientist emerges like a butterfly from a chrysalis.

CHAPTER 7.   S PROCESS AND MY JOURNEYMAN'S RATING

A few weeks after our watershed discussion in his office, Willy dropped by my office before going to lunch to invite me to go with him next day to Robinson Hall in order to attend the Astronomy Research Colloquium. His friend Jesse Greenstein was to discuss an especially interesting topic. We walked in next day and sat down in the front row. Beginning his presentation, Greenstein commented that the "Kellogg contingent" had shown up, and that we were most welcome. I glowed inside. Being officially attached in research to Willy Fowler was a comfortable feeling. Jesse Greenstein was an extraordinary gentlemen, a man of charm and sophistication as well as erudition. I liked him at once and did so for the remainder of his life, which would be lengthy. Jesse first showed some spectra of the light from old stars obtained at the Palomar 200-inch telescope. The smooth light intensity plotted as a function of wavelength was punctuated by many dips of intensity at specific wavelengths. These were atomic absorption lines in that star's atmosphere, and the depths of the intensity dips reflected the abundances in that star's atmosphere of the elements causing those absorption lines. Most of those absorption lines were much weaker than in the sun, revealing that the heavy element abundances in that star were about ten times smaller relative to H and He than they were in the sun. This was my first example of what astronomers call a "metal-poor star". What metal-poor means is that the elements heavier than hydrogen and helium, the two lightest elements, were much more rare than in solar gas. The heavier elements, called "metals" by astronomers without implying that they physically behave as metals do, comprise 2% of the sun's mass, but only 0.2% in the star that Greenstein was displaying.

Greenstein described a picture that could make sense of such an observation. One need but assume four things: that the elements were created in stars, as Hoyle had advocated; that the initial hydrogen and helium of the interstellar gas in the galaxy was slowly augmented in heavy element content by the creation of new elements by nuclear reactions in stars, because the new heavy elements that were created in past stars were ejected into the interstellar matter when those stars died; that when new stars are born, they are born from average interstellar gas; and lastly, that astronomers see only the surface composition of each star, which remains the initial composition of the gas from which the star was born. Assuming that picture to be correct, late forming stars will have more heavy-element content than early forming stars. That is, "young stars", those born recently from the current interstellar gas, would have a greater abundance of the heavy elements than do "old stars"—just the opposite of what the idea of element creation in stars might at first glance seem to suggest. Greenstein was implying that his example star had formed from interstellar gas at an early time when its heavy-element content was tenfold smaller than today. His talk explained that this was another piece of evidence that the heavy-element content of interstellar gas was increasing with time, and that this star formed so much earlier than the sun that its abundances were at that time much less than they were later when the sun formed. This was, he added, another example of a general science called nucleosynthesis, which consisted of quantitative evaluation of each of the four assumptions that he had enumerated. From here on I will use the term *nucleosynthesis* to mean all interrelated questions of science that are united and explained by the idea of creation of the chemical elements within stars, with special emphasis on the nuclear processes that act within the stars to synthesize the new nuclei. *Nucleosynthesis* is this entire theoretical system rather than the single idea of creation of elements in stars.

"How long ago was that star born?" I found the courage to ask amid that audience of knowledgeable astronomers.

"It is hard to say because of the difficulty of pinning down the ages of old single stars," Greenstein answered, "but from its luminosity and color I estimate it at about eight billion years old." That would make it almost twice as old as the earth! What I was hearing excited me as no reading in a book possibly could have. But in truth my main emotion was relief that my question had not made me look ridiculous. To some extent I had yielded to my old weakness, bluffing, of pretending to understand more than I did. But Greenstein's answer and Willy's sideways glance of approval told me that I had hit the mark this time. I began to see myself more as scientist and less as student. Science feeds on the asking of questions.

But Greenstein was not yet finished. He pointed to the absorption lines of the element barium as being an exception to the general trend. Their strength in his metal-poor star was even greater than in the sun. Why did this absurd contrast with other heavy elements exist in the otherwise metal-poor star? This required the element barium to be some twenty times more abundant relative to other elements in that star than it was in the sun. What he was demonstrating was my first example of an *overabundance* of a specific element. The words *overabundance of barium* in Greenstein's star describes his finding that the ratio of the abundance of barium (Ba) to that of other heavy elements was at least twenty times greater in that star than is the analogous abundance ratio in the sun. This seemed very strange indeed. One element, Ba, is overabundant in a star where all others are underabundant. Was nature simply being whimsical? Taken at face value his data showed that the abundances of the elements did not all increase in lock step, but that some, as Ba in this case, had made a big increase within a single star. Greenstein concluded on that very point,

"This must be the *s* process acting in this star and bringing its products to the surface where we can see them". Wow! What a lucky day for me, to hear from an expert astronomer new evidence for ideas relevant to the *s* process, ideas that would motivate much of the next decade of my life. I walked back with a spring in my step.

This colloquium by Greenstein had alerted me to a set of ideas that I would need to understand in order to become a player in nucleosynthesis theory. *Nucleosynthesis* deals with abundances of the elements and the distinct isotopes of each element. Greenstein's star and the sun had differing abundances. When one speaks of the numerical abundances, one may refer to a specific star, or perhaps to the interstellar gas, or maybe to the entire content of a galaxy. Amazing! Was I correctly hearing that distinct objects in our universe are made of distinct mixtures of the chemical elements? That demands an explanation.

The abundances are not universal, but sample a specific object. To avoid the obvious fact that the entire galaxy must have overwhelmingly more total atoms of the element silicon, say, than does any single star, astronomers devise an abundance scale that compares equal masses of differing samples, so that one can distinguish the abundance compositions of the samples rather than their total numbers of each element. Since hydrogen is overwhelmingly the most abundant element, astronomers compare the chemical compositions of stars by considering an average sample of each that is chosen to contain 10^{12} H atoms, one thousand billion hydrogen atoms. That relatively large number of H atoms allows the much smaller abundances of the heavy elements to be expressed by numbers of convenient size. A very abundant element, sodium (Na) for example, on that scale then has the abundance Na=1.6 million atoms (per 10^{12} H atoms) in the sun; the heavy metal gold (Au), by contrast, then has the solar abundance Au=7.1 atoms (per 10^{12} H atoms). The qualifier *per 10^{12} H atoms* is thereafter understood, and does not need to be repeated. The choice of sample size makes conveniently sized numbers for the abundance compositions and enables those compositions to be defined by tables of numbers. Such tables of numbers allow separate samples to be easily compared. I seized that concept and would never let it go.

A related concept that I had to deal with was *precision*. The abundances were not handed down on stone tablets! They had to be won from nature by careful observations. The precision of an abundance number in any object depends upon how accurately astronomers can measure it. At the time there existed considerable uncertainties in these numbers, but as astronomical instruments and computers would evolve in the decades to come the precision was steadily increased. Furthermore, when astronomers measure a stellar abundance of an element from the strength of an absorption line in the star's spectrum, they measure the combined abundances of all stable isotopes of that element. The isotopes of an element differ only in atomic mass, and they do so by virtue of the numbers of neutrons contained in the nuclei, each of which has the same nuclear charge, making them isotopes of the same chemical element. It is the charge of the nucleus that determines the chemical properties of the atom.

These concepts were very familiar to me as a physics student in Kellogg Laboratory. My own experiment had concerned the rate of destruction in stars of the mass-13 isotope of carbon, ^{13}C. All carbon atoms have nuclear charge of six, for it is the six protons in the nucleus that makes them carbon, but the ^{13}C isotope has atomic weight 13 because its nucleus contains seven neutrons, in contrast with the more abundant isotope of carbon, ^{12}C, which has atomic weight 12 because its nuclei contain six neutrons. What I had to grasp was that the theory of nucleosynthesis concerns the abundances of the differing isotopes of each element, whereas astronomers measure in stars the sum of the isotopes of an element. The isotopic composition of an element can not be expected to be the same from star to star, and this variability would be confirmed in the decades ahead when astronomers developed instrumental techniques for measuring also the isotopic composition of some major elements. For this reason nucleosynthesis theory concerned itself during my student years with the isotopic composition measured in solar-system material. The solar system provided the only sample in which the isotopic abundance ratios could be measured in terrestrial laboratories with high precision.

Furthermore, I learned that for many elements astronomers

could not measure their abundances even in the sun. If the stellar absorption lines for that element were too weak, or if they lay in wavelength outside those that penetrated the earth's atmosphere, the solar abundance could not be measured. Therefore we took the isotopic composition of the sun from the isotopic ratios measured in a sample of earth or of the meteorites. These seemed to yield identical isotopic abundance ratios, so we took that to be the isotopic composition of the sun. And lastly, the abundances of many heavy elements could not be measured in the sun at that time. By carefully measuring element abundances in the meteorites, chemists were able to show that one type of meteorite, called C1, seemed to always give the same abundances as those as those that were known in the sun, so those meteorites were taken to yield the best solid sample of solar heavy elements, and gave the values of the chemical abundances for all of the heavy elements. A review paper[1] about solar abundances published in 1956 by Hans Suess and Harold Urey gave fertile stimulus to the growth of the theory.

Understanding the preceeding paragraphs is the key to understanding my scientific life. Therein lie ideas that became as much a part of me as the Iowa farms. They arrived like a *news flash* during Professor Greenstein's bold seminar. By having to master all of this, it became clear that the abundances must be won from nature by careful observations. Because confidence in nucleosynthesis depends on agreement between observed abundances and those yielded by the theory, I was forced to come to terms with the varying precisions with which the abundances were known. At first an annoyance, I soon saw this as being the nature of science itself. We believe ideas to be correct when their consequences agree with measured numbers. Ideas are proven wrong when their consequences disagree with measured numbers. Not a year passed but that some abundance was determined with higher precision, making it an exciting time to determine the impact of that improved abundance on the theory.

My scientific life became the breathless interplay of the 1950s between three essentials for nucleosynthesis science; namely, how well are abundances known, how well can nucleosynthesis theory predict their expected abundances, and how good is the nuclear data that governs their calculated abundances in the stellar theory. This was an intensely exciting fabric of interplay to me, so that I became just as motivated in exploring all three aspects as I had initially been in dealing with the nuclear data. I believe that it was my emotional urge to explore the origins of things that made me so motivated in this quest. One thing is sure; I would love science forever. The spirited interplay that built nucleosynthesis theory had won my heart.

Changing what is known

During summer of 1960 I finished my experiment on the Van de Graaff accelerator. During this crucial period I sat up nights until about 4:00 AM recording the alpha-particle counts at the exit of the magnetic spectrometer. I sat before the bank of vacuum-tube circuits that controlled the operation of the accelerator, making adjustments as needed. The ^{3}He ion source in the van de Graff dome was functioning beautifully. Because of the small counting rate at the exit slit of the spectrometer, I needed a lot of beam time to obtain a statistically significant number of alpha-particle counts at each magnetic-field setting. I found it possible to run the Van de Graaff without competition during late night hours; so I put those hours to steady use.

Willy's hunch had been right. The energies of the alpha particles liberated by the nuclear reaction showed no sign of the previously reported state of ^{14}N near 7.60 MeV of energy. Had that state existed, the ^{13}C isotope would have been destroyed rapidly in stars. My result showed that its destruction rate was small, however. This meant that the large abundance of ^{13}C that exists naturally on earth could have been simply the consequence of the hydrogen-burning nuclear reactions that provided the stellar power. This made the ^{13}C abundance an early success of nucleosynthesis theory.

Two noteworthy things for my personal growth as a scientist followed. Tommy Lauritsen, called me to his office, just two doorways west of Willy's along the same plaster-walled interior corridor in Kellogg Laboratory.

"Clayton, have a cigar", Tommy offered, extending his box of Roi Tans.

"Don't mind if I do", I replied, exaggerating for sake of the occasion. Soon smoke billowed from my own.

Tommy Lauritsen, the keeper of human knowledge of energy levels of light nuclei, examined my data for only fifteen minutes, then rose and led me from his office to the master drawing of the ^{14}N energy levels, posted on the corridor wall not five feet from his doorway. With red pen he drew a wavy red line through the state shown on that energy-level diagram at 7.60 MeV and offered me his congratulations. I surely blushed. The drawing was then sent down to Vic Ergott to be redrawn with the incorrect state removed[2]. I was dizzied by the realization of what I had done—*changed the state of human knowledge*. The feeling is like none other known to me. It is the life blood of the scientist. At a boundary of human knowledge, I had just made a change. It was a transforming experience, a refining fire, one that epitomizes the PhD degree in a world-leading academic institution. He who has had this experience will never forget it.

I now had the responsibility to write a research paper for review and publication. My article was submitted and published[2] in *The Physical Review*, the leading organ of physics knowledge. The Kellogg professors emphasized that research is never done until it is published. It avails nothing for me to privately know something unknown to the world. A thing is *known* only when published, so that anyone can know.

Meeting psychological fear

At about this same time, our first child, Donald Douglas Clayton, was born. Mary Lou struggled, tentatively, fearfully, with the pains of labor, and after a few hours the attending physician at the Huntington Hospital in Pasadena suggested a caesarian section. Mary Lou wanted that. Her own mother had from Mary Lou's youth repeatedly warned her of the pain of childbirth. Following Mary Lou's birth, her mother had thereafter slept in a separate room from her father, vowing to never again experience the "inconceivable pain" of childbirth. Not surprisingly, her father, Louis Roland Keesee, had established a relationship with another woman, one whom he married three decades later following the death of Mary Lou's mother. Now, in this hospital crisis, my own confidence was severely shaken. I could not think straight. Instead of immediate sympathy, my first reaction was to realize that Mary Lou was not to be like my own mother, who had delivered naturally four times, starting with myself in her bedroom in Iowa. Uncertainly I agreed to the doctor's suggestion, and Donald was delivered within the hour on January 14, 1960. I struggled thereafter with this surprising and unexpected shock to my self image. I tried to sympathetically embrace my wife, to console her, if consolation were even an appropriate word, to share joy over our baby boy. I tried to pretend that this small detail did not matter. In my mind I knew that it should not matter. But I was not strong enough. My emotional development had not been sufficiently complete. The modest abdominal scar glared huge and red in my brain. And if our marriage had suffered from youthful naivete, from our own fearfulness over sex, it was, for better or for worse, crippled on that very day. It would take eight years to admit that defeat. I do not doubt that feminists will despise me for this weakness. But what can I say? It happened. And if I could not admit it then, I can now.

Author and wife Mary Lou expecting, December 1959

The Feynman Lectures

We were fortunate to become friends with several of the Caltech faculty. One was Matthew Sands. Tall and lanky, longish straight hair parted on one side, smoking cigarettes, laconic, he was a sort of no-nonsense Texas-cowboy kind of physicist. One year earlier I had attended his lectures on electron theory, which amounted to a combination of classical Electricity and Magnetism, relativity, retarded potentials, radiation by accelerated electrons, and the motion of electrons. This course was designed for readying students for research in Caltech's electron synchrotron, but I welcomed the chance to study the topic despite beginning research in Kellogg. It was a fine course, one that would become useful in my later career in astrophysics. Sands had at the same time become concerned that the Caltech undergraduate physics curriculum was very old fashioned for a school of Caltech's modernity. He worried about the same problem at Caltech that I had experienced more extremely at SMU. Out of the blue he offered Mary Lou a job. She had quit school teaching after Donald's birth, but wanting work for fewer hours she had begun technical editing services when work of that type was offered to her. Knowing this, Sands wanted her contractual help with a new series of lectures for physics majors that Richard Feynman would begin in fall of 1961. Her experience at technical editing, her love of English, and the fact that I was a Caltech physicist made her a convincing choice in Sands' mind. Mary Lou accepted. I was very enthusiastic about this. The Ford Foundation provided the money that hired her. As a result we began work as part of the team on *The Feynman Lectures in Physics.*

Richard Feynman was a legend at Caltech and in the world. I had attended Feynman's lectures in advanced quantum mechanics, partly to learn more but partly just to be in class with this legend. His presentations were exceptional in their teaching skill, showing Feynman as indeed a genius, one who also loved to help students understand what they were learning. His presentations of advanced topics showed a desire to bring everyone along with him and to point out the stumbling blocks. His Caltech image went beyond physics to bongo drums, topless bars, and lock picking during the Manhattan project. We students revered him for his solution of quantum electrodynamics and for the *Feynman diagrams* that we all learned to draw in order to illustrate mathematical perturbation theory. Everyone was confident that he would soon be awarded the Nobel Prize (as happened in 1965) for this formulation. Because *The Feynman Lectures in Physics* became one of the celebrated intellectual efforts in the history of physics teaching, it is worth telling how it was done in the trenches and how Mary Lou and I helped.

Twice weekly, Feynman would present his lectures to Caltech undergraduates and to a host of onlookers, faculty, postdocs, and graduate students in the second-floor lecture hall in Bridge Laboratory. The undergraduate students enrolled in the class were entitled to the front two rows. They must have been really pumped to see the entire campus turn out for the lectures intended for them. The reader may wish to reread the initial sentence, because that outpouring of advanced audience has been unprecedented in my experience. Feynman's lectures were recorded on tape (supported by abundant note taking by others). The tapes were transcribed quite literally by a typist, who listened to the tape and tried to type what she heard. That typed transcript then came to Mary Lou. She would read it while also listening to the tape and attempt to convert Feynman's spoken word into prose. That was our contribution.

It is no insult to genius to point out that this translation to prose was difficult. Feynman was the first to realize that the spoken word and the written word differed greatly--especially Feynman's spoken word. He like to interject things like "You see!", or "And all we did was this", followed by scratches on the blackboard or waving of arms. I made it my own priority to attend Feynman's lectures, and doing so was one of the great experiences of my life. Never was fundamental physics so clear as it was while Feynman was explaining it. My contribution, strictly informal inasmuch as I had no official capacity with the project, was to help Mary Lou by answering many questions that she had about his intent at certain points on the tape. Often it took a physicist to choose the right word, or even to understand the word when Feynman had used it. The typist often could not recognize a word. We spent a lot of time at that, time that sometimes annoyed me because it interfered with my other goals, but time that sometimes thrilled me when I understood what his exposition and explanation had achieved. Many have written on this great event in physics, which produced books and ideas that were, and are, unlike anything that the human race had done before. Here was a genius attempting to make all of physics clear to Caltech undergraduates. It was a creative act of very high order. He tried to make the underlying ideas of physics correct but understandable, and without advanced mathematics that freshmen had never seen.

We had no access to Feynman personally during this time. Mary Lou could not ask him what he had meant. That would not have been appropriate. So she asked me. Feynman had gone up the mountain. He was "in a zone". He was working full time on the lectures; it was up to the team to preserve them. Matt Sands was almost always Mary Lou's point of contact, though Bob Leighton also was contacted from time to time. She would take him what she had put together, and he would take her product on to others, armed with his own notes from the lectures. Together they tried to blend prose and presentation. I had very little to do with all of this; I was just involved continuously in that one peripheral way. I listened to more of the Feynman voice on tape than did almost any other physicist, trying to reconcile what was on the transcription with what was on the tape, and to make sentences of it. Often my help to Mary Lou was only to clarify a word; "He said *entropy*" for example. After forty-eight years, I still hold a great fondness for that adventure. It remains a high point in my life in physics education. My three bound volumes are among my precious possessions. They inspired me to resolve that I too would, if given the chance, make extra effort in my own physics presentations during my own faculty career.

Family Life

We resided for two years at 583 N. Mentor Avenue owing to Donald's birth and to the birth of our second son, Devon Charles Clayton, on May 24, 1961. Devon's birth, also in Huntington Hospital, was, of course, also by caesarean section. In the N. Mentor single-story complex, ours was the front cottage within a Spanish styled horseshoe of modest two-bedroom cottages. Our rent there was waived in return for my being on-site handyman, contact point with the owner about conditions, and rent collector.

Early in 1960 I finished the other half of my PhD thesis, the mathematical formulation and solutions of the *s* process for heavy-element nucleosynthesis. Fowler and I submitted our paper "Neutron Capture Chains in Heavy Element Nucleosynthesis" on 15 July, 1960 to *Annals of Physics*[4]. I had just turned twenty-five years old and our first son was just six months old. This paper became one of the landmarks of nucleosynthesis, causing Fowler to claim "the *s* process is now a solved problem." This was an exaggeration in that the physical circumstances and details of its operation in stars was unknown for decades, and to some degree still is. My PhD thesis defense had to be postponed until August 1961, however, because Fowler was off for another year in Cambridge. Feeling that we should celebrate during the 1961-62 academic year, we seized an opportunity to reside at 729 Lakewood Place, just one block south of the intersection of California Street and Lake Avenue. Lakewood Place is a curved tributary of South Lake Avenue, from which it bows outward to the west but rejoins one block further south. Its houses lay on the west side of the curved street, its east side being a small parklike island separating it from Lake Avenue. We leased this house for a year from a UCLA economics professor, Ed Rada, who lived in Pasadena despite teaching in Westwood. This boggled my mind, but such was California. The location was princely for Caltech. I easily and quickly walked to Caltech each day, just four blocks to the east along California Street. It was a very handsome house, in Spanish style with stucco outer walls and a red tile roof. This made it exciting for us as a place to live, our first venture into the post PhD world of paid employment for doing physics, and therefore a bit of a status symbol for us. Its two floors included three bedrooms upstairs and a small room off the main bedroom that we used for the baby bed and crib for Donald and Devon.

Devon was just two months old and Donald sixteen months old when we moved in, high in hopes but struggling in our intimacy.

With my *s*-process project finished in 1960, Fowler again took sabbatical leave in Cambridge, one of the loves of his life. His affection for it helped imbue Cambridge with a magical significance in my own mind, one that would blossom during my own seven-year period there with Fowler and Hoyle, an epoch still six years away. During Willy's year away, Tommy Lauritsen asked me to post new findings in nuclear astrophysics onto the Kellogg bulletin board so that faculty and students could keep up with developments during Fowler's absence. This was a good assignment for me, causing me to integrate several topics into my own growing awareness. In retrospect I find it humbling that Tommy would ask me to do this, for it is clear to me how green I was in general matters of what would become my life's work. When one becomes the world's leading expert in a scientific topic, as I had unquestionably become on the *s* process, one does not automatically become knowledgeable in the many larger science topics that are impacted by it. I do not think I did a very good job in keeping Kellogg up to date; but no one complained.

Tommy and Marge Lauritsen were among our guests at more than one grand hard-drinking dinner at our house on Lakewood Place. Mary Lou was a hostess extraordinaire. We fancied ourselves contributors to Caltech's image for hard playing and hard working. To some degree these parties helped hide from ourselves an advancing disintegration within our marriage. Mary Lou began to smoke far too much and lost weight. She became depressed after Devon's birth, to the extent that I sometimes found it necessary to go home at mid day to change his diaper and clean him up. That crisis lasted but three months; nonetheless, our views of each other deteriorated further. I find this all quite sad, especially so because of the guilt that I now feel in thinking that, had I been more maturely purposeful about serious issues of marriage, I should have been able to help her more. I tirelessly offered help with chores, but these lay in the realm of *assistance* rather than of those deeply personal realms that are so decisive for well being. I honestly judge myself a failure as a husband. No rationalizations will be offered. I was too immature and to preoccupied with myself. It is also sad that the many good antidepressants available today were not available then, because they surely would have helped Mary Lou's condition.

Calculating the *s* Process

What precisely had I achieved scientifically during graduate school? The answer has several parts requiring some technical discussion. It is also subjective, involving one person's opinion on what he accomplished. It is in part the hope of tangible results of high significance that inspires scientists to Herculean efforts. So indulge my views.

In the first place, my experiment on the three-million-volt Van de Graaff accelerator showed, contrary to earlier evidence, that the mass-13 isotope of carbon would not be rapidly destroyed during hydrogen fusion in stars. Performing that experiment and evaluating the physics of Bethe's carbon-nitrogen cycle in which ^{13}C is created from ^{12}C, the lighter more abundant isotope of carbon, endowed me with lifelong appreciation of experimental nuclear astrophysics. I gained the confidence to get in the ring and box with the heavyweights of thermonuclear power in the stars. By the same token it preserved ^{13}C as not only an abundant nucleus in theory as in fact, but also a nucleus toting an excess neutron that can be liberated in *s*-process settings by reacting with hot helium ions. The significance of that fact grew as decades passed, when it would become increasingly clearer that ^{13}C is the major neutron source for the *s* process in stars.

Secondly, my work provided early support to Fowler's strategic goal of making Kellogg Lab a hotbed of nucleosynthesis theory. Understand that very little nucleosynthesis theory had been attempted in Kellogg Lab when I joined that effort in 1957. That newness is not widely appreciated, even within the science community. The common assumption being that, because the new review paper (*B²FH*) coauthored by Fowler was to become quite celebrated in nuclear astrophysics, Kellogg Lab must have been at that time quite active in nucleosynthesis theory, just as it had been in the laboratory measurements of nuclear reaction rates for the stars. But such is not the case. I had been the first research student taken on by Fowler to work on nucleosynthesis *theory*, as opposed to *laboratory* measurements of relevance to nuclear astrophysics. I was Fowler's wedge for creating a Caltech presence in the theory of the origin of the elements. Without that transformation, the theory of nucleosynthesis would remain Hoyle's theory, a product of Cambridge University. Willy wanted that leadership for Caltech. Oblivious myself to such issues of institutional reputation, I did help to make nucleosynthesis theory a Caltech specialty.

Thirdly, my creation of a quantitative description of the time-dependent abundances produced by neutron captures in the stars was a major step in the transition within Kellogg Lab. My work made it possible to calculate numerical answers for the *s*-process abundances starting from iron, a process that is responsible for the natural abundances of about 120 of the stable isotopes of the elements heavier than iron. I showed that these were created in low-mass stars as a byproduct while they are fusing helium into carbon. The quantitative theory, in contrast to simply its phenomenological ideas, was such a numerical success that the origin of our elements in stars was an idea now beyond doubt. What my Ph.D. thesis calculated was the temporal evolution[5] of *s*-process abundances. That new approach to the *s*-process abundances sought those superpositions of the time-dependent solutions that I had found that could, when summed together, reproduce the solar abundances. It was in those superpositions that the astrophysical circumstances lurked.

This opportunity was my good fortune and its success made my subsequent career possible. I had been given the chance to become a pioneer of nucleosynthesis by giving better definitions and time-evolving quantitative treatments to the static outline in (B^2FH). The structure of thermonuclear nucleosynthesis in stars, which had been introduced by very important papers by Fred Hoyle in 1946 and 1954, had not been improved upon until Fowler, with his inimitable infectious enthusiasm, created a bandwagon for nucleosynthesis theory. His first step had been publication in 1957 of the review paper by Burbidge, Burbidge, Fowler and Hoyle (B^2FH). This review achieved iconic status by being cited *pro forma* by future works as a catchall for the idea of nucleosynthesis in the stars. It came to be called simply *B²FH* among astrophysicists, the square of the *B* standing for both Burbidges, and began a trend in journals by which influential works could be cited *by acronym*. *B²FH* had been submitted as a review paper for the ideas of nucleosynthesis in stars, albeit with considerable new material concerning the synthesis of the heavy elements. A *review* paper is one submitted to a review journal to summarize the state of a research field. It had done that profoundly well for the elements heavier than iron; but, as I slowly realized[3] during the following decade, did not review Hoyle's picture for the origin of the common elements lighter than iron. This weakness went unnoticed as nucleosynthesis developed. *B²FH* became a science icon, as well as the default citation for new researchers entering this field. Fowler described how Ed Condon accepted it for *Reviews of Modern Physics* without peer review, causing Fowler to quip, "Those were the days".

My primary research accomplishment had been making the s process something that could be calculated. To share what this means requires some technical discussion. The *B²FH* treatment of the s process had focused on nuclear correlations between abundances in solar system material (earth and meteorites). It had not presented a calculable theory. It described a simplified correlation between the abundances of the isotopes of the elements and the nuclear properties of those isotopes. It did so by assuming the existence during s-process nucleosynthesis of a steady-state situation in which the abundances did not change with time[5]. They sought the values of abundances required in order that their values would not change during passage of time in a fixed environment. Each isotope was to be created at the same rate that it is being consumed, and that balance implied a specific value of the abundance of each isotope. That balance also required that every isotope in the chain be destroyed at the identical rate. One single destruction rate would characterize each of the 120 s-process isotopes. These were very restrictive assumptions, and *B²FH* had made them for both the s and the r processes. The mnemonics s and r stand respectively for the *slow* capture of free neutrons in the star and the *rapid* capture of free neutrons in a separate stellar setting. Owing to the assumption that the abundances were not changing with time, their treatment could not address larger astrophysical questions. In the real world abundances do change with time. Their description, by assuming that each isotope was destroyed during the *s* process at the identical rate, required that the abundance must be inversely proportional to a purely nuclear property, the neutron-capture cross section. Those isotopes having smaller capture cross section must have larger abundance, and conversely. This inverse correlation was approximately correct over a small range of atomic weights, as my work confirmed it should be, but not correct over a large range of atomic weight. The inverse correlation of abundance with neutron-capture probability was also not new, although posterity tended to attribute it to B^2FH because they clearly expressed an astronomical context for it.

The time-dependent nature of nucleosynthesis was acknowledged by B^2FH only by another physically incorrect assumption; namely, that the actual rate of destruction of each isotope would decline smoothly with increasing atomic weight (instead of maintaining a constant value). They drew a cartoon-like smooth curve to illustrate that expectation. That sketch was not based on any calculation, as many believed. It was a hunch. They did not address whether this global decline happened in individual stars or was the net product of galactic growth of element abundances. Like much of B^2FH, one could not calculate abundances with their description.

Because observations by Margaret Burbidge, by Jesse Greenstein, and by others showed that *s*-process barium was enriched perhaps 20-to-50-fold in some red giant stars, Fowler had asked me if I could build a quantitative theory of the *s* process that would show if fifty-fold enrichment of barium is reasonable within a single star. Large overabundances could not be produced by merely postulating equal destruction rates for each isotope. Rather it required transmutation during time of some abundant seed nucleus to atomic weight A=138, the atomic weight of barium. Realizing this, Fowler asked me if neutron irradiation of iron, which was much more abundant than barium in the stars, could produce a large overabundance specifically of Ba. That was the speculation stated in B^2FH. We discussed that this would require a time-dependent *s*-process theory that could compute the spreading in atomic weight of the seed abundance as it moved toward larger atomic weight. I had welcomed that challenge, thankful to this day for Willy's having taken me along to Jesse Greenstein's colloquium.

In later years I explained how I generalized the *s* process by an analogy to the populations of insects. One can ask what the populations of insects would be if they exist in a steady state, that is if their populations are to remain unchanged as time passes. That requirement requires that the birth rate of each species be exactly balanced by its death rate. Only then can the population be unchanging. In that case one can calculate what the abundance (population) of every insect would be. It would be proportional to the birth rate of each insect (number born per year) and also proportional to the average time that each insect lived when all causes of mortality are taken into account (natural death, predation, disasters). When ecologists observe insect populations, however, they find something quite different. Many species go through dramatic population cycles— two slim years followed by a year of decided overpopulation, say, or in the case of the cicada, a seventeen year cycle! To understand these things, population biologists had to consider the time-dependent situation. They had to discard the simple idea that birth rate balanced death rate. They had to consider how the changing environment and the changing population of one insect influenced that of another. That sophistication was momentous for population biology. This is similar to what I achieved in my formulation of the *s* process (and soon thereafter of the *r* processes) of nucleosynthesis.

I had succeeded in answering Fowler's questions. My thesis devised new analytic solutions of the transport and spreading of nuclear abundances as time passed, giving the Ba overabundance produced by neutron irradiation of iron as a function of time. The answer was that Ba overabundances as large as 10,000 could in principle be achieved from such irradiation, but only if the neutron exposure had occurred for precisely the optimum time. But for other durations of irradiation the Ba abundance varied by huge factors. A second problem for the optimum neutron exposure was producing too many of those heavy abundances having atomic weight A>130. So such special history was ruled out. I was able to demonstrate that no single neutron exposure could approximate the solar abundances. The iron in galactic history had been exposed to varying numbers of free neutrons, irradiations that took no single value but instead covered a range of exposures[4]. I showed that this range had the requirement that the fractions of iron that had been irradiated had been smaller and smaller for increasingly larger values of the neutron exposure. And most iron had not been exposed to neutrons at all.

As a result I showed that the destruction rate of each isotope was not constant as a function of atomic weight, as B^2FH had assumed, but had regions of near constancy between the so-called magic numbers of nuclear structure, N=50, 82 and 126. And the destruction rate declined steeply in the vicinity of each magic number. It soon became apparent that the natural abundances showed exactly that same "ledge-precipice structure", as I dubbed it[4]. I had shifted the emphasis to the distribution of neutrons to which Fe has been exposed during the natural evolution of stars. By requiring a distribution that could produce the solar s process abundances in each red-giant star, I concluded that the Ba overbundance was about equal to 50, just as required by astronomical observation. Willy was ecstatic with these new insights, bragging about my work to others, "It's a solved problem"

With that proud claim by Fowler, my apprenticeship ended. In the language of the guilds I had earned my journeyman's rating and could take up the trade on my own. The Germans call their thesis advisor *Doktorvater* (Doctor Father), because that apprentice relationship is as meaningful and formative within a science lifetime as the father relationship is within life itself. Nor did our relationship dissolve thereafter. Indeed, it intensified after my Ph.D., but the relationship became one of colleagues rather than of master and apprentice. Fowler deserves substantial credit for what I achieved, because it was he who asked me key questions and inspired me to find their answers.

Despite the success of my work we realized that more secure evaluation of s process nucleosynthesis would require better neutron-capture cross sections for the isotopes along the s process chain. To this end Willy sent me in 1962 to Oak Ridge National Laboratory as an invited visitor, so that I could serve as stimulus for an experimental program by Dick Macklin and Jack Gibbons. Macklin and Gibbons did indeed open a new systematic chapter in neutron cross sections with their experimental setup. I visited that program again in 1964. Their review paper[6] on their findings was published in 1965 in *Reviews of Modern Physics*. I had been very pleased when their acknowledgements thanked me for my help. It was the first occasion on which a scientist thanked me for help on his problem, and it left a lasting impression on me, drawing me more deeply into the spirit of science. Dick Macklin remained my lifelong friend and colleague on the *s* process, and Jack Gibbons later become science advisor to President William J. Clinton.

As I neared the end of my stay in Kellogg, Willy hoped that we could similarly elevate the steady-flow *r* process of B[2]FH to a time-dependent solution that could address supernova dynamics. The rabbit-out-of-a-hat-star of Kellogg parties, Phil Seeger, magician par excellence, took the lead on this by building improved semi-empirical formulae for the masses of nuclei that participated in the r process, a crucial need for any *r*-process calculation. Willy and I collaborated with Phil to calculate the first time-dependent *r* process. We traced the evolution (in fractions of a second) between short-time solutions limited by the duration of the free neutron flux, and the long-time solutions characterized by cycling caused by fission of the heaviest nuclei. Our 1965 paper[7] was another pathbreaker, setting the standard for two decades for *r* process calculations.

Frank McDonald at SMU had been proven right. Graduate school had changed my life by opening the opportunity to have a career of scientific research. My gratitude for that will not die. I was also very lucky to have been Willy Fowler's first graduate student in the theory of nucleosynthesis. Fowler saw that nucleosynthesis had a great future as a discipline of physics research. He proposed to the National Science Foundation in 1960 to form an NSF sponsored research team in this subject. That grant was approved, so that after my thesis defense in August 1961 I was offered one of four postdoc positions that this grant would support. My appointment as *Research Associate* allowed me to remain at Caltech for two more years of exciting investigations. My work during that period featured the formulation of the *r* process with Fowler and Phil Seeger and my unexpected discovery of the radioactive clocks for nucleosynthesis.

The age of the chemical elements

It is a measure of a good new theory that it immediately finds new applications and explains misunderstood facts. My treatment of the *s* process certainly had that property. The most glamorous of these was my discovery of radioactive clocks for nucleosynthesis. Radioactive clocks had been used in geology to determine the ages of rocks and even of the earth itself. The central idea is that when a radioactive element decays it transmutes into a stable element, uranium decaying to lead, for example, and if this occurs within a rock one can tell how long the rock has existed by measuring the buildup of the stable daughter within the rock. Claire Patterson at Caltech had measured the age of the earth by determining how much the isotopes of the heavy metal, lead, had been built up by the decay of the radioactive uranium on earth.

My new radioactive clocks based on the *s* process were able to measure a quite different age; namely the age of the elements themselves. The nucleosynthesis theory explained how the elements were created but it did not explain *when*. By using the *s* process I could determine what fraction of the daughter abundance had been created by radioactive decay of its parent[8] after nucleosynthesis began. As in the common hourglass, if one knows the rate at which grains of sand drop into the lower glass one can determine the time when the hourglass was inverted from the amount of sand buildup in the lower glass. My discovery of this method enabled me to demonstrate that nucleosynthesis had begun 13-17 billion years ago, three to four times the age of the earth.

"Willy", I exclaimed bursting into his office, "I have found a radioactive chronometer for nucleosynthesis!" I had spent several days drafting my initial calculations so that I could show him.

"What *are* you talking about!" he exclaimed, looking up from his work. I showed him how I had been able to calculate that half of the abundance of an osmium isotope had been created by decay of radioactive rhenium. After a long exchange based on this new idea, Willy congratulated me with sincere admiration.

"How did you find this?" he asked. "Fred and I searched high and low for a long-lived radioactive nucleus that we could use."

"You probably assumed that it would not be possible to determine what percentage of the osmium was the result of radioactivity and what percentage was the result of nucleosynthesis" I suggested, and he agreed. "Don't you feel that we should write a paper on this?"

"*You* should write a paper on this, Donny boy" he replied. "This is your discovery. If *we* write the paper people will assume that it was my discovery. But first, submit an abstract for a talk at the American Physical Society meeting, and go present it there. The Navy will pay for your travel!"

As long as I live, I will never forget that just and selfless act by a great man.

My discovery won instant fame. This proved to be important to me, by the way, during my job search for the years after I would leave Caltech. When I was scheduled to present my method and results at the *American Physical Society* meeting at the Sheraton Park Hotel in Washington D.C in April 1963, the science press requested a press conference. *Wow! Me give a press conference.* I have saved their telegram requesting my appearance. In the press room I passed out a sheet recapping my talk and answered questions to a bustling press corps. To one reporter's question I replied that my methods had ruled out any possibility that all of the elements were created at the same time; and to another I explained that my method could be made more accurate in the future by improved measurement of several nuclear properties. I was separately interviewed by famed science writer Walter Sullivan, whose report appeared in the *New York Times*. A story entitled "Happy Birthday" appeared in *Newsweek*. Front page stories were printed in the *Washington Post* and *Washington Star*. I became immediately well known to the physics world.

My s-process research had stimulated this discovery in an eye-opening way that shows how one stumbles onto discoveries rather than plans them. Bear with me for a few technical details. While decomposing the solar-system abundances into their *s*-process parts and their *r*-process parts, a project to which I had turned, I noticed that the mass-187 isotope of one very heavy metal, osmium, the 76th element of the periodic table, was too abundant for the theory. The ^{187}Os abundance was about twice as great as it should have been. I at first feared that the *s*-process theory was not working. Trying to resolve this I looked up the properties of the mass-187 isotope of rhenium, the 75th element, a neighboring heavy metal. I was stunned to see the answer immediately. The nucleus ^{187}Re, which was a product of the *r* process rather than of the *s* process, was in fact not stable, but was radioactive and decayed very slowly to ^{187}Os by radioactive decay. The ^{187}Re halflife was very long, 40 billion years. This halflife was longer than the age of the earth or even of the estimated age of the universe! Nonetheless, I reasoned that if the age of the elements was something of order three times the age of the earth, a sufficient fraction of the ^{187}Re would have decayed to ^{187}Os within the interstellar gas of our Milky Way galaxy to have approximately doubled the ^{187}Os abundance therein. That solved the problem. In my 1963 *Astrophysical Journal* paper[7] explaining this discovery, I repeated my Washington statement that the preliminary age was 15 billion years. The elements were roughly three times as old as the earth itself. "Happy Birthday" indeed!

Why does this topic fascinate people? It speaks to the origin of our material world by showing, first of all, that an origin existed. The elements have not always existed. It shows that some cultural "just so stories" are false. The notion that the elements and the earth were created at the same moment is shown to be demonstrably false. An alternative idea, that the elements have always existed, is also shown to be demonstrably false. Yet another idea, that the entirety of the heavy elements was created at any single moment of time is also demonstrably false. This is exactly what the scientific method consists of; namely, measurments that show a previous belief to be false. The history of science is in fact a huge rubble pile of false and discarded ideas. Like the refining of gold, scientific knowledge is the residue of improved ideas that meet all challenges. In this case, the creation of the heavy elements was demonstrably spread out over a very long time. Science had again tightened the noose on popular beliefs. Many people were forced to relinquish their comfortable images of what has occurred. And if the creation of the elements did indeed have a beginning several earth ages ago, why did it begin then? Why not earlier, or why not later? What was happening in order for this first birthday for the heavy elements to have such a date? Today we can see that the big-bang origin of our universe provides the natural answers, and that my methods had pointed to the need for a *Big Bang*.

SINS in Kellogg

The responses to Willy's job listing of two-year NSF Research Associateships were outstanding: Richard Sears came from Michigan to build computer models of the sun, with an eye on predicting the rate of arrival of neutrinos from the sun; John Bahcall came to work on the weak nuclear force in astrophysics, especially electron captures in the sun and stars; and Icko Iben came from a teaching position at Williams College to build a computer program for automated computation of advanced stellar evolution. We all began our positions together in 1961. Digital computers had just arrived, and they changed our tools more than I could imagine. Realize what fun those next two years were, working with those three and with Phil Seeger, with Willy riding shotgun. One of us had the bright idea (I think it was John Bahcall) to name the team project *Stellar Interiors and Nuclear Synthesis*, which could be known by its acronym *SINS*, much to Willy's delight. Our weekly *SINS* seminar became well attended. Even Richard Feynman attended occasionally, as he did when I gave my talk announcing the radioactive chronologies. When in response to two of his penetrating questions I finally said, "You could be very good at this", Feynman's face first looked shocked, then his eyes narrowed in a wide grin as the audience burst out in laughter.

CHAPTER 8.   VENTURING OUT ON MY OWN; TEXAS ONCE AGAIN

I had not sought to move back to Texas. I did not feel that my heart was in Texas. Caltech and life in southern California had opened my eyes to a more free-thinking mindset than I associated with Texas. I wanted that open mindedness. But as my two-year postdoc at Caltech was drawing to a close without a subsequent job in hand, I began feverishly to seek academic positions. I could not predict my future, but at this time I certainly wanted to remain in academic life.   A life of scholarship was the life for me. So I talked to everyone I could about academic positions. I discovered that I really should have started this process earlier. Stanford offered some chance associated with nuclear physics, but they seemed wary of nucleosynthesis as a research specialty. I tried the University of California at San Diego, because Willy thought that their excellent program in meteorites might provide a suitable home for study of the origin of the elements. I was invited to give a talk there, and that was well received, but they were really seeking someone to experimentally pursue the origin of the meteorites and of the solar system. Although it was a good time generally for faculty expansion at universities, my nucleosynthesis specialty did not fit easily into existing departments. No university save Caltech and Cornell could point to a nucleosynthesis expert among their faculty. But the biggest problem was that I was too late beginning my search for many openings.

The Department of Space Science, Rice University

My frequent tennis partner at Caltech, Curt Michel, who had been a physics undergraduate at Caltech, told me that he had applied to a faculty search at Rice University, which was just then forming a new Department of Space Science. Rice was not Caltech, but it had an excellent reputation.

"What do they mean by *space science*?" I asked.

"I guess it will be pretty broad," Curt replied. "I asked the Chairman, who is the only one there now, and he thought it could be anything we want it to be if associated with science in space-- but not space engineering."

Curt's plan was to become one of the scientist astronauts, so an appointment in this new department would be ideal for him. He would focus on near-earth physics, the kind that would be relevant for astronauts in space. That would include the earth's magnetosphere, the Van Allen radiation belts, the aurora borealis, the solar wind and the magnetic storms caused by its interaction with the earth's magnetosphere—things like those. Curt would become a recognized expert in those topics. The department's name, Space Science, was, though vague by academic standards, precisely right for NASA relationships, and the proximity of Rice to the Johnson Space Center was perfect.

"Do you think that the origin of the solar system and of its chemical elements would fit in, or would that be too esoteric?" I asked.

"I think it would fit; but call Alex Dessler, the Chairman, and ask him. Basically I think he is deciding such things now."

Curt could not clarify the goals of this new department because they had not been defined. So I searched out the ad in *Physics Today,* the monthly magazine of the world of physics. The ad mentioned hoping to explore the origin of the solar system, which I saw related to the origin of the elements and the history by which they came to exist within the newborn solar system. I telephoned Alexander J. Dessler, first chairman and in fact the only faculty member currently in place in the department. He told me that he had already hired a specialist in scientific satellites, Brian O'Brien, from Van Allen's group at the University of Iowa, and that he planned to hire Curt Michel. Dessler was receptive to the science perspective that I outlined and he vigorously encouraged my application. He promised to speak to President Kenneth S. Pitzer about the possibility of a fourth faculty member for 1963-64, the first year of existence of the Department of Space Science, inasmuch as he had been given authorization to hire only three.

Because it was almost April I informed Dessler that I would soon travel to the Sheraton Park Hotel in Washington D.C. (April 1963) to present my discovery of Rhenium-Osmium cosmochronology by a contributed talk, and also respond to a request to a press conference there. Dessler leaped at that! He asked for full details of this press conference. I described the Telegram that I had received asking me to speak to the press. He took that information to President Pitzer as part of his proposal for a fourth faculty member. They checked about the press conference by a telephone call to the Press Officer of the American Physical Society. The upshot was a telephone call to me at the Sheraton Park Hotel during the meeting, offering to me the newly authorized fourth position, Assistant Professor of Space Science. My salary would be $8,300 for the nine-month academic year, with augmentation at the same monthly rate for research during the summer.

I experienced strange *deja vu* returning to Texas. To do so would retrace my father's path, and thus mine as a young boy. My father had ventured from Iowa to Texas to build his life. Leaving Iowa at age five had left within me deep unrecognized scars. And now I ventured from Caltech to Texas to try to build my life when I really did not want to leave Caltech. Leaving Caltech was hard, and I was unsure that I could make it on my own. Such feelings were not uncommon.

My parents were delighted at the news that our relocation would be to Houston. That seemed comfortably nearby. Our Pontiac Firechief made its last drive across the southwestern deserts, to

Dallas, where we settled in for a weeklong reunion with my family and with Mary Lou's parents. During the middle of that week my father sat down with us to disclose one of the many forward looking, optimistic acts of generosity and love that so characterized him.

"I think you might want a newer car to begin your faculty position at Rice," he said.

"But the Pontiac still works quite well," I countered, already sensing the familiar inner cramp at the thought of giving up part of my life, one that he had given me. I did not know why I was that way, fearful of losing something precious. Was he also remembering my five year-old panic when relinquishing our initial DeSoto after its drive to Texas?

"Well, I found a really nice Buick LeSabre at a dealer downtown," he carried on. "Let's go have a look at it."

We drove down to Harwood Street in downtown Dallas, and there on its used-car lot was a two-year-old 1962 LeSabre Sport Coupe, a desert-sand-colored two-door beauty with a light brown roof. Its two doors were each huge, bestowing a rakish but elegant look and allowing easy access to the rear seats behind tip-forward front-seat backs. I was smitten, and traded the Pontiac Firechief on the spot. My father beamed, knowing that that is what he would have done. For $3000 it was a good deal. I was to love that car for twenty years. And, yes, it was indeed nicer for a new Assistant Professor of Space Physics. My father always had an instinct for looking one's best, one's most convincing. I tried to adopt that habit in my own life.

My parents invited us to leave Donald and Devon, then ages three and two, with them while we sought a place to live. After driving the new Buick to Houston Mary Lou and I met Alex Dessler at Rice University. He showed us our initial suite of four offices on the second floor of the geology building, assuring me that plans were in the works for a new space science building in a couple of years. He recommended buying a home in Meyerland, where he resided, but I was not ready to buy. I was much happier renting a nice one-story furnished home on Georgetown Street in West University Place, just a mile west of Rice University. This setting was emotionally very comfortable for me. It reminded me of my childhood home on Sherry Lane, and West University Place resembled University Park in Dallas in many ways. Throughout life I would love homes that reminded me of my childhood.

The Course and the Book: *Principles of Stellar Evolution and Nucleosynthesis*

Alex Dessler, my new chairman, scheduled a second meeting for the next afternoon at which we discussed the teaching of classes in space science. Rice University had at that time no undergraduate majors in space science. What we needed were courses suited for beginning graduate students. I proposed to offer a course in the physics of the structure and evolution of the stars. Work with Richard Sears and Icko Iben at Caltech had made me very interested in that subject, but I hardly dared hope that doing so could be my academic assignment for the next three years. I was ecstatic when he enthusiastically endorsed my suggestion. The truth is that Dessler was not at all sure what courses would be most appropriate because he was not sure what our department would become. Its title, Space Science, had been chosen to ride political enthusiasm for what Houston and the Manned Spacecraft Center might become, so it was appropriately noncommittal about topics of study. Alex then stimulated my ambition by expressing his hope that we could produce a series of textbooks for space science. I resolved to write one because I had discovered by attempting to learn this subject myself that there was a serious textbook vacuum emphasizing the physics of stars and nuclear astrophysics.

During 1963-66 I taught that course and produced a typed set of notes for a textbook entitled *Principles of Stellar Evolution and Nucleosynthesis.* An embarrassing aspect of this lay in my proposal to record my lectures on tape and have them typed as a rough draft of the book. That is how Feynman did it. The result was worthless and was discontinued after two months. In the first place, I was no Feynman. His magisterial view of physics had no parallel in my tentative and halting efforts to decide what to include in my approach to the stars. In the second place, what I said in class was not even a crude approximation of what one should write. This wasted time and money, causing me to blush at my vanity.

On the other hand, having to develop physics treatments of the ideas and to decide what to include for graduate students was pure gold. It became my driving motivation for the eventual product. We enrolled eager graduate students, and the course drew enrollments of six-to-ten students annually. The large enrollment resulted from students having only four classes offered, because Dessler had been able to get administrative go-ahead for the existence of the department that recognized that a single course from each faculty member was both appropriate and adequate in so new an academic discipline. I found that the questions by these eager students and the need to create physics problems that illuminated the principles of physics guided my expositions. Of the many original problems that I designed, which were constructed to address the physics of pressure within the stars, the physics of energy flow, the physics of the opacity of stellar matter impeding radiation flow, the physics of nuclear power generation capable of keeping stars hot, the physics of structural stability, and the physics of nucleosynthesis, many found their way into my textbook. The result was a textbook designed to actually help students understand the things that they would need to know and that were not well explained elsewhere. I was not tempted to write a monograph intended to impress my peers. The record shows that this was a brilliantly successful project for me.

Although I and our sons were settling happily into our Georgetown Street home, Mary Lou began planning almost at once to find a different house. I discouraged that, wanting to live without distractions for the next few years in order to flesh out my Rice career plans. Emotionally I was happy in a setting that reminded me of my happy childhood home; but Mary Lou moved to the beat of a different drum. This house also resembled her childhood home on Purdue Street in University Park, but that was a home from which she emotionally wanted to escape. Her parents had a sterile marriage and uninteresting daily lives. At the time I had no understanding of this difference between us; but I did understand that our own marriage tended to slide toward sterility. Our lives were clearly exciting, but our intimacy lacked comparable fire. So, fueled by her zeal, we moved in 1964 to another rental house of different character, this one south of Rice on Maroneal Street, near the Shamrock Hotel. It was a traditional two-story having an English architectural character. I agreed that I did like that house better, despite my resistance to moving, and took to it enthusiastically myself.

First Rejection by a Referee

I experienced at this time the realization that scientific referees, even famous scientists, may err in the face of a good new idea. I was writing a short review paper[1] "Chronology of the Galaxy" for *Science*, the weekly journal of the *American Association for the Advancement of Science*, when I became concerned that estimated ages of the globular star clusters[1] exceeded 20 billion years, whereas my ideas on radioactive chronology placed the beginnings of nucleosynthesis near 13 billion years ago. Sitting in our department colloquium listening to a talk about the solar wind the idea hit me: *the sun is now less massive than it once was.* The stars of even lower mass than the sun that are today the most massive of those remaining within the globular clusters might even have lost a larger fraction of their initial mass than has the sun. Their winds might have been stronger than that of the sun. The entire paper danced before my imagination. The mass of helium fused per second at the stellar center is roughly proportional to M^4, the fourth power of the star's mass. Because it produced central helium faster in the past when its mass was larger than at present, the buildup of helium, which is the aging process that astronomers measure, would make the star appear to be older than it actually is. I envisioned a simple calculation that isolates that effect, performed the calculation, prepared a figure to illustrate the result, and sent it off to *Astrophysical Journal* on March 25, 1964, one week after my 29th birthday.

The report of the referee was not received until after a long and significant auto trip east with Mary Lou during June and July. We first visited Oak Ridge National Laboratory, where I had begun two years earlier to consult with Richard Macklin and Jack Gibbons on the topic of their measurements of neutron capture cross sections for nucleosynthesis. Then we drove to Tilton, New Hampshire, where a Gordon Research Conference, *The Physics and Chemistry of Space,* would take place. That too assumed great significance for my later career, because in Tilton I met many leaders of the study of meteorites. So it was not until late July that I first saw the report of the referee on my submission to *Astrophysical Journal.* I could not believe my eyes when reading the referee's conclusion that such an important finding could not be published on the basis of a simple hand calculation. I made revisions of the only suggestions that actually had merit and sent my draft back to the editor, S. Chandrasekhar, with a letter saying that the referee had not understood why the simple calculation was not only adequate but preferable, and how the paper should now be clearer to him. S. Chandrasekhar, a brilliant astrophysicist who would share the 1983 Nobel Prize with Willy Fowler, sent the paper and the first referee report to Martin Schwarzschild at Princeton for adjudication, the standard procedure by editors whenever an author argues that his paper was given an unjust appraisal by a referee. Schwarzschild was himself a celebrated astrophysicist and the reigning expert on stellar evolution, but his report lamely agreed with the first referee. My paper was rejected.

Getting really hot now, I telephoned Chandrasekhar at the offices of *Astrophysical Journal.* "The referees are wrong", I said firmly.

"I know", Chandrasekhar replied. I paused. Had I heard him correctly?

"Do you agree with me?" I asked to be certain.

"I do", Chandrasekhar replied.

"Well, what can I do to resolve this impasse?"

"Send it back to me with another letter", Chandrasekhar replied, "and I will publish it."

He did, without further refereeing.

My paper appeared in the November 15, 1964 issue[2]. Chandrasekhar is remembered as a very strong editor, who read every paper submitted and who actively played a role in shaping this journal. I had found a lesson of great value. This experience gave me resilience when a decade later I fell into tremendous rows with poor referees of even more important works by me.

The paper itself, despite failing to maintain high importance as years pass and new data appear, had those characteristics that would reappear later in many significant publications. It provided a short discussion and calculation of a result of physics that had been overlooked by predecessors. It presented the simplest possible calculation and emphasized the most important effect for clarity and relevance. It was what I call "an interdisciplinary insight", one that emerges from my awareness blend from several differing specialties of science. In this case I needed awareness (1) of cosmology and the apparent age of the universe; (2) of astronomical technique for determining the ages of star clusters; (3) of the physics of the structure of stars and how they age; (4) of the data then emerging on the solar wind, a flow of ionized gas away from its solar source. As one of only a handful knowing each of these areas, I was able to put two and two together. Such interdisciplinary insights would primarily occur to me alone and be published as a single-author paper. These papers would invariably encounter referee resistance by reaching outside of accepted frameworks. "Stellar winds and the ages of old clusters" was the first example of such an occurrence in my life, and, significantly for me, the first example away from the intellectual ferment of Caltech, my scientific womb. I share these thoughts here not in an attempt to toot my own horn but rather to clarify my own sense of grasping for independence from Caltech and how I would achieve it.

Astronomy with Radioactivity: Gamma Rays

During my first academic year at Rice we recruited another new faculty member, Robert C. Haymes. This portended very good luck for me. For his thesis from NYU Bob had been detecting neutrons in the earth's atmosphere, and he arrived with the idea that he might detect gamma rays arriving from outer space. This change of thrust was supported by the faculty, although some favored instead continuing work on atmospheric neutrons. One fateful day Bob approached me with a stimulating question:

"Does stellar nucleosynthesis produce sources of gamma rays by ejecting radioactivity?"

This was a pregnant question. I described a theoretical idea of supernova explosions that produced transuranic elements via the *r* process, which by its very nature must be explosively ejected from the sources in order that the *r* process work properly. Radioactivity from the main line of nuclear fusion reactions, on the other hand, was thought at that time to decay within the star's deep interior, rendering it unobservable. I happened to be preparing a paper defining the dynamics of the *r* process for publication at that very moment, continuing the project started with Fowler and my Caltech classmate, Phil Seeger, at Caltech; so I gave Bob a rundown on its ideas. I described how Burbidge, Fowler and Hoyle had even speculated that the transuranic radioactivity produced by the *r* process might be the cause the slow, roughly 60-day, decline of the optical light after supernova events, as first recorded by Chinese astronomers in the year 1054. Burbidge, Fowler and Hoyle suggested that a newly created isotope of the element californium was responsible. The transuranic elements Neptunium, Plutonium, Americium and Curium, elements not found on earth but that are known from nuclear properties to be created by the intense neutron burst that is the *r* process, subsequently decay to isotopes of the elements Uranium and Thorium, which are found in terrestrial rocks and which also decay but over much longer geologic times to the heavy metal lead. Among my notes at Caltech I had gone so far as to calculate the energy input today to the *Crab Nebula*, the remnant of that 1054 explosion, owing to the energetic alpha particles that are emitted into the nebula by such radioactivity. I had hoped to explain the hints of apparent acceleration of the speed of expansion of that remnant, as suggested by time-lapse photography by astronomers; but I gave it up as probably too small. What had not yet occurred to me was to calculate the gamma rays given off. Gamma rays pass easily through dilute matter, so they could not heat the remnant, but by the same token those gamma rays travel at the speed of light through the interstellar matter of our galaxy to us. That very evening I did a hand calculation at home to show that if the amount of radioactivity was as great as would be required in order to qualify as a sufficient energy source for the 60-day decline of the optical light it would also be capable of producing observable gamma rays at earth. This excited me, perhaps providing the new direction that I felt I should seek. .

I gave this news to Bob Haymes and asked him if he could detect a rate of arrival as small as one gamma ray arriving per square meter each second. That rate of arrival per fixed area is called *the flux of gamma rays*. That was the expectation that I had calculated the night before. I was astonished that Bob thought that his detector could do that, so I asked for a description of his detector system. When he did so it seemed that it might just work! But my calculation needed more careful bookkeeping of its details. In particular, these gamma rays would arrive having very specific energies determined by each isotopic nuclear decay scheme. Each radioactive isotope produced gamma rays having their own characteristic energy. So from my class I enlisted my first graduate student, a young man named Wade Craddock. I taught him how to read from reference books the detailed gamma-ray-line decay schemes emitted following each decay, how to read the halflife of each decay, for they differed from isotope to isotope, and how to use the laws of radioactive decay to then determine the numbers of each gamma ray emitted per second, and to do this calculation for various ages of the remnant. Wade worked zealously in response to the chance he had been given. Within a month the calculation was done, then checked by repeat calculations until I was satisfied. The largest flux from the *Crab Nebula* came from a mass-249 isotope of Californium, ^{249}Cf, having halflife 351 years, just less than half the age of the supernova remnant that is the *Crab Nebula*. We submitted our paper[3] to *The Astrophysical Journal* on January 18, 1965. It opened a new area of astronomical spectroscopy at just the time that Bob's balloon-borne detector would soon stimulate the field of gamma ray astronomy. I would pursue the hope for gamma-ray-line astronomy for four decades. Clearly in this work I could not have conceived it with being technically aware of Bob Haymes' expectation for his gamma-ray telescope.

Robert C. Haymes and the author with the flight gondola
for his new gamma-ray telescope.

There was a fly in this ointment. From my work on this at Caltech I had already come to the conclusion that the so-called *californium hypothesis* that had inspired it was a false hope. Burbidge, Fowler and Hoyle must be wrong. The californium hypothesis was the proposition that ^{254}Cf was synthesized abundantly by the *r* process intense neutron flux and would, with its 60-day halflife, generate the light from the young but aging remnant so that that light might also decline with a 60-day halflife. All of this was a lot to suppose, but this idea by celebrated pioneers of nucleosynthesis was regarded by many contemporary writers as a foundation stone for the reality of nucleosynthesis, whereas it was in fact only an interesting phenomenological suggestion. Before my paper with Craddock was submitted I went through those old numbers again and was again discouraged. If all supernovae ejected that much *r*-process matter there would exist in nature one-hundred-times too much uranium as well as other *r*-process nuclei. To make this idea work restricted the *r* process to occur only in some rare type of supernova rather than in all, one so rare that it might occur only once in 10,000 years. But the number of observed supernovae revealing that two-month decline in the emitted light was observed to be more common that that. So I wrote up my reservations and included it in our paper[3] on page 193. This is what scientists do, to express our reservations about our own work as we display our results. This tacit agreement is an essential part of the scientific process. A scientific paper should not be a sales pitch. Furthermore, I saw no easy way within a supernova remnant following a month of its rapid expansion for the kinetic energy input from the spontaneous fission of ^{254}Cf to be converted into light. Despite praise given to me for the stunning idea of this paper, I privately thought, *No. This is not the explanation for the light curves, and we will not find these gamma ray lines from the Crab Nebula."*

On a more positive note a powerful new idea was set loose. If the correct ideas and the correct telescopes for gamma-ray lines could be designed, detecting radioactivity could test or confirm the theory of nucleosynthesis. That theory was, although already believed to be true by most astronomers, woefully short of experimental confirmations. For that reason alone the halflife coincidence between many supernova light curves and that of ^{254}Cf had elicited much overly enthusiastic admiration. But I now realized that positive detection of radioactivity shortly after a stellar explosion could confirm unequivocally that the event had synthesized new nuclei. Radioactive nuclei are the only demonstrably *new nuclei*. Any stable nucleus could be either new or thirteen billion years old or any age in between. But a radioactive nucleus can not be much older than a few times its halflife. Detecting radioactivity in interstellar matter would prove that nucleosynthesis was happening today, not in some distant unknowable past. This simple new idea had beauty and power. Having myself now given birth to this idea I was determined to pursue it for better hopes than those that had been offered by the flawed californium hypothesis.

Oak Ridge and Gordon Research Conference

The auto trip with Mary Lou to the northest during summer 1964 contained four significant moments for my life—two entirely personal and two scientific. It was a time of maximal stability in our lives, one in which affection flourished. Sailors must have this feeling when after months on stormy seas they land on a calm tropical island, relax in the sun and eat coconuts. We seemed to have recovered from the bad two years that had followed Devon's birth in 1961. The boys were excited to spend three weeks shuttling between their pairs of grandparents in Dallas, so Dallas was our first stop with our 1962 LeSabre. The three weeks together without them released Mary Lou in understandable ways so that they count among our happiest times. We had not seen the southeastern states, which provided our route to Oak Ridge, Tennessee. We were especially moved while leisurely surveying the Vicksburg battlefield, my first taste of civil-war history save for the slight remarks passed down from my great grandfathers Clayton and Kembery, who had both served.

Oak Ridge was our first science destination. Willy Fowler had sent me on my first visit in 1962 to Oak Ridge National Laboratory in order that I might assist the experimental program by Dick Macklin and Jack Gibbons, which was designed to measure the neutron-capture cross sections of nuclei having high importance to the s process. They had, like I, employed a Van de Graff accelerator, but in a most ingenious way. Pulses of protons lasting only seven nanoseconds (7 billionths of a second) struck a lithium target and released neutrons from the $^7Li(p,n)^7Be$ nuclear reaction. The mean energy of the neutrons could be set by the energy of the proton beam. In this way they had obtained a pulsed beam of 30-keV neutrons, just the neutron energy range needed to measure the capture cross sections for heavy nuclei during the s process in stars. Macklin and Gibbons were now in the writing phase of their review paper on their findings. My only contribution was that of looking over their findings and making some suggestions. But secretly I was thrilled, because I regarded the s process as my baby. With so many new measured cross sections the shape of the s-process σN

curve was now evident, and it revealed clearly the ledge-precipice structure that I had predicted from my PhD thesis. We had a fine social time there, a highlight being invitations to all to the home of Joe Fowler, director of the physics division at ORNL. I there proposed that in honor of Willy we should drink some martinis, but only Joe and I carried through. Their pathbreaking paper was published[4] in 1965 in *Reviews of Modern Physics*. I was and remain proud of their thanks to me in its conclusion. This had been a unique time for me, to also have the chance to contribute to its experimental underpinnings. I will never forget those two visits. J. H. Gibbons later become science advisor to President William J. Clinton. R. L. Macklin remained for four decades a leader in experimental *s*-process studies.

Continuing our drive north we spent one night at the Sheraton Park Hotel in Washington D.C. so that I could show Mary Lou where the 1963 American Physical Society meeting had been held and where the press conference on my findings had occurred. Then our first visit to Boston, where thanks to Mary Lou's planning we had lobster dinner at famed Durgin Park restaurant. We paused at Concord to visit our first revolutionary war site. Our destination was the Tilton School in Tilton, New Hampshire where Mary Lou and I shared a dorm room for the week of the Gordon Research Conference *Chemistry and Physics of Space.* How quaint it was to debate such a huge topic by this relaxed setting of world experts. Today any of dozens of topics on the chemistry and physics of space could itself engage an entire weeklong conference having more participants than did these small and unpretentious meetings held under the umbrella of the Gordon Research Conferences. Willy Fowler had sent me in 1962 to the first of these as well, and not only did I want to return to the second one but was to remain grateful to Willy for exposing me to this community, which was at that time dominated by chemists studying the meteorites. Meeting so many of the great leaders of that field unwittingly prepared me for a dramatic foray into meteoritics that I would make ten years later. These two conferences were precursors to the annual meetings in March at the *Johnson Manned Space Flight Center* after the return in 1969 of samples from the Moon. For the first time I met Harold Urey, Edward Anders, Robert Pepin, Robert Walker, Heinrich Wänke, John Wood and many, many more who would become very significant for me later.

A Whiff of Iowa Sod

The fourth significant moment of this trip took place at the Kembery farm in Iowa, the home farm of my early years. We had routed our return trip back through the Midwest to visit with grandparents now living in Greenfield, with Aunt Elva and Uncle Art at their farm, and with my cousins. Mary Lou had once visited Iowa with my family years ago, prior to our marriage, young and romantic, finding our sex on that soft soil between the corn rows that provided our only privacy. When I parked our car on the road in front of the Kembery farmhouse of my earliest years, some tightness gripped my chest. I arose from the car and approached the front gate, beside the pine tree that I had so often climbed. I gazed with tear-filling eyes at grandma's farmhouse, where someone else now lived. I could no longer just go on into my farmhouse.

I looked at the corn crib, at the barn, and my eyes sought out the cowpath along which they had always returned. I could see that Grandma's garden still existed behind the house. I could hardly speak with Mary Lou, who wondered how long I would just stand there and stare. I was suppressing sobs, an embarrassment to a 29-year old. My departure from this place had indeed been forever. There was no way back, no way to make it all right. This was what I had feared at age five; and I had been right. That fear lived on in me, as I could now not deny, as did some homesick loss that had inspired that fear.

On that long drive back to Texas I tried to explain myself to Mary Lou, tried to share a grief that I could not name. How can you grasp that amid all the excitement and positives of my life in science, those minutes standing there with tear-drenched eyes were the most important moments of my summer. I trembled at the power of my memory. And when we drove the quarter mile down the hill to the farm of Uncle Herman and Aunt Annie Rechtenbach, as I had done so many times twenty-five years earlier, my joy at seeing them again was choked with tearful sobs. And when back in Dallas we first saw Donald and Devon, saw Donald's joy and saw Devon's caution, his unwillingness to run into out arms, his doubtful glances as if to ask "Are you really back?", I was at first surprised, then disappointed, and then overwhelmed with a sense of sympathy for what he must feel. I ran to Devon and swooped him up into my arms and loved him, trying to make his loss right.

Time-Dependent *r* Process

The paper with Phil Seeger and Willy Fowler on the dynamics of the *r* process was submitted just after our return from Iowa. The description of the *r* process in the famous 1957 review paper by Burbidge, Burbidge, Fowler and Hoyle (*B²FH*) had postulated for the sake of the exposition that the creation rates and death rates of heavy nuclei within the intense neutron flux were equal, thereby putting the abundances of those nuclei into a steady state, unchanging with time. That simplification enabled them to state the main ideas of the *r* process quantitatively. However, they had not actually been able to compute more than a rough estimate for the abundances in that steady state despite the simplification. Our paper, submitted to *Astrophysical Journal* on July 14, 1964, utilized years of work by Phil Seeger to create a new prescription for the unknown masses of those nuclei involved in the process, all of which were quite radioactive and not even known in the laboratory. His improved semi-empirical formula for these masses enabled us to calculate the evolution with time of the abundance distribution resulting from the irradiation of iron. It illuminated not only the shapes of the abundance distributions but also the sequence of those shapes from one key time step to the next. The evolution was very rapid, befitting an exploding stellar core within a supernova. Our figure[5] 12 showed dramatically the change of the shape of the abundance distribution at successive moments 0.44 seconds, 1.77 seconds, and 3.54 seconds after the free-neutron density was turned on[5]. These results showed the likelihood of many *r* processes. Figure 13 showed the same distribution of abundances that would occur if the free neutrons remained present for a long time (10 seconds or more). This was the first truly modern treatment of the *r*-process idea, so that our paper set the standard for subsequent works that would have to address more difficult aspects[5] of the process.

In a nice touch we used the new neutron cross sections measured by the Oak Ridge program to calculate a theoretical fit to the solar-system σN curve for the *s*-process abundances, so that by subtraction we were able to use this new technique to determine what the observed *r*-process abundances actually are. This too was a major step forward, one with which we were well satisfied and one that has been repeated by subsequent workers. What exceeded our grasp was the knotty question[5] of how the explosively expanding matter would cool and how the neutron density would simultaneously decline, so that we were unable to follow the transition from the supernova interior to extrastellar space. This remains a knotty problem four decades after our paper.

I was beginning to first feel my wings of independence at about this time. I became confident that I could work on more problems than those that derived from my Caltech experience. This feeling, halting at first, was very important to me. A difficult period in the research life of any PhD from any leading institution is to find his own voice out in the provinces, so to speak. For two or three years I felt the issue of my independence from the Caltech setting was seriously in doubt as I mostly tried to extend my thesis work to new details. Then I discovered a new voice in an area that I called "extreme astrophysics"—temperatures and densities so high that new phenomena within nuclear astrophysics appear. With a grant from the Air Force Office of Scientific Research I hired my first postdoc, Peter B. Shaw. He and his wife Karen moved to Houston for three years of that support and we became good friends. In 1965 we published in *The Physical Review* a paper[6] with Curt Michel on photo-induced beta decay, a heretofore unknown effect at high temperatures where the density of photons is large and can *cause* beta decay to occur. And in 1967 we added another new process[7] by which Coulomb deexcitation of an excited nucleus caused by scattering of fast nuclei could increase the rate of deexcitation of nuclear excited states. Although these investigations did not spur practical applications because their effects were a little too small at temperatures reached within stars, they justified a new sense of pride and independence within me. Frosting on that cake would occur a year later when I was invited to deliver the weekly Caltech physics colloquium to explain these new processes, placing me on the other side of the fence from my graduate-school years.

Author at the construction site of the new Space Science Laboratory at Rice University, 1967.

A moving episode took place prior to my move back to Caltech in fall of 1967. As space physicists our department faculty members were invited to present scientific lectures to the astronauts at Johnson Space Center. This was appropriate in the sense that Rice University had been involved in obtaining the land on which the Johnson Space Center had been built and that in return NASA had been quite supportive of President Pitzer's plan to create a new Department of Space Science there. That plan had created my faculty position at Rice. I would always feel very fortunate at that confluence of historic events.

The goal of our faculty presentations to the astronauts was to create a familiarity within the astronaut corps of certain fundamental understandings in space science; so I chose to deliver two lectures on the physical structure of stars, with emphasis on the sun. Edward H. White was in these lectures in spring 1966, and I spent leisure time chatting with him poolside at a NASA pool party for participants. He was muscular, in great shape, friendly and modest, a quite impressive guy, and I liked him. We chuckled together that my lecture on the sun's structure would hopefully not be of much use during his next flights. An Air Force Lieutenant Colonel, he had spent 3-1/2 years in Germany with a fighter squadron, flying F-86's and F-100's before becoming a test pilot, and he was genuinely interested in my father's barnstorming days in Iowa. We swam a few friendly laps together before getting a drink from the bar and talking more. Ed had been chosen with the second group of astronauts in 1962 and chosen pilot of *Gemini 4*, during whose orbits of the earth in June 1965 he had become the first American to make a spacewalk, 21 minutes outside the capsule. I valued this contact. It raised my enthusiasm for astronautics. So it was a dark day when shortly after I took a leave of absence from Rice University to return to Caltech for the 1966-67 academic year, Ed White was killed by an Apollo spacecraft flash fire during a launch pad test at Kennedy Space Center, Florida, on January 26, 1967. Virgil Grissom and Roger Chafee also died in that fire.

CHAPTER 9.   GETTING THE CALL

Substantial portions of my scientific life jelled in two of the oldest European university towns. I resided in both for similar seven-year periods of difficult and exciting work, and then withdrew. I feel mystical attraction to this coincidence but try to avoid sentimentality. These European life segments changed me in many ways. Both affiliations would involve life plans that would grow stronger as their relationships progressed, and that would then collapse, suddenly, even shockingly.

The first seven-year epoch occurred in Cambridge, England, where I resided each summer during 1967-74, also during spring 1967 and during fall term of 1971. Two full-year leaves of absence from Rice University liberated much of that time. Ten years later, 1977-1984, the second seven-year epoch occurred in Heidelberg, Germany. I resided there each summer for eight years and twice took yearlong leaves of absence from Rice University, the first as Alexander von Humboldt Senior Scientist Awardee and the second as Fullbright Fellow during my second sabbatical year. During both European epochs I poured myself into creative work and into the cultural and architectural histories of the towns, beautiful middle-sized towns dominated by eight centuries of student life and by the psychological power of medieval roots. On both occasions I pursued romantic family aspirations, and both ended in ruins.

These annual pilgrimages would never have happened had not Willy Fowler persuaded me to return from Rice University to Caltech for the 1966-67 academic year. Fowler was pursuing a dream of his own, and I count myself fortunate to have been part of that dream. Fowler had already embraced Cambridge as the love of his life during two sabbatical leaves there (1954 and 1962). In 1966 his enormous energy and enthusiasm focused on his vision of Fred Hoyle's planned *Institute of Theoretical Astronomy* in Cambridge. That institute was a structural manifestation of Hoyle's plan to establish astrophysical leadership for England in Cambridge. Fowler badly wanted that plan to succeed, for it opened his dream of becoming the Englishman that he would so have loved to be, and that he even considered becoming.

The Cambridge years came over the horizon in the form of a totally unexpected letter[1] from Fowler's Caltech office, dated February 14, 1966. After some preambles it launches in paragraph two into two points in ways that for accuracy's sake I quote *verbatim*:

> *Second, I have heard rumors that you are writing a*
> *book on nuclear astrophysics. As you know, Fred Hoyle*
> *and I are also doing one. If you are doing so, how about*
> *getting together? We have some stuff already written on*
> *H-burning, He-burning up to and including e-process.*
> *I would guess you have s-and-r-processes and nuclear*
> *cosmochronology well in hand. It sounds like a natural*
> *hookup to me. In regard to details-- credit, financial, and*
> *otherwise-- it would be alphabetical, CFH, and the sales*
> *and royalties would certainly be more than three times*
> *otherwise with the H!*

> *Third, is it not time for you to have a sabbatical? If*
> *so, how about coming back for a year. The atmosphere*
> *here is terrific, especially with the fight between the local*
> *quasarists and the cosmological quasarists. In a year we*

*could finish the book. Come back! Bring Mary Lou and the
boys. If Rice won't pay, we will.*

> *With all the old pepper,*
> *Willy*

That was some letter! Although Willy and I were close, he had not telephoned me to feel out his ideas. What this suggested to me was that his mind was made up, and he thought it more likely to succeed by surprising me dramatically with a formal offer by mail than by negotiating by telephone. With regard to its two issues it said to me that he attached sufficient importance to his plan to simply extend a sweet offer from the outset. The book, Clayton, Fowler & Hoyle, would be sure to secure my fame and future. It also occurred to me that he sensed that I was too far along in my own book project for them to catch up, and that if they did not have me they would not be able to get their own done. "If Rice won't pay, we will" spoke to Willy's legendary determination to make the best things happen without leaving them to chance. My having "s-process, r-process and nuclear cosmochronology well in hand" seemed to acknowledge that I had moved all three topics past their rudimentary outlines in B^{2}FH. He knew that by recognizing my accomplishments in this letter, by welcoming me as partner in the select circle of founders of the theory of nucleosynthesis, I could not easily decline.

I spent a full week on an emotional high, weighing my conflicts. I could not put it out of my mind that the current flood of attributions to *B^{2}FH* for all matters of nucleosynthesis theory could be supplanted by references to *CFH*. In terms of intellectual stature that was a heady prospect. But I was troubled that my own book was almost finished. Willy rightly anticipated my reluctance to relinquish that lead. Furthermore, I wanted to publish my book as I had conceived of it, not as an advanced monograph on nuclear astrophysics. I regarded it possible that as a textbook it would have more lasting value than would an eventually dated assessment of the status of nucleosynthesis theory, even though the *CFH* version could be quite good. I was also worried about several sketchy parts of B^2FH that I felt sure needed first to be clarified before any book could assume lasting utility. On the other hand, I did want to return to the excitement of Caltech. Mary Lou definitely wanted to return to that excitement!. It gave her more chances to be the social vertex that lay in her nature. Another concern: I was nowhere near the required years of service for an earned sabbatical from Rice University, and Willy certainly knew that fact without need to figure. I was in my third year on the Rice University faculty, not my sixth. All these thoughts reverberated through my brain for days. But within days I had decided to decline their offer to coauthor the book if it meant holding back on my own.

On February 18 my secretary at Rice typed a three page letter in reply. My letters always seemed to expand into documents. Such is my proclivity for elucidation. I had also decided not to telephone Willy to discuss his invitation. Better would be to answer in kind. Rather than declining his offer I decided to explain why my book was not really in competition with what they wanted to write. Quotes from my reply letter included the following statements:

"...but first let me explain the nature of the book that I am writing. For three years here I have taught a one-year graduate course in the physics of stellar structure, evolution and nucleosynthesis. I have been working on a parallel book. It is specifically designed to be a textbook in the physical principles of the subject, complete with well chosen problems for the student; but it is not an advanced monograph."

I then described its seven chapters, which were, I claimed with some exaggeration, almost finished. In truth I still had much text to write. "I regard them as horses of different colors. I believe both projects are worthwhile", I wrote. I admitted that I was eager to return to Caltech "to dunk myself back into the mainstream". I wrote that the Sloan Foundation had contacted me about a Sloan Fellowship and that I could probably share salary with Caltech. But still, I had declined his book offer, no question about that, and so I put the letter into the mail with some trepidation. I knew that many would question my wisdom. Then I waited.

On February 28, 1966, Willy's typed reply from his Kellogg secretary arrived. I had been sleeping poorly. I ripped it open.

> *Dear Don:*
>
> *It is clear that you are far along on your textbook and should push it to conclusion. FH and WAF will just have to get down to work on their book!*
>
> *My best,*
> *Willy*

I was relieved. Willy had simply accepted my conclusion, and agreed with it, at least on the face of things. But what a terse reply. It seemed to withdraw the offer of also writing a book with them. I wondered what he really thought of my judgment. In no time I learned that I had gained his respect by my decision. During the following week he telephoned me to outline his goals with great clarity. They really excited me much more than the first letter had done. Rather than trying to remember the details after 30 years, I will quote the letter that his secretary typed to me on March 3, 1966:

> *Dear Don,*
>
> *This will confirm our telephone conversation of last week. You will be most welcome, any time, for any period, and with full or partial stipend. Just let me know what you want--I am your humble servant, Sirrah.*
>
> *The best plan would be for you to finish your textbook, spend a few months in Europe for the good of your soul, and then come here to work with Fred and me on our book, so long delayed and so little along. This is <u>not</u> a requirement--you can come to do whatever you d--n well please (as long as you bring Mary Lou!).*

Actually I think the book would go pretty fast if some one person could give it his undivided attention. What is required is to update CFHZ, CF, SFC, FH etc...You could put nucleocosmochronology where it belongs....All sounds great if there was just a spark plug.

In regard to the Europe trip, I have a second thought. I will probably spend the summer of 1967 at Fred's Institute in Cambridge. If you were here for academic 1966-67 we could finish the thing off in Cambridge, and you and Mary Lou could have your time abroad under the best of all possible circumstances.

My best,
Willy"

That is exactly what we did. I accepted the invitation to return to Caltech, which in turn led to the beginning of my seven years in Cambridge. Those years would change my life; but Willy never would get his book. We found too much that needed doing, as I had anticipated. Of some historical interest in Willy's reply dated March 23, 1966 was, "I have recommended to Carl Anderson that you receive an appropriate appointment with C.I.T. for the period September 1966 through August 1967 with one month vacation at a stipend of $7,500. The extra $500 is in lieu of travel and dislocation expenses for your family." The corresponding letter from Carl Anderson, Chairman of the Physics Divison, followed on April 19, 1966. It is some measure of the subsequent inflation that we felt that amount to be quite reasonable. It was constructed intentionally to cover about half of my academic (nine-month salary) as well as half of three months of summer salary from Caltech; in turn, Rice University paid 5/9 of my academic-year nine-month salary, which was $13,700 in 1966-67, plus an equal share of the summer-month salary. Although that total seems small today, it was just in the range where young families felt that their compensation had moved upward out of the minimal range. Most jobs earned near $10,000 per year in 1966, whereas my total, counting the extra salary that academics can earn in the summer if they have the grants to pay for it, was almost $18,000.

Climbing the academic ladder

Another escalation of my salary had taken place during the previous year. It was abetted by a competitive offer from the University of Texas. Their astronomy chairman, Harlan Smith, had hoped to persuade me to move from Houston to Austin; and that opportunity held considerable appeal owing to the large and excellent astronomy department there. Their offer, an Associate Professorship, with promise of continuing tenure, forced Rice University to either match it or wish me well. Alex Dessler, our Chairman, ascertained by conversation directly with President Kenneth Pitzer that Rice would indeed match the offer, enabling me to make the decision on whatever merits I saw. The good luck for me was my promotion to Associate Professor in either case, with attendant increase in pay, and on a somewhat accelerated schedule in comparison with the norm. In a close call, I had decided to remain at Rice University. But the episode taught me two significant things about academic salaries. The first is that because they will grow slowly in the absence of any pressure, competitive pressure was a useful way to move one's salary ahead. All top academics learned this during the 1960s, when competition for good faculty was strong among universities that were then expanding their faculty numbers. Much the same thing became the rule in executive pay during the decades to follow. It is today a rule of business that large pay increases come from marketing one's skills to other companies. I also learned that it is the pay increases early in an academic career that really count. A $2000 raise applies to *every* subsequent year, not just the next one. A $2000 raise for me in 1966 presaged perhaps a $100,000 raise of my career earnings. My early raise had a benefit analogous to that of early contributions to retirement programs. They count far more than later contributions.

It was from this period of my life that I began to feel myself to be among society's affluent members. I was 31 years old and earning more than most workers. I speak not of wealth, that entirely different status of which we can but speculate, but rather of the comfortable conviction that we received a fair portion of society's wealth. For example it is no coincidence that Mary Lou and I would purchase our first house when we returned to Houston after the coming year away. That next year would also mark the beginnings of intellectual and geographic mobility for us, another mark of affluence. Job travel in the mid 1960s was but a glimmer of the mushroom of activity today. We paid attention to our money carefully; but the worries that made us so frugal at the beginnings of my academic career at Rice, already with two small children and some debt from graduate school, began to be displaced by the conviction that we would share the best of American life. This has proved to be the case, and I express a lifetime of gratitude at being well paid for pursuing intellectual frontiers. There exists no greater wealth than this--to be paid generously for pursuing creative efforts that one would choose to pursue even if wealthy. Golf enthusiasts might liken this to earning handsomely by playing the professional golfing circuits. And I believe this to be both fair and wise. The good salary comes not from having a Ph.D. degree. That degree is no meal ticket today, and it was not then. The good salary reflects the value to society of having talented people teaching our best college students while simultaneously demonstrating intellectual frontiers to research students. It is the intellectual excellence of our college faculties and their hard working lives in pursuit of truth that inspires students to graduate with the confidence to forge their own frontiers.

The letter of appointment dated May 3, 1966 arrived from Caltech President Lee A. DuBridge:

> *"I am happy to inform you that the Board of Trustees, at its meeting on May 2, 1966, appointed you Visiting Associate of Physics for nine months, effective September 1, 1966, or on arrival, with a salary of $800 per month. This action was taken on the recommendation of the Physics Division and with the approval of the Provost and the President."*

I was and still am proud of having had this appointment, because the position, Visiting Associate, is at Caltech one carrying faculty status, and not to be confused with the appointment of Research Associate, which is at the postdoctoral level. Nonacademic readers may have noted that the appointment involved three letters. That is not unusual in academia, although two more commonly suffice. My three "appointment letters" were from the Professor directing the research area, the Chairman of the Division, and the President of the University. A similar procedure had attended my appointment at Rice University.

Weighing Possible Fame

My decision about the writing of the books on nuclear astrophysics was sufficiently momentous that it may be worth explaining why it was so momentous and how it turned out. Nuclear astrophysicists everywhere were aware that throughout the 1960s Fowler and Hoyle were the world's most famous twosome in astronomy. They became legends. Willy Fowler, with whom I remained personally very close from 1957 until his death in 1993, later won the Nobel Prize in physics for his work. Fred Hoyle, with whom I would become personally close, was then the greatest of all theoretical astrophysicists since World War II, and later won the Crafoord Prize in astrophysics, the "Nobel Prize" in fields not designated in Alfred Nobel's will. This superstardom was at its zenith in 1966 when I was called upon to make my decision. To not accept their invitation with whole-hearted enthusiasm entailed considerable opportunity risk on my part. Weighing in on the negative side, however, was my thought that the field was not sufficiently well developed for an advanced treatise. I had grown to be aware of the sketchiness of their 1957 *Reviews of Modern Physics* paper, a publication that was to become universally cited by other authors desiring to encapsulate the beginnings of stellar nucleosynthesis with but a single reference. I was aware that much of their review paper would soon have to be supplanted by better formulations. I was also nervous about being lead author of what would amount to an updating of their 1957 review paper. My words here are intentional, admitting that I was already aware (as was Willy) that their celebrated 1957 paper would eventually be seen as a timely review rather than as a source paper. Its lively description of neutron-capture nucleosynthesis processes was its highlight; but my papers had already moved beyond that description. My awareness from being a part of the development of the theory of nucleosynthesis in stars positioned me to see how much yet needed to be fixed up. And in this I turned out to be correct. During the 1966-67 year I would be totally immersed in creating a quasiequilibrium theory for the nuclear burning of silicon within the interiors of evolved stars, cleaning up thereby the sketchy B^2FH attempt at their so-called "alpha process". Clearing that up would be our major achievement of the impending year. That work was very satisfying but shoving any talk of the book further into the background. We were frankly too busy to get to it.

It is possible, I suppose, that had I coauthored in 1967 the best book that we three could write, perhaps putting our new results into that book rather than into journal papers, perhaps I would today be more commonly thought of among the formulators of the theory of nucleosynthesis in stars. Friends have suggested so. But it is not an important issue. There were very few "founding papers" for nucleosynthesis in stars outside of Hoyle's two pioneering publications in 1946 and 1954, and, in veins related more to stellar evolution than to nucleosynthesis, the great hydrogen fusion papers by Hans Bethe and the resonant quasiequilibrium helium-fusion reaction by Edwin Salpeter. Cosmochemist Hans Suess had seen the neutron-capture systematics in the heavy-element abundances, and B^2FH dressed that up nicely but stopped short of an actual calculation of s and r processes. Given those works, I believe that all the rest had to follow. This is the way science is, and it does not demean the subsequent insights to observe that had they not been published by the historical leaders they would have later been published by someone else. Science is the hunt, not the answer. We are hunters after the true description of nature. I was among these following the great new idea, slightly later chronologically than Cameron and Fowler but contemporary as far as discovering the correct formulations was concerned. Indeed, looking back over it all, I now sometimes describe myself as a "clean-up hitter", a specialist in finding formulations mathematically and physically that stand the test of time. I am content, *ecstatic* actually, that my record of this has been among the best. But it is the fun of it that really counts. All of these subjective issues attended that gut-wrenching decision made by me during spring 1966.

William A. Fowler, the author, and Maarten Schmidt in Fowler's Caltech office during winter 1966.

One final thought along these lines highlights their relevance. Namely, I believe that my career productivity was ignited by return to Caltech in 1966. Despite the marvelous seven years of my first sojourn there, 1956-63, and despite my luck with three wonderful constructions for nucleosynthesis, I was in 1966 struggling to find my own voice. Without the inspiring seven years with Willy, Fred and Cambridge that I was now beginning at Caltech, I might well have sunk back to a workaday career that remembered its first moments as its finest. This is after all the case for many of us, that the great flaming contribution that we can offer in our youths from the gardens of Valhalla is not so easy from the provinces. For this I am eternally in the debt of Willy Fowler, just as surely as I had been for his introducing me to nuclear astrophysics in 1957. Without Willy I would have been an entirely different person.

As far as my own textbook is concerned, I believe that my thoughts on that also proved correct. I will later describe how I got it published. It became a standard textbook of astrophysics education for three decades, not only because I did a pretty good job but also because such a book was badly needed for astrophysics education. Despite many regrets in my life, I am glad that regret over this book issue did not arise. I had indeed made the best decision. But I was very fortunate at subsequently having a career that could be spoken of in the same breath with theirs. Had I been less successful I might surely have forever harkened back to the day that "I almost wrote a book with Fowler and Hoyle". I was also much relieved at the time, and now even more so, that my reluctance did not interfere with the unforgettable friendships that I was forging with these two very great scientists. Fortunately, Willy's enthusiastic urgings in his letters and the excitement that 1966-67 was yet to bring us had removed all doubts over friendship. I really loved him for that.

Weighing Defense Science

At about the same time I received a letter from William G. McMillan, Chair of the UCLA Chemistry department. Its first sentence read,"This is to inquire as to your availability for and possible interest in participating in the third session of the UCLA/ ARPA *Defense Science Seminar* to be held 1-26 August, 1966." The Advanced Research Projects Agency, under a contract to UCLA through the Department of the Navy, had supported two prior efforts in 1964 and 1965 to educate promising scientists in the science problems of national defense. My recent selection by the Alfred P. Sloan Foundation for a Fellowship alerted them to include me in the invitation list. Because it encapsulates something of these times I quote the third paragraph of that letter:

The classification and consequent inaccessibility of much of Government and Defense Science work, together with the unusual nature of some of the requisite scientific specialties, constitute a natural barrier which threatens to impair the availability of new talented scientists to the government. The UCLA/ARPA Defense Science Seminar is aimed at helping to develop a new generation of responsible people, knowledgeable in the scientific-technical problems of Defense and Government, as a base from which might be drawn new technical committee members, counselors and even occasionally Government administrators. Two accelerating trends lend growing urgency to this endeavor. First, those scientists currently called upon for such work largely got their start during World War II and are now beginning to retire from the scene. Secondly, it is clear that science will play an increasingly important role in national and international affairs and that the vanishing race of Government advisors must not only be resupplied but must in fact be considerably expanded to meet the need."

The contract offered consulting pay for the month and living expenses. It concluded with a questionnaire and asked me to determine my reply within a week. It was an outstanding opportunity to be educated by experts, so I accepted, but not without a few misgivings. I had become an implacable foe of the war in Viet Nam. I had participated in antiwar meetings on the Rice campus, in which students and faculty expressed openly their concerns. My conviction that we were exceeding our national interests in dangerous ways tended to produce an antigovernment feeling. Many of my friends did become antigovernmental. However, I quickly sorted this out as irrelevant, being a problem in national political judgment rather than of technical support. I was, and am, patriotic to the core. The aims of the seminar were clearly justified and well thought out. We had planned to go to California in any case, so I agreed to arrive for the month of August at UCLA, where the formal briefings would occur. I completed the necessary papers for the *Secret DOD* clearance.

So it was that in July 1966 Mary Lou and I again drove across the southwestern deserts to southern California. Our wonderful 1962 Buick Le Sabre sport coupe had replaced the previously wonderful 1954 Pontiac Fire Chief; and there was a strong sense of *deja vu*. Donald and Devon were now six and five-year old boys in the back seat. But we hauled no trailer this time. I mailed boxes of books, especially my textbook manuscripts, and we shipped a few boxes of clothes. The motels along the way had noticeably upgraded during the decade since our first such drives. They were now air conditioned, as was our Buick, and best of all, the motels had swimming pools. Each day we looked ahead to stopping early at a nice motel pool so we could relax and swim in the late afternoon heat. These would be long remembered within our family, even after its tragic breakup. Donald in particular would long refer back to his favorite motels, and look at the photographs that we had snapped in order to record each. As usual over physical matters, Devon was more diffident. He liked to just splash around while Donald experimented with breathtaking springboard dives to his water-treading Daddy. And Mary Lou just grooved.

Our drive did not take us directly to UCLA, however. Willy and Ardy Fowler had invited us to join them during their summer vacation in San Clemente, California. They had rented a house for two weeks directly on the beach. What Willy always understood was how to combine fun with a running start on work. After an overnight stop to visit the San Diego zoo, whose fame was rapidly spreading, we finished the small drive up the coast highway to San Clemente. The sun was still high, and Willy and Ardy were sitting out on the beach in front of the house. Their daughters Mary and Martha were playing in the sand. Willy proposed martinis all around and began his explanation of the windbagging techniques that we were sure, he promised, to love. We listened with all the old pleasure at Willy's enthusiasm. We watched one Fowler demonstration of his windbagging technique, then retired to dinner. It was very exciting to go over our plans for the year ahead, and to hear of Willy's hopes for Fred Hoyle's new Institute.

This began a seven-year period of continuous contact with Fowler, during which we became closer as friends than had been possible 6-9 years earlier while I was his graduate student.

Fowler was then an athletic 55 year old, and his demonstrations of windbagging technique on the following day illustrated his zest. Taking a large pillowcase, which he first got wet in the surf and then held up with its aperture open to the wind, he raced into the headwind into the surf. The fully inflated pillowcase was then wrung shut quickly at its neck. Struggling with it deeper into the surf in order to catch the breakers at the right point, Fowler suddenly turned and leaped his round girth onto the middle of that inflated sausage and paddled madly in the direction of the breaking wave trailing him. Frequently he caught it right and had an exhilarating ride to shallow surf. My old photos show Fowler riding the end of one such success onto the sand. We all tried it, save for Ardy and Devon. Donald's few successes were not for lack of trying often. Devon preferred just to wade in to a point of seeming danger, then leap in self protection when the wave arrived. By 11:30 we had all had it, when Willy called me aside. Opening his small thermos, he poured us dry martinis. "Having a martini before lunch is my greatest pleasure," he explained, "but only on vacation."

An interesting August 1966 in the Defense Science Seminar at UCLA reveals something of the times. Many technologies were openly presented by experts, such as antisubmarine detection, rocketry, warheads, and ionospheric signal propagation. Daylong excursions from the secure UCLA site took us to selected arsenals to witness weaponry, to missile silos to inspect the intercontinental warhead delivery system, and to the Air Force Systems Command within the NORAD mountain. In all of these things, the technology and science was combined with the gee-whiz tours to help us understand the problems of defense and to entice us to form some consulting relationship down the road. It happens that I never did this, but many physicists did consult with defense technology. I did not object to it, but an opportunity of interest to me never developed. I think I was just not sufficiently interested to encourage such a relationship. To participate we drove from Fowlers in San Clemente to Santa Monica, where the organizers had rented a bungalow in the *Star Dust Motel* on Wilshire Boulevard for the entire month. These bungalows, behind the motel in front, were one-story cottages having a small kitchen and two bedrooms, so our little family made it comfortably through the month. It also had the required swimming pool. I might add that in 1966 this western end of Wilshire Boulevard was a bit down in the tooth, not at all the high-rent district that we today associate with that famous street. Here was the schedule and speakers from those four weeks.

1 August: Defense Research; Missile Technology

2 August: Strategic Systems (Polaris, Poseidon, ICBM)

3 August: Nuclear Weapons Effects
 (Fireball, EMP, Communications)

4 August: Command, Control & Communications

5 August: <u>Field Trip</u>; Vandenberg Air Force Base

8 August: Nuclear Weapons Physics (my favorite day)

9 August: <u>Field Trip</u>: Defense Atomic Support Agency Field Command, Albuquerque

10 August: <u>Field Trip</u>: North American Defense Command, ENT Air Force Base

11 August: Radar & Reentry

12 August: Ballistic Missile Defense

15 August: Civil Defense

16 August: Counterinsurgency Operations

17 August: Army Weaponry & Biological and Chemical Warfare

18 August: ARPA and Viet Nam

19 August: World Hot Spots (NATO, Southeast Asia)

22 August: <u>Field Trip</u>: Picatinny Arsenal, New Jersey

23 August: Anti-Submarine Warfare

24 August: Naval R&D and Communications

25 August: <u>Field Trip:</u> US Naval Ordnance Test Station

26 August: Arms Control; Security Debriefing

These set an exhausting pace, even for a 31 year old. Many days did not end until long after nightfall. I saw rather little of Mary Lou, Donald and Devon during the month. I tried to share what our schedule called for; but sharing the knowledge gained was out of the question. Mary Lou needed some adult company after looking after the boys all day long. Donald needed to get me into the swimming pool. Devon needed his story time, and cajoling out of his habitually grumpy demeanor. Although I remember it all reasonably well, I remember most of all feeling that, as interesting as it had been, the time was just not worth it. It seemed like a month's investment of time for about ten hours of useful information. Of course, others felt differently about that, depending on the value they attached to the first-hand visits to key sites and the personal briefings from significant military personnel. Had I moved into government service, probably I would have seen the time as more valuable. That was what the program hoped for from many of us, and which did not come about in my case. The biggest benefit to me was an improved view of military technology needs and an increased sensitivity to the subtleties of the cold war. Those understandings enriched my subsequent life and, I believe, made me a better citizen. But two decades later I would actively oppose President Reagan's "Star Wars" defense ideas as dangerous and unworkable, which may not have been the outcome that many engaged in this seminar would have wished. Truth was, my heart was committed to nuclear astrophysics. That was all that I wanted to think about.

Home Again in Pasadena

When at August's end we packed up again and moved to Pasadena I carried with me a letter dated April 26, 1966 from Mrs. A. Jaget in the Caltech Office of Residence in Throop Hall. She advised us to visit her office when we arrived and to obtain a listing of nearby residences. It is interesting to read the price guidelines that we could expect for housing within the Caltech area of Pasadena, an area that is, by the way, a very nice community and is today regarded as expensive.

Furnished Houses Or Apartments: 1 Bdrm....$ 75 to $120

2 Bdrm....$135 to $175

3 Bdrm....$175 to $300

Unfurnished houses were somewhat less; however, we came seeking a furnished one. We visited many on her list, but the real jewel was not on her list. It was found for us by Ardy Fowler. A fine remodeled old house at 1361 E. Del Mar now contained as centerpiece to a grouping of apartments stood just one long block north of the Caltech campus. We snapped it up, for $300 per month, nicely furnished two bedrooms on two floors. From there I would enjoy daily walks down either Chester or Holliston Avenues to my office in *Kellogg Radiation Laboratory* on the California Boulevard side of the Caltech campus. The pleasantness of walking to work inspired my resolve to try to live within walking distance of Rice University when we would return to Houston 12 months later. We also enjoyed the nearby park, between Del Mar and Cordova on the other side of Chester Avenue, where the boys spent many a happy hour when they needed to be out of the house. That was often. It was also at that park, at this time, that I began a tentative foray into jogging. I have never stopped since. It was a modest beginning, ten minutes spent before breakfast in about five laps around that park. The activity was to grow into a friend for life, sometimes supporting me through darker moments.

Any family moving to a new location faces the question of school. Donald was 6 1/2 and Devon 5 years old when we settled in at our Del Mar address. Our previous seven years of life in Pasadena had only prepared us enough to know what the best school option for advantaged children was. By *advantaged children* I mean children who have been culturally advantaged rather than wealthy. Academic families produce school-age children that are ahead of society's norms, and our family was no exception. We wanted Donald in the first grade and Devon in kindergarten with other advantaged families. Times were changing from my own years in school when University Park Elementary had been good enough, even very good, where all families were advantaged, when social work and remedial schooling were not necessary within advantaged communities. This was in a time when society did not really know about disadvantaged children. But the late 1960s became times of great social turmoil on many fronts, and public schools began to face a much wider range of preparedness among first graders. We checked with old Caltech friends who confirmed that *Polytechnic School* remained the best choice. Located at the corner of California Boulevard with Wilson Avenue, across the street from the corner of the Caltech campus and from Caltech's Paddock Athletic Fields, *Polytechnic School* had grown originally because of the large numbers of well prepared Caltech children. So we entered the world of private schools, rare 31 years ago but now threatening to take over from public schools. We had to foot the tuition bill, but it was within our reach.

Polytechnic School proved a wonderful choice. For one thing its medium sized classes, 20 in first grade and 22 in kindergarten, in a small school having good discipline, enabled Donald and Devon to adjust immediately and confidently. For another thing, it was truly a neighborhood school. Everything was within a block of the Caltech campus. It was a shining little world, well justifying the slogan on my city map from the Pasadena Chamber of Commerce: "PASADENA, The Most Distinguished Address in the West". It *was* a wonderful city. As often occurs, road planning later cut through the heart of northern Pasadena with a major freeway, damaging many good communities that helped make Pasadena special, including that area near Orange Grove Boulevard and North Mentor, where both Donald and Devon had been born in 1960-61. But that freeway would come later, and in 1966 Pasadena felt whole and lovely. Individual portraits of each of Donald and Devon's classmates are aligned on the respective 8 x 10" photographs of both the First Grade class and the Kindergarten class, along with their teachers, Mrs. Seaman and Miss Hawkins, and the Headmaster Mr. Stork, and an inset photograph of the inner courtyard of the school.

Mary Lou did not seek employment. I was grateful for that, because I hoped for a simple and efficient family life. I confess that I wanted her to fit into my plan, my special year of opportunity; and she did. Two children beginning school and a new residence made a powerful combination. Moreover, we planned only to remain until April. Our plan was to then depart for Cambridge. This excitement brought other family demands. Travel plans had to be worked out. Contingency for the boys' early departure from *Polytechnic School* needed some explaining. The school was understanding on that need, provided that we would enroll Donald and Devon in school in Cambridge when we arrived. So that needed thought. Through all this excitement I wanted nothing more than to think about nuclear astrophysics. And that is pretty much what I did, thanks to Mary Lou's decision to be a homemaker for those seven months. In all the years of our marriage, I remember this year as our best. Unquestionably the quality of that life, for me and for our two boys, rested on her decision to be a homemaker. Regrettably, it would not hold thereafter, for Mary Lou's was a restless spirit.

The Nucleosynthesis Watershed

Daily discussions with Willy Fowler explored our course for the year. We agreed on the top priority, which was a better description of the detailed mechanisms and pathways by which a gas of hot silicon nuclei would transmute to iron and nickel. It was already evident that the most abundant isotope of silicon, ^{28}Si having atomic weight A=28, had to be a very abundant product of nuclear burning in stars. After the stellar center has become hot enough for oxygen nuclei to fuse with other oxygen nuclei, the resulting "oxygen burning", as it was called, produced ashes that were overwhelmingly ^{28}Si. As stars always do, that ^{28}Si center becomes squeezed to higher temperature and density by the gradual compression. Stars get hotter as they shrink, even though they simultaneously radiate heat energy during the process. This slightly paradoxical behavior is the result of gravity's work during compression, as pioneers of the theory of stellar structure had explained. So dominant is that principle, moreover, that I had featured it in the foundation physics section of my course notes, now gathered in binders on shelves above my office desk in the Kellogg ground floor (the ground floor with respect to the road entering the Caltech campus to the rear of Throop Hall). My book's bound notes bore the title *Principles of Stellar Evolution and Nucleosynthesis*, a title that the eventually published book would also bear. However, my notes were also stuck on the question of the transmutation of silicon to iron. Further progress of the science of the evolution of the presupernova star and of the composition of the burned ejecta depended on a quantitative understanding of that transformation. Willy and I agreed that it needed to be solved.

In their massive 1964 paper on supernova physics in *Astrophysical Journal*, Fowler and Hoyle had described only the way that transmutation began. A. G. W. Cameron also had it in his *Yale Lecture Notes*. It was not possible to heat ^{28}Si nuclei enough to fuse with others in a direct path to the ^{56}Nickel nucleus (^{28}Si + ^{28}Si $\Rightarrow$ ^{56}Ni). The charged repulsion between the two silicon nuclei was simply too great. Before sufficient temperatures to overcome that repulsion could be reached, the ^{28}Si began to disintegrate in the heat. This impasse differed from earlier epochs of nuclear fuel. When successively heated, helium ashes of hydrogen burning fused with other He nuclei, carbon ashes fused with other C nuclei, and oxygen ashes fused with other O nuclei. The ashes of one fuel became the next fuel, and the fuels burned by fusing with their own kind. But for silicon it had to proceed differently. Silicon *melted*! Small fragments were chipped away by the heat. This much had been correctly reasoned by Fowler, Hoyle and Cameron. Fowler and Hoyle had published a good description of the pathways by which hot silicon broke down. Primarily the energetic photons in the heat chipped off a proton from the ^{28}Si nucleus, followed by alpha particles being broken off of the consequent ^{27}Al nucleus, and continuing until the original ^{28}Si nucleus had been broken down into seven ^{4}He nuclei.

What was lacking was the behavior of the entire ensemble of nuclear particles in the hot gas. As we discovered, the back reaction of that system of free alpha particles on the decomposing silicon slowed its rate of decomposition. For example, the reaction ^{24}Mg $\Rightarrow$ ^{20}Ne + ^{4}He, which was a step in the silicon melting, was opposed by the inverse reaction utilizing the free ^{4}He nuclei that were maintained within the gas by the high temperature. The system as a whole would have to be taken into account to describe how the nuclear gas was to transmute from a gas of silicon nuclei into a gas of iron nuclei. Statistical physics required that that transmutation had somehow to occur. But how? In talks I sometimes offered analogy with melting ice; namely, we know that warmed ice turns into water, but how is that transition to be described? The answer would determine not only why iron became the most abundant of all heavy elements but also why mass A=56 became its most abundant isotope.

The fuss over iron comes about by virtue of it being the most abundant of the common metals in the universe. It is the most abundant of all elements having atomic weight greater than 28 (silicon). That glaring fact demands an explanation if the contents of our natural world are to be interpreted as the result of natural history. A glance at a graph of element abundances reveals the degree to which iron in the universe outnumbers all of the common metals. It also reveals that although ^{57}Fe has on earth only 2% of the abundance of the more abundant isotope, ^{56}Fe, iron is so abundant as an element that the ^{57}Fe isotope is more abundant than the elements chlorine and potassium combined! Three of the isotopes of iron (masses 54, 56 and 57) are counted among the most prominent isotopes of any element. Fowler and I would show during this year that two of its isotopes (^{56}Fe and ^{57}Fe) have been derived in nature from the radioactive decay of nickel isotopes of the same atomic weights outside of exploding stars; but this gets ahead of my story.

This sharp abundance peak at iron was the feature that Hoyle had seized upon in 1946 as a pointer to the physical process by which the heavy elements had been created. Observations of other stars affirm the universality of this pattern. Iron is among the easiest elements to be measured in the spectra from stars by astronomers. It had led Hoyle to develop the theory that the elements had their origins within stars, rather than in some early dense universe. In his 1946 paper, Hoyle had explained the sharp abundance peak at iron as a property of thermodynamic equilibrium occurring at a temperature so high that the nuclear reactions were in equilibrium. Each nuclear reaction then proceeds at the same rate as its inverse. This situation bears analogy to saying that if one cools water enough ice will result. But Hoyle presumed that nuclear properties of ^{56}Fe nucleus itself (and also the nuclear properties of the other iron isotopes) were germane to the situation. He took it that the excess numbers of neutrons within the equilibrium bath took on appropriate values for the existence of abundant ^{56}Fe as the dominant abundance. The ^{56}Fe nucleus, however, contains four more neutrons than protons. The real truth would prove even more fascinating.

During our 1966-67 work Fowler and I were led to the remarkable idea that this very abundant chemical element was created not in its own chemical form but that of radioactive nickel progenitors. That idea is novel in attributing the natural abundance of any element (or isotope thereof) not to its own nuclear properties but to those of a radioactive parent that subsequently decays to it. This idea was not in itself original; but the existence of the astrophysical argument supporting it using natural evolution within the stars was new. Science grows not on the mere statement of an idea but on the demonstration that the idea works naturally.

The specific idea identifying the parent for the most abundant isotope, ^{56}Fe, had appeared first in post-WWII Germany. It arose from nuclear-structure physics and the description of the spin-orbit forces in nuclei that validated the newly advanced shell model of the nucleus. The purpose of those researchers had not been the origin of the nuclei. They were designing the paradigm called *the nuclear shell model* for understanding the spacings and properties of the excited energy levels of the atomic nuclei. It was in this context that one of the coauthors of that Nobel Prize effort first recognized very special properties of the mass A=56 isotope of nickel, an isotope that decays radioactively to the stable A=56 isotope of iron. The isotope ^{56}Ni is called "doubly magic" by nuclear physicists because both its number of protons (28) and its number of neutrons (28) are members of the sequence of so-called "magic numbers", those specific numbers of nucleons that have extra strong binding energy in the nuclear shell model. Otto P. Haxel, one of the coauthors of the evidence for strong spin-orbit forces in nuclei, seems to have been the first to state that ^{56}Fe is so abundant because of the doubly-magic nuclear properties of its isotopic progenitor, ^{56}Ni, rather than because of the properties of ^{56}Fe itself, as Hoyle introduced at about the same time in England. That idea would require that ^{56}Fe be first created as ^{56}Ni, which decays radioactively to ^{56}Fe. My later interviews with Suess, with Haxel (see photo <1997 Otto Haxel and Clayton>), and with Hoyle (see photo <1974 Hoyle and Clayton in Hoyle's Cockley Moor home>), conducted over the following decades, confirmed that Haxel was the first to express that idea. The astrophysical reasons for this had to await our work now underway in 1966; and the observational demonstration of its correctness had to wait four decades. The path is twisted by incorrect conclusions on which celebrated astrophysical interpretations were made[2]. This issue became the focal point of our work at Caltech in 1966.

The confusion over ^{56}Ni ejection was due to the lack of a good picture for how, when, and why the elements were ejected from the exploding stars. A hindrance to this had been the problem of the fusion (called "burning" by astrophysicists) of ^{28}Si into heavier nuclei, the very problem that Fowler and I now agreed to attack. And it was also in 1966 that temporary confusion was accidentally introduced by an ambitious general research program conducted by A. G. W. Cameron and his students at Yale University. Following a decade of research in nuclear physics, Cameron had burst into scientific attention in 1957 with his own treatment of Hoyle's ideas of nucleosynthesis in stars. He argued that the abundances built up in the slow static burning of the star were largely irrelevant; instead, the shock wave started by the central explosion would so heat the stellar matter as it blasted through it that furious burning would alter all of the abundances in a matter of seconds. He was linking the nucleosynthesis with the dynamic process of ejection. It was a very bold view, one that is today believed quite relevant for silicon-burning matter explosively ejected from the stars. But a description of that Si-burning process would first be needed. On the question of the *e*-process production of ^{56}Fe, Cameron simply followed Hoyle's 1946 production of ^{56}Fe as itself. Adding to the confusion, Cameron and his students at Yale then published a calculation that put all nuclear reactions onto a new computer; and their first result in 1966 erroneously concluded that ^{54}Fe, not ^{56}Fe, was the most abundant nucleus formed during silicon burning. When that first paper by them crossed Fowler's desk he called me from my office shared with Bob May in Kellogg's ground floor to his own, handed me their preprint, and said with puzzlement, "See if you can figure out what is going on here."

The so-called "alpha process" of B^2FH was all that they offered in their famous *Reviews of Modern Physics* exposition concerning this important transmutation. It was vague at best, wrong in thrust. I found it essentially useless by this time. But it was at this time I read Hoyle's 1954 paper and was stunned. Hoyle's description was much clearer than the B^2FH alpha process. Hoyle had made a brilliant observation; namely that the binding energies of alpha particles in A=4n nuclei declined with increasing mass above ^{28}Si. He realized that the consequence would be that hot photons would break alpha particles out of nuclei faster as atomic weight increased, so that nuclei between ^{28}Si and ^{44}Ti would be decomposed in a downward flow toward ^{28}Si. I then confirmed by calculation that nuclei between silicon and iron would be disintegrated by the heat faster than would silicon itself. This presented a conundrum. If the silicon nuclei recaptured proton and alpha-particles fragments that had been liberated from other silicon nuclei, those resulting nuclei heavier than ^{28}Si would be quickly disintegrated. At first glance this seemed an impasse, as if matter were dammed up at ^{28}Si. After a few days I recognized the meaning of that situation. It reminded me of the *r* process, which we had calculated in our 1965 paper with Phil Seeger. In that process heavy nuclei capture neutrons rapidly, only to lose them even more rapidly in the heaviest stable isotopes, again because of the high temperature. The consequence of the "damming of the flow is that a partial equilibrium is set up, an equilibrium in which neutron capture comes into balance with neutron ejection. The same seemed likely for the silicon-to-iron region. Protons and alpha particles would be ejected from a whole ensemble of nuclei at essentially the same rate that they were being captured by that same ensemble of nuclei. Despite this flash of insight, I became temporarily stuck when trying to summarize the system for silicon burning. While I was describing my picture to Fowler, he reached into his file cabinet and handed me a few sheets of paper. They were handwritten notes by Caltech graduate student, R. A. Wolf, presenting a simple solution to this very situation.

Aided by a key idea in Wolf's notes, Fowler and I reread Hoyle's idea and afterward reasoned that the entire abundance region between silicon and iron would exist in what I called a "quasiequilibrium abundance distribution". In quasiequilibrium all nuclear reactions involving protons, neutrons and alpha particles with nuclei lying in mass between silicon and iron proceed at almost exactly the same rate as their inverse reactions, just as they would in a complete nuclear equilibrium. Nuclei would be created and destroyed by reactions that almost exactly balanced each other; but a tiny residual deviation from exact equality exists because of the slowness of the decomposition of silicon. I described the whole system of nuclei between silicon and iron as "just waiting", in the spirit of r-process waiting points, "waiting for something to change". The total number of nuclei from A=28 to A=56 was decidedly not in equilibrium with the number of free neutrons, protons and alpha particles. The total number of ^{28}Si nuclei decreased only very slowly. That abundance of ^{28}Si nuclei was the feature that slowly changed with time during silicon burning, shifting the entire abundance distribution with it as it did. The "waiting" was for the abundance of ^{28}Si to be reduced. I called it "quasiequilibrium"

Excitedly I reworked Wolf's suggestion, adding a few improvements making it more complete. In a small three-man seminar with Dave Bodansky in Willy's office I presented the solutions and suggested that we make graphs of the abundance distributions that would result. We constructed a scheme for doing this in collaboration with Bodansky, who was on sabbatical leave from The University of Washington. Bodansky was a nuclear physicist eager to enter nuclear astrophysics. He programmed most of the computations on the early Caltech computers. These computations consisted, as did most of that epoch, of self-consistently evaluating many analytic functions simultaneously; but they were not treatments of the time-dependent differential equations, as Cameron's group was attempting. Our formulation clearly identified ^{56}Ni as the most abundant product and therefore the natural parent of ^{56}Fe. There existed no chance that ^{54}Fe would remain the most abundant isotope. Something was wrong with Truran's and Cameron's paper. Our method worked, and the results were astonishing to the eye. They revealed a good match to the natural abundances of almost all of the most abundant nuclear species between ^{28}Si and ^{56}Ni. We had before us an excellent match between a theoretical calculation and the facts of nature. Willy was very excited and, after opining that we should submit this finding[3] at once (January 1967) to *Physical Review Letters*, quipped, "We will have to put this into our book!"

What our full quasiequilibrium solution accomplished was the complete description of a smooth and continuous transition of ^{28}Si into ^{56}Ni. It predicted the abundances of every isotope between atomic mass A=28 and A=62, as well as how they changed during the transition of ^{28}Si nuclei into ^{56}Ni nuclei. Those intermediate abundances resembled a bridge hanging between A=28 and A=56, where the mass A is a proxy for distance along the bridge and the abundance is the height of the bridge, which is high at both ends, though not equally high, and sags greatly to a lowest level halfway in between. But the bridge of abundances was not smooth, but possessed jagged ups and downs along its general trend. What was stunning was the remarkable degree to which those jagged ups and downs resembled the shape of the natural abundance distribution. This pulse-quickening discovery suggested that our quasiequilibrium process had much to do with the natural abundances of the chemical elements. A more precise understanding of why a few of the isotopes seemed to be missing would have to wait until I took on a special research student after my return from England.

James Truran, Cameron's student at Yale, at my request shared printouts of a detailed computer calculation that he had made with Cameron that agreed with our results but that had been calculated by their technique. Truran had been able to put all of the nuclear reactions onto a computer and to follow in time the consequent changes to the abundances. These two works, taken together, created a turning point in the history of nuclear astrophysics and the origin of the chemical elements. In the first place they convinced the world that abundant iron was so abundant owing primarily to its synthesis in supernova explosions as radioactive nickel. This overturned the interpretation of the *e* process that Hoyle and Fowler had championed since 1957. It showed the limitations of Hoyle's pioneering effort and also that of the B²FH review paper of 1957. But it marked the beginnings of the modern approach in two other ways that are equally significant. In the first place our papers established a kind of duality for looking at nuclear reaction networks. The attempt to analytically understand the behavior of a large and complicated nuclear system, as exemplified by my approach, was carried to a new level and was brought into concurrence with the approach by Truran and Cameron based upon numerical integration on computers of the nuclear reaction-rate equations for the abundances. The science that achieved this was a model of healthy collaboration. Modern papers of high impact now carefully join these dual views, stating the conditions under which a simpler analytic result will be valid and sufficient and those for which detailed numerical integrations provide the only reliable path. In the second place, these works also marked an historical shift away from the burning during the static phases of stellar life to the abrupt and violent burning, typically only seconds in duration, that occurs when the strong supernova shock from the rebounding core propagates outward through the stellar mantle. Attention was shifted from the nuclear burning under constant conditions to that which occurs during the process of getting the matter out of the stars.

Robert May on his trajectory and Willy on his

During Fall 1966 I became well acquainted with my office mate, a quite enjoyable and remarkable man destined for fame. Robert M. May also held the position *Visiting Associate in Physics* in the Kellogg Lab. Bob was on leave from Sydney University in Australia. Our shared office in Kellogg presented me with the opportunity to describe to him a novel idea in nuclear reactions. I argued that the nuclear reaction $^3He + ^3He \Rightarrow ^4He + 2\ ^1H$, which is important in the sun, was so forcibly suppressed by the Coulomb repulsion keeping the two ^{3}He nuclei apart that, instead of interacting in the usual nuclear sense in which the nuclei approach so close to one another that they partially overlap, the reaction was completed by the tunneling of a neutron from one ^{3}He into the other. The tunneling of the transferred neutron converts the other ^{3}He nucleus to a ^{4}He nucleus and leaves the remaining two protons free. Winkler and Dwarakanath, and later Bacher and Tombrello, were actively engaged in measuring the rate of that important solar reaction at higher energy in Kellogg Lab at that time. I showed Bob my simple calculations, and explained their consequences. If the probability for the reaction to occur could be made greater at the low particle energy of the solar center, the proton-proton chains would be shifted more toward PPI and away from PPII[4]. My suggested mechanism could do that, in which case the flux of neutrinos from the decay of radioactive ^{8}B within the sun, a solar flux that Raymond Davis was tooling up to measure in the Homestake Mine in Lead SD, would be smaller than expected. Bob was intrigued and said that he knew a better way to do the calculation. He explained his formulation to me, which I very much admired, and we did the calculation and eventually submitted the paper to *Astrophysical Journal* after my return to Houston[4]. Bob did the theory expertly and rapidly, and I was very impressed at the results, which showed that the probability should indeed become increasingly significant at lower energy than at the higher laboratory energy used the Van de Graff accelerators. This energy mismatch was always the problem when measuring nuclear reaction rates for the sun, leading me to explain that mismatch carefully in my textbook notes[4].

In our paper Bob and I pointed out approximations that we had made in our formulation and held out hope that an even better treatment might give even more dramatic dominance to this mechanism at low energy. Willy was ecstatic, saying once to me, "This is the greatest thing in stellar reactions in a long time." His enthusiasm was overoptimistic; but it did give me a nice glow. I predicted to Willy a great career in nuclear physics for Bob May. How was I to know how short it would be! Rather soon Bob had a brilliant computational idea in evolutionary biology which made him very well known. When I next saw him 18 years later he was a Professor of Biology at Princeton University. Even Princeton could not hold him, and he accepted a Royal Society Professorship which he divided between Oxford and London, becoming science advisor to her majesty's government and one of the most recognized names in all of science.

While my work on radiogenic origin of iron was underway I stopped one afternoon in Willy's office. He was at his desk pouring over tables of numbers and formulae that I recognized to be evaluations of thermonuclear reaction rates. He was checking the numbers, being sure they were right. He picked up his Keuffel and Esser slide rule and slid the cross hairs to the product Z_1Z_2, then moved the slide bar to divide by the cube root of the temperature beneath the crosshairs, then slid the crosshair again to insert a factor that represented the square of the electric charge. He had only glanced up at me, to let me know to wait quietly. When done, he compared his number with the printed entry in a table that Barbara Zimmerman had constructed by programming the new Caltech computer.

"By God, she got it right!" Willy exclaimed. The computer programming, which was a modern thing to do, was being correctly done.

"Let's go play softball", I suggested. The World Series was just over. "There is a Kellogg pickup game at 4:30". Willy was a great fan of baseball. But his atypical answer left a deep impression.

"Don, I can't. I've got to get these tables done." Then solemnly, "This is my life's work. I've got to finish it before we leave for England."

I had never heard from him such a serious statement of personal purpose. To simultaneously recognize both the significance and the urgency of his goal touched me. He was recognizing that his summary evaluations of nuclear data for thermonuclear reaction rates, of which these data tables destined for his *Annual Review of Astronomy and Astrophysics* article were the latest example, would be the work for which he would be remembered, even as it turned out, by a Nobel Prize. But he was also acknowledging that he had to finish it now or risk loss of focus in the eventful year ahead.

Richard Feynman and my Textbook

During October 1966 an excited celebrity rushed into my office. "Hi. I'm Dick Feynman", he said. I had to smile. Everyone knew who Richard Feynman was. "I hear that you have a textbook on stellar physics", he continued. "I wonder if I could borrow it. I have been reading Chandrasekhar's book and Schwarzschild's book, and can't get anything out of them." What he meant was that those books did not put physics first, whereas Feynman was unable to not put physics first. I was honored, scared, and dumfounded. "Let me explain", he continued. "I have been asked by Hughes Aircraft to give a series of three lectures on the stars, and I'm having trouble finding a source that explains the ideas physically." It happened that Hughes had offered him $5000 for those three lectures, celebrity bagging for Hughes and a handsome offer. He had accepted and now grew impatient at not knowing how he wanted to approach the material. With significant insecurity I handed the great man a copy of my typed notes and text for the book. He thanked me and glanced through the structure of my notes, then departed. I must have sat there for 30 minutes, ruminating on this emanation from the gods. Richard Feynman had been a hero to me since graduate school. How could it be that *he* sought *my* help? Would I be humiliated by the simple explanatory material that I had handed to him, telling him that it was really for beginning graduate students? "That's just what I want!" he had countered.

About three weeks later he burst into my office again. "Clayton, this is great stuff. You have to publish it", he said, handing my text to me. "It's very good".

"Well I want to", I said, "and it's time to began seeking a publisher. I am not sure anyone will want it"

"Tell you what," Feynman interjected. "I'll call an editor, Brad Bayne at McGraw-Hill and tell him to ask to see it".

I thanked him sincerely and he left. Within days my phone rang. Brad Bayne was calling from McGraw-Hill in New York. He told me that they wanted to publish my book! This amazed me because they had not even seen it. And, furthermore, he wanted to fly to Los

Angeles to bring me a contract to sign. Unbelievably, that is what happened. We had a friendly time, meeting once with Feynman in his office, and I signed a contract well beyond my dreams. McGraw-Hill was at that time a temple of physics publishing. I still had my silver spoon. This story explains my thanks to Richard Feynman expressed in the Preface to the published edition of my textbook.

Fowler's *Stellar Interiors and Nuclear Synthesis* group was very active on many fronts. He had won a generous National Science Foundation grant to complement the Office of Naval Research support that Kellogg Lab had long enjoyed. It supported many postdoctoral research associates. Some of them remain especially vivid in my memory. Kip Thorne, a former superstar Caltech undergrad now fresh from his Princeton PhD, was presenting a lecture course on general relativity, which I attended rigorously, while he was also starting several projects in relativistic astrophysics with students and other research associates. Kip was very energetic and is today one of Caltech's famous professors. Robert V. Wagoner was working on nucleosynthesis in the Big Bang. John Bahcall had been promoted from a Research Associate to Assistant Professor, and he was working with Giora Shaviv on models of the sun to test for its expected neutrino emission. When I had left this SINS project after my final year as Research Associate at Caltech (1962-63), astrophysics was but a modest sideline of the laboratory for nuclear physics; but now during my return in 1966-67 it seemed the main thrust of the laboratory. And its research was branching out from nuclear astrophysics into relativistic astrophysics.

CHAPTER 10. BRANCHING OUT IN ENGLAND

During the winter of 1966 Fred Hoyle arrived in Pasadena, an event that created considerable excitement for me personally and within the Caltech environment. This was my first opportunity to talk with Hoyle. His previous visits belonged to Fowler. After a lengthy conversation about the establishment of quasiequilibrium during silicon burning he became a convert to its new picture. He thereafter supported the radiogenic origin for iron as radioactive nickel. The original e process was now dead, although the statistical-mechanical idea lived on, but now characterized by only a small neutron excess incapable of yielding ^{56}Fe directly. Hoyle urged us to run with this new quasiequilibrium idea, but he would be busy with new goals of his own. Hoyle's main scientific effort during that winter was the idea of nucleosynthesis during the *Big Bang*, even though he was not convinced that the universe actually began in such incredible manner. He and Fowler had begun calculations with Robert V. Wagoner, a postdoc from Stanford, of the detailed nuclear reaction paths that would occur as dense matter expanded according to the cosmological rate given by Einstein's equations for a homogeneous universe. The bringing of their famous paper to conclusion by them made 1966-67 a banner year for nucleosynthesis theory. Both the quasiequilibrium origin of radiogenic iron and the origin of ^{1}H, ^{2}H, ^{3}He, ^{4}He, and ^{7}Li in the *Big Bang* were determined.

A well known photograph of the four of us was staged in Willy's office <1966 Fowler, Clayton, Wagoner, Hoyle>.

Planning the trip to England in spring 1967. From left, Robert V. Wagoner, William A. Fowler, Fred Hoyle, and the author. Taken at Caltech between East Bridge and Sloan Laboratory.

Hoyle's other goal featured plans for his new institute. By all odds the most exciting development for me was Hoyle's speaking to me about wanting to establish a group of core regulars in nucleosynthesis for the anticipated five years existence of the *Institute of Theoretical Astronomy* in Cambridge. That was the name Fred chose. He hoped that I would agree to do this, whether we went forward with our book project or not. I accepted without even talking it over with Mary Lou, as Willy had accurately forewarned us of Fred's plan. Mary Lou loved the idea. Hoyle's flexibility over the monograph on nucleosynthesis relieved me, because after the excitement of the past months I doubted even more strongly that it could materialize. Hoyle had too many irons in too many fires. Fred was happy that the Burbidges, Fowler and myself had already committed to the project, as would Wal Sargeant, who was central to a different plan of Fred's. Bob Wagoner agreed to come for the first summer. Photographs of four of us at this time were staged in Willy's office <1967 Fowler, Wagoner, Hoyle, Clayton >. Hoyle had gathered five years of financing from several sources, as described in his autobiography.[1] It would thereafter be reviewed by a university board that would evaluate its effectiveness with an eye toward continuation. As to my finances, he promised that I would soon receive a letter from the Assistant Registrar appointing me *Visiting Fellow* for 1967 at the institute for a 1200 pound stipend. I did not know how much that would buy, but I was confident it would suffice because, having been selected an Alfred P. Sloan Foundation Fellow with $20,000 for my use from them, I planned to use that money for the travel expenses for several years and to live in Cambridge for four months on the 1200 pounds. That worked out quite well.

Those older than age seventy may remember how special it was in the 1950s to travel to Europe. I was not aware of anyone in my 1953 High School class at Highland Park who had done so, whereas today in such an affluent school district almost all have been to Europe with their parents, perhaps more than once. Affluence in the 1950s seems to me to have had a different meaning for parents, few of whom aspired openly to visit Europe. Wealth was not so opulent, save for the very few very rich, and most families at Highland Park strove to better their worldly conditions and those for their children by practical concerns. At Caltech graduate school my good friend during 1957-61, John Domingo, had been to Switzerland for a year on some kind of student exchange; and because of that we regarded him with some awe as a holder of deep secrets of sophistication, although John would laugh at such a thought.

It was therefore still a matter of some glamour that Mary Lou and I were to travel to England in spring 1967 with our two sons on the Cunard Line's *Queen Elizabeth*. Fred Hoyle returned to Cambridge in March and Willy and I quit our work in March to organize this adventure. My friendly banker in Houston, Bob Robertson, had agreed to loan $3000 in order that we could purchase in advance through Volkswagen a new red 1967 beetle convertible for delivery to their dealership in St. John's Wood in London. That was a typical sum for a new car of that class in 1967; and we concluded the transaction from Pasadena. Mary Lou hosted one final dinner party at our Del Mar Avenue apartment, this one for Willy and Ardy Fowler and for Richard and Gweneth Feynman. Gweneth had experienced a childhood similar to Fred Hoyle's, from the Yorkshire Pennines, so that confluence provided the favorite topic for us at the dinner. We hung on her every word, trying to anticipate what England would be like. But having Dick to dinner was the really great privilege, and he was as always brilliant in conversation. We had been engaging in friendly talks ever since I loaned him my draft copy of *Principles of Stellar Evolution and Nucleosynthesis*. He was very entertaining and gracious on a number of topics. But unquestionably the most interesting aspect of that dinner insofar as physics history is concerned was Dick's needling of Willy. It began at a moment, the immediate cause for which I am unable to remember, and continued for fifteen minutes or so, during which it became apparent that Dick regarded Willy as somewhat of a plugger, not really impressive as a physicist but good at getting things done. This needling discomfited me, not least when Dick implied with a wink to me that I, like he, recognized Willy's true pedestrianism and that I would rise above it. I had not the courage to correct him by admitting that I too am a plugger, one who admired Willy beyond words and what he had been able to achieve. Because of my anxiety during this conversation it is a singularly clear memory, my only memory of a Feynman action that was beneath him. Maybe it had been the wine. It was the singular goals of our upcoming trip to Cambridge that had pulled us together with such glamorous company on an otherwise very merry evening.

My books and papers were boxed and shipped to Rice University. My family climbed into our 1962 Buick LeSabre Sport Coupe for its return trip to Texas, this time to my parents' house in Dallas, from where we were ticketed to fly to New York. The mirth and excitement of it all made spring 1967 one of the best of times for Mary Lou. She was ebullient planning our good times, and tireless researching worthwhile things to do. She possessed sensitive cultural and historical antennae, a very good thing for me as a physicist. She could hardly wait to see a Shakespeare play on its home turf. The first step, however, was to spend four days in New York, which neither of us had ever seen, before sailing.

Her happiness spilled over into our intimacy, creating the best of our dozen years together and leaving me hopeful for a better future. At a level of expense that seemed reasonable at the time, we booked four nights at *Hotel St. Moritz* on Central Park South. My photographs from that visit emphasize our walks in Central Park, visits to famous places in the city and our ferry ride to the statue of Liberty, which we climbed, all four of us, with only Devon making cranky remarks with his unforgettable young wit. Our friend from my graduate school days, Henry Ruderman, accompanied us to the statue, inasmuch as he was living in New York for his job at Brookhaven National Laboratory.

On the fourth day in New York we taxied to the Cunard docks on West 42nd Street at the Hudson River. The hustle and bustle of passengers turning their bags over to Cunard staff for delivery to their state rooms remains vivid in my mind. A Cunard photographer recorded us on the entry plank.

My family boarding the original Queen Elizabeth for our voyage to Southampton, England. April 1967. Front are Devon (left, age 6) and Donald (right, age 7); rear the author and Mary Lou Keesee Clayton, my first wife.

Donald, Devon and I were each in coat and tie, a measure of the sense of grandeur that sailing ships evoked in 1967. Mary Lou had purchased new suits for each son, handsome indeed, for this and for our transatlantic dinners, which we understood would be dressy affairs and for which we requested the first sittings in order to maintain a better family schedule. We dressed for every dinner of that five-day crossing. They were a highlight of each day, in part because our table's assigned waiter assumed a friendly chattiness with Donald and Devon, each of us enjoying his London accent. We saved one dinner-menu folder, the one for Saturday, April 29, 1967. The fish course featured choice of Poached Salmon, Cucumber and Hollandaise Sauce or whole Lemon Sole, Meuniere. The featured entree was Rock Cornish Game Hen en Casserole, Chez Soi, although grilled entrees and a ham joint were options. Recommended wines were 1951 Barsac, Reserve de Mousquetaires, or 1961 Brauenberger Reisling (of which I ordered a bottle). Wines were then unknown to me. It seemed magical. Many a wine lover will admit, as I now do, that a long love affair with wine began at just such a special point in life that one remained willing thereafter to suspend disbelief over the arcane postures of oenophiles.

Sailing on the original *Queen Elizabeth* was a unique moment in time, so much so that in the 21st century I have yet to meet a single person who has sailed on her. Her stately old-world appointments and carved wood trim seemed already old, as indeed they were. She had been renovated after service in World War II and we did not know at the time that she was at the end of her career. In mid-ocean the ship passed the *Queen Mary* on her way to New York, and passengers flocked to the railings to watch the passage. *Queen Mary*, with her three smokestacks, and *Queen Elizabeth*, with her two huge ones, tooted loud signals to each other, surely more for the mirth of their passengers than for nautical reasons. The *Queen Mary* too was destined for imminent decommission.

On the fifth cruise day we sighted land and docked at Cherbourg, then departed again after unloading many passengers and turned back across the channel for our docking at Southampton. By this time the afternoon shadows were quite long. We watched the slow procedure from the railings, again dressed in coats and ties. After disembarking we had only to board the boat train to London, to where our bags had been checked. This trip ended long after sunset, so we had yet to see any countryside. Victoria Station was where we found our bags in an area for Cunard passengers, all in order. It was also there that I was grateful for the VIP treatment that Fred Hoyle had arranged for us. He had hired a Cambridge driver with instructions to take us to the home that they had booked for our stay. His meeting card read "Professor Clayton", held before his own coat and tie. Tired but enchanted, the London lights faded into the pitch black of the old A10 as I watched the occasional lights of English villages pass by. We encountered only a few autos on the entire trip after leaving greater London.

White Cottage and Kathleen

It was almost 10:30 p.m. when the driver turned onto Storey's Way and then into the drive at number 15, *White Cottage.* The boys, ages six and seven, were exhausted. Barbara Hoyle was there to greet us and to introduce us to our housekeeper, Kathleen Bilham, who had lived in the house as housekeeper for two decades. Kathleen showed us to our rooms, where we deposited some bags, then asked if the boys would like a cup of Horlicks. I had no clue, so I said yes, one for us all. As we sat having our first taste of a warm English traditional sustenance, Devon, who was trying to be grown up, elegant, and stay awake, spilled his Horlicks cup onto his new suit. Slowly, as consciousness dawned of his great gaff, tears filled his eyes and he cried embarrassed exhausted tears. We led him up to bed, in his own sweet room after reassuring him that it all was OK and could be cleaned. Finally, after we were totally enchanted by the "cottage", Barbara Hoyle turned to us anxiously and said, "Oh, I do hope it will be all right", meaning the house and housekeeper. She had found it for us herself.

It will be difficult for the reader to credit my astonishment on the next morning. With morning sun playing on the window curtains, I rose to draw them back to expose my eyes for the first time to an April day in England. I must have gasped at the slight haze that hung over the rear garden, which ran back 200 feet or more to a simple wooden fence at its rear. Behind that fence lay a large green, which I learned was the Trinity Hall Sports Ground. The central grass walkway pierced topiary hedges cut in forms of large fowl atop cubic hedge bases. It seemed instantly the most beautiful sight I had ever seen. The central square of the rectangular garden was a green shortgrass lawn, whereas the portion nearer the house in front of the topiary was a mixed English flower garden, and the rear portion behind the hedge-lined central square was blossoming fruit trees and strawberry beds. It seemed a country house rather than one in a busy market-and-university town. At that very instant I bonded with England, or perhaps with a dream of England, one that burst upon my senses with a magical grip. I stood at least ten minutes, Mary Lou rising at last to see what I gawked at. Perhaps that garden view, like so many in the seven years to follow, touched my childish imagination, filled with folk tales, nursery rhymes and legends. Perhaps it recalled as suddenly as a sharp pain the beautiful, well loved view of Grandma Kembery's vegetable and flower gardens behind their Iowa farmhouse, a view that I had deemed forever lost except within my yearning heart. But there through wispy morning mist it lay! To my astonishment I had similar experiences throughout England during the 1960s, truly then an idyllic emerald isle. It was *Hobbiton, in The Shire*, and the English were hobbits! Within weeks I was having fantasies, such as wanting to move to England and to raise my own boys there rather than in Houston or California. It is no exaggeration to admit that this fantasy haunted me for the next three decades, until reality asserted itself through my veil of yearnings.

Rear view of White Cottage, Cambridge, from its garden.

The house had belonged to Mr. Freeman, a recently deceased manager at the Tolly-Cobold Brewery. His daughter, Pamela Turfrey, had decided to lease it out while deciding the future of the property, and we were the benefactors of that waiting period. We paid 100 pounds per month for this privilege. Kathleen Bilham had been his housekeeper, and now she took my family under her wing and taught us the English ways, molded somewhat by her Yorkshire upbringing. She took us to town, to the central market, where with string bag and infinite patience we waited through queues at the baker, the fishmonger, the greengrocer, the butcher, and other specialized shops. I lost whole mornings doing daily shopping but found it so enchanting that I did not care. Those grocers are mostly gone now; the old Cambridge town is gone, lost to boutiques and tourist businesses. How fortunate I was to have loved the town in the 1960s, before its nature changed. The Cambridge of the 1960s could be described by the tradition "town and gown". *The town* was simple people, merchants and others who kept life functioning, who lived in town while serving the populace. *The gown* was the university, the college Dons walking purposefully from one task to another, and the students in their own gowns on their bicycles, being treated like noblemen by each college staff. Storey's Way, like other old-fashioned streets, was occupied by people of the town, each tending his own garden, living as neighbors, whereas the Storey's Way that I would revisit thirty years later had become an expensive bedroom community for business entrepreneurs and occasional professors who spend relatively little time at home, and whose gardens grow weeds because gardeners had become scarce. I embraced the town of the 1960s with little awareness of its fragility, assuming that proper Englishness would maintain it. An emotional analogy from my childhood can be seen in the transition from family farms in Iowa, which I had taken to be the immutable and the true way, to the agribusiness of today's industrial farms.

With Fred in IOTA and in the Scottish Highlands

The walk from *White Cottage* to the *Institute of Theoretical Astronomy* (hereafter *IOTA*) made a pleasant interlude--along Storey's Way to its right-angle turn, where the red pillar box (mailbox) stood, right into the rear entrance of *The Observatories'* gardens, and another hundred yards past *The Observatories*. I became acquainted with several of the owners along Storey's Way, especially Andy Anderton, then about eighty years old, who invited me to golf at the Gog Magog Hills course after learning that I had been a golfer. Fred's new institute was behind construction schedule, however, so the modest American contingent found other places to work. For myself it was a small wooden hut in a field of sheep that lay between the Observatories and Madingley Road. The hut was charming. It was shared with Willy Fowler and with Bob Wagoner, who came for that first summer.

219

The author (left) at age 32 with Willy Fowler, outside the hut in the field of sheep on the grounds of Cambridge University Observatories. This hut was our first office in Cambridge prior to the construction of Hoyle's Institute of Theoretical Astronomy.

I was so enchanted with the settings that I accomplished rather little that first summer other than writing the paper on silicon-burning quasiequilibrium. The stroll was so easy and so safe that seven-year old Donald appeared daily at the hut, despite my admonishments against his bothering me at work. The Burbidges lived and worked in Churchill College just next door to *The Observatories*, and I believe that Wal Sargeant worked in *The Observatories*. That was it--Fred's American contingent, as it would be for the next six years--Fowler, Geoff and Margaret Burbidge, Wal Sargeant, Phil Solomon and myself. We were the regulars that Hoyle hoped would provide continuity to IOTA's programs for six summers of his directorship. I walked that walk about four times daily during my first two summers in Cambridge, as we arranged to retain *White Cottage* for a second summer by finding that my Rice colleague David Hellums (Professor of Chemical Engineering) and his family would lease it during school term for his sabbatical in Cambridge. They too became immensely fond of Kathleen Bilham, who, with her dog Turie, was inseparable from the house--at least for the time being.

Our family finances were modest but quite adequate. In January I had received the letter at Caltech appointing me Visiting Fellow for the year at a non-pensionable stipend of 1200 pounds. This 300 pounds per month for that four months covered all of our local expenses. The pound was quite strong in 1967, $2.8; but more significant was how much a pound could buy in England in 1967. Our magnificent house was 100/month; a pint of ale was 17p, about the same as a loaf of bread. Other foods scaled accordingly. So although the pound now buys little in England, it was a solid and significant sum in 1967. Americans who lived there in the 1960s agreed that living in England seemed very inexpensive, although that is no longer true.

During this 1967 summer Fred Hoyle invited us to go climbing in the highlands of Scotland. This was a special love of Hoyle's, one of those human passions that may exceed reason. We agreed at once to go, as did Willy and Ardie Fowler and Geoff and Margaret Burbidge. Fred booked for us all at the Loch Duich Hotel, with its view of famed Eilean Donan Castle <1967 Fowler, Clayton and Hoyle at Loch Duich>. We departed Cambridge early in our red Volkswagen convertible in order to bed-and-breakfast our way north. We were totally charmed by one B&B that we found in Hawes, in Wensleydale, Yorkshire. The entire dale is a postcard. Beguiled, we visited the local dairy and bought local cheese and bread, which we ate while watching fields of young lambs. Owing to the town's name, however, our favorite B&B on this journey was in middle Scotland in a town named Spittle in the Street. At Loch Duich we were charmed by the cozy little hotel rooms, where with extra cots my family of four slept in one of the larger rooms. Over the next two days I joined Fred and Willy <1967 Hoyle and Fowler on hike> on two hikes to the tops of 3000 ft mountains, slogging through bogs and pausing during intermittent rains <1967 Hoyle and Clayton on hike>. Others stayed at the hotel or went sightseeing. It was very jolly, but more than that, the comradeship and talk of astrophysics with Fred and Willy so hooked Willy and myself that, unknown to us then, this was to be the beginning of six years on the Munros (those mountains above 3000 ft in height). In this long summer daylight, from 4:00 am to 10:00 pm, we talked no more than two-to-three hours of astrophysics, but they seemed immense discussions in which we returned without preamble to a topic raised hours earlier, or indeed days before. It was the first occasion of my life to be at lengthy leisure with great scientists, and I would need pages to describe my growing sense of feeling comfortable at that particular frontier of science.

We said not one word of the proffered book. They knew. They could see for themselves that I had been correct. The summer was too busy, getting Fred's Institute running, taking long lunches with pints of real ale at country pubs in Madingley, Coton, or Grantchester, finishing the lengthy article for *Astrophysical Journal Supplements* on quasiequilibrium during silicon burning, helping host the first of Hoyle's annual summer conferences on astrophysics. In the breathless pace all knew that events moved too fast for a book on the state of knowledge. By contrast, Fred and Willy knew that I was examining page proofs and queries from my publisher, McGraw-Hill, for *Principles of Stellar Evolution and Nucleosynthesis.* I even showed them the simplified version of the quasiequilibrium that is established during silicon burning that I was to include within it. Even as we parted for my flight back to Houston for the beginning of the fall semester at Rice University we spoke animatedly of our plans for next summer. Always in high spirits, Willy joked, "And we'll finish the book!" He knew that we had too much on our plates. Our enthusiasm is characteristic of research humming along. One derives immense energy from the sense of accomplishment and from anticipation of the next tasks to be undertaken.

CHAPTER 11. THE ASTRONOMY OF RADIOACTIVITY

Important family decisions pressed when my family and I returned from London to Houston in September 1967, just prior to the beginning of classes. To allow search time for a residence we stayed in a long term residential hotel near Texas Medical Center. For the first time we felt justified in looking at homes for sale rather than for rent. I had just been promoted to Associate Professor and my nine-month salary was increased to $15,000. Donald and Devon were both of required school age by this time, and from conversations we had come to favor Poe Elementary School in the Southampton neighborhood just north of Rice University. Our great luck was that an elderly couple at 2346 Tangley Street, rather central in that block between Greenbriar and Morningside, had decided to sell and move. It was perfect! After our summer in Great Wilbraham we were relieved to find another home that all four of us could love. At $23,500 it was easily within reach by mortgage. It was ours. Within a month we took possession and the boys were in nearby Poe Elementary. The school and district reminded me so much of University Park Elementary that I was extremely comfortable with this. I purchased bicycles for the boys and coaxed them toward that decisive moment of bravery when I would lift my hand from the bicycle while running alongside on Tangley Street. There is no substitute for the experimental physics of learning that the bicycle is much more stable while rolling than at rest. I would learn some things relevant to myself during the decades to follow from the sentimental lifelong attachment to this house made by Donald, who was seven years old when we moved into the Tangley house. Repeated evidences of that attachment during the next four decades of Donald's life stimulated me somewhat to revisit my own sense of loss of my childhood homes. A curious cultural coincidence was to be the featuring of Tangley Street as the set in the movie *The Evening Star* which was the sequel to *Terms of Endearment,* the final part of Larry McMurtry's Texas trilogy, coming after *Moving On* and *All My Friends Are Going to Be Strangers.* Larry McMurtry and I had become friends in 1965 when both of us were faculty associates of Brown College of Rice University and he was teaching creative writing. His easy going speech was just perfect for his appearance and writing style. Several Rice faculty members provided character models for characters in Larry's trilogy.

Looking for Radioactivity

Bob Haymes was taking graduate students into his research program in gamma ray astronomy at Rice University. Bob's design of a shielded sodium-iodide crystal as detector[4], pointed by an active collimation shield that defined its angular aperture, was very successful and was copied by many subsequent instrument makers. Because the scientific objectives for detecting radioactivity in the galaxy relied on nuclear astrophysics and nucleosynthesis, Bob naturally asked his research students to enroll in my course, *Stellar Evolution and Nucleosynthesis.* Now returning from my year away at Caltech and Cambridge I found a robust enrollment awaiting me. Two of these students were to become among the most significant colleagues of my career. Several were students in Bob Haymes's program. One of these, Jerry Fishman, was destined for great success in this field. His attendance in my class led to one of the important serendipitous stimulants to my research career. And Stan Woosley, who became my own research student, taught me more than any other student of my career. Both later earned fame as scientists.

Several issues about gamma-ray astronomy need to be understood. Firstly, it means different things to different people. Gamma rays arise from many different physical mechanisms giving them differing ranges of energies. The gamma ray is akin to light and X rays in that all are fundamental quanta, particles with no rest mass, each being an excitation of the electromagnetic fields. Each is a bundle of electricity and magnetism tied into a knot by the laws of physics. Those natural processes that create the most energetic quanta differ from those that create less energetic ones. Gamma rays are the most energetic of these quanta, those that are broadly speaking too energetic to be useable in an optical system. Optical systems rely on the wave properties of the quanta. They focus quanta, or diffract them into members having differing wavelengths. Gamma rays are too energetic for that.

Secondly, the flavor that my research was to give to gamma-ray astronomy at Rice University was a focus on radioactivity. A radioactive nucleus has too much internal energy, with the result that it is excited by extra motions that are not required. It may shed this extra energy by creating a quantum of electromagnetic energy, a gamma ray. The gamma rays created in this way possess energies in the range 0.1 million eV to about 10 million eV. The eV is the energy range of ordinary light, so a million eV, written MeV for short, is a million times more energy than a light quantum.

Thirdly, if Haymes' gamma-ray astronomy detector was to detect such gamma rays from the galaxy, it is necessary that the galaxy contain radioactive nuclei. My own scientific thrust focused on the processes that maintained radioactive nuclei in the galaxy. Because each radioactive nucleus is characterized by a halflife, that time during which half of a collection of them will decay, simple reasoning shows that they should all be gone today. Consider that the main radioactive nucleus that our work would expose was ^{56}Ni, having halflife of 6 days, and that the average age of a stable nucleus is about 10 billion years. Clearly, no ^{56}Ni nuclei remain unless some were very recently created, within the past few weeks. My work focused on the creation of that "new ^{56}Ni". Finally, the radioactive nucleus must emit a gamma ray if it is to be detected from afar.

On the heels of publication with Bodansky and Fowler of our

quasiequilibrium description of silicon burning, the nucleosynthesis process that creates the ^{56}Ni, I naturally gave that process a lot of emphasis in the nucleosynthesis portion of my course during spring semester of 1968. I created several problems to assign to the students and explanations that made their way into my textbook. One day I was describing the nuclear reasons for a most fascinating transition on which much was to depend; namely, at atomic weight 40 and below, the most abundant A=4n isotopes are all stable against radioactive decay and are found on earth. These include ^{28}Si, ^{32}S, ^{36}Ar and ^{40}Ca, which are respectively the most abundant isotopes of silicon, sulfur, argon and calcium; but at atomic weight 44 and above, the most abundant A=4n isotopes are all radioactive so that they are not found on earth. These include ^{44}Ti, ^{48}Cr, ^{52}Fe and ^{56}Ni, which are respectively the most abundant species at those atomic weights, but which subsequently decay by positron emission (positive beta decay) to stable atoms ^{44}Ca, ^{48}Ti, ^{52}Cr and ^{56}Fe. Those decays have become today the celebrated sources of the most abundant isotopes of titanium, chromium and iron. The reason for this transition between stable nuclei and radioactive nuclei is a subtle nuclear conspiracy involving the law of nuclear masses, which are determined by details of nuclear structure and which also are influenced increasingly by the Coulomb repulsion within the larger nuclei. The mutual electrical repulsion of the positively charged protons renders their binding energy less than that of the neutrons, which experience no electric force. Consequently the most stable nuclear structures are those having more neutrons than protons. For example ^{56}Fe is more stable than ^{56}Ni, which has equal numbers of protons and neutrons, so that ^{56}Ni must decay to ^{56}Fe. But the silicon burning quasiequilibrium does not have enough excess neutrons to create ^{56}Fe, so it must create radioactive ^{56}Ni instead. I became fond of saying in class, in colloquia and in writings, "Iron is a radiogenic element." And for analogous reasons the abundances established by the silicon-burning quasiequilibrium are of stable nuclei in the lower half of atomic weights and of radioactive nuclei in the upper half.

I had been explaining all of this in Space Physics 551. After this class one of Haymes' graduate students, Jerry Fishman, approached me to ask,

"Do those radioactive nuclei emit gamma-ray lines?"

"Probably, but I never looked" I replied. "Let's go check it out."

My mind raced as Fishman and I walked back toward my office in our newly constructed building, Rice University's *Space Sciences Laboratory*. Because these nuclei were explosively ejected by the thermonuclear explosion, the radioactivity within the rapidly thinning ejected mass would rather quickly be able to be seen from the outside. This might be the idea that I had long awaited, the idea better than the *Californium hypothesis*.[2] Jerry Fishman, on the other hand, was very aware of my paper with Wade Craddock because that paper was serving as a motivating target for the attempt to detect gamma-ray lines from radioactive nuclei, even though I had made known my skepticism about the Californium hypothesis upon which it was based. Haymes's detector was mounted on a gondola[4] that dangled from the high-altitude balloons that were Bob's space vehicles in the 1960s. One must get above the atmosphere to detect gamma rays from astronomical objects, because the gamma-ray photons, gamma rays being simply photons but photons of very high energy, interact so strongly by colliding with the atoms of the earth's atmosphere that they can not reach the ground. The gamma rays collide both with the atomic electrons of the air atoms and also with the concentrated charge of the atomic nuclei, in the presence of which they transform into a electron-positron pair, a particle and its antiparticle. The experimental situation is very rich, and Jerry Fishman was busy learning all that Bob Haymes had to teach him on that subject.

In my office I took Viola's compendium of nuclear energy levels and decays from my shelf and turned impatiently to ^{56}Ni. There it was, loud and clear. Almost every decay of ^{56}Ni was accompanied by gamma rays having special energies that identify the ^{56}Ni decay. At the blackboard I calculated that if a supernova ejects a mass of ^{56}Ni that I estimated to be equal to $1/10^{th}$ of the mass of the sun[1], that object would be detectable by Haymes' gamma-ray telescope as gamma-ray lines of energies 0.847 MeV and 1.24 MeV. My feet practically left the floor! Jerry was also excited by our discovery.

"Would you like to work on this with me?" I asked him. "It will definitely make a publishable paper."

"I sure would", Jerry replied.

So it was agreed. I loaned him Viola's book and asked him to check for detectable gamma rays from all of those nuclei between atomic masses A=44 and A=62 that were prominent in the quasiequilibrium. And I gave him a copy of my paper with Bodansky and Fowler so that he could read more of its details. In those pre-internet days that now seem as if from the dark ages, the only way to share a paper not yet out in the journal was to give one a copy of it. Fortunately I had several copies from Fowler's mass-reproduction system that mailed out preprints bound in orange covers, preprints that he, in a typical Fowlerism, named *Orange-Aid Preprints*. So Willy had given me five copies, and Jerry got one of those.

As it always was with me, my mind could not quit thinking about this discovery during that evening. At dinner I shared it with Mary Lou and the boys. Later I focused on a special peculiarity of great discovery as it occurred several times during my life. Why had it taken me so long to think of this, needing Jerry's question after class to prod me? Why did it take me so long despite the fact that I had been seeking this moment since my paper with Craddock in 1965 that first brought the idea of testing nucleosynthesis by detecting gamma-ray lines into print? I had begun work on the quasiequilibrium in winter 1966 at Caltech, and I had continued it through the Cambridge spring and summer of 1967. I had convinced not only Willy but also Fred Hoyle that their *e* process was wrong, that iron and chromium and nickel were synthesized as radioactive progenitors during silicon burning without the large number of excess neutrons that their *e* process required. Why had I not taken one more step to this new and beautiful opportunity? At the time I felt somewhat inadequate for having taken so long. Later I realized that new scientific ideas happen like that. The discoverers who always seem so brilliant to the world at large have in fact pounded their heads against a puzzle for a long time before finally discovering the intensely satisfying resolution. The human brain, at least my own, focuses on the ideas that motivate daily activities. My science life experienced so much excitement in the 1966-68 time span that my brain was always focused on the next task of its flow. My attention had been diluted by so many issues that were not science but rather my scientific life. I had not been sitting around for two years asking repeatedly where I would find the new and better hope for testing nucleosynthesis with gamma-ray lines. There are always so many new questions. It is only in retrospect that I, and I suspect almost all discoverers, can say, "I should have seen it sooner." Inventors may make a new invention in a flash, but scientists conceive a new idea only slowly, burdened by the weight of skepticism that is the scientific method, burdened by having to understand all that has gone before in order to understand what is truly new.

Jerry Fishman and I worked fast. We had all of the decay schemes mapped out within two weeks, and I had turned my attention to the question of whether the gamma rays can get out of the supernova so that they might be detected from outside. While the quasiequilibrium is being established after the heating of the supernova event, the matter lies deep within the structure of the expanding postsupernova star. The overlying layers of supernovae are completely opaque to gamma rays, vastly thicker than the earth's atmosphere which is itself sufficient to prevent gamma rays from reaching ground. The ^{56}Ni halflife is only 6 days, but it first decays to ^{56}Co, which is the nucleus whose decay to stable ^{56}Fe is the one accompanied by the 0.847 MeV gamma ray. The ^{56}Co halflife is 77 days, so even 77days after the explosion half of the initial ^{56}Ni still remains alive in the form of ^{56}Co. After 154 days, one quarter of the initial supply would remain. These exciting ideas and numbers encouraged my optimism. The surface layers of the supernova expand into the near vacuum at speed near 10,000 kilometers per second, so the amount of mass along a line of sight might decline sufficiently rapidly to allow escape of gamma rays. Each interior layer expands more slowly than the surface in proportion to their distances from the center, so the quasiequilibrium radioactivity might, I reasoned, expand at 2000 km/sec. This is still very fast. But supernova models were primitive in 1968, so that I felt that this Type II supernova was too much in doubt. Fortunately, most of the ^{56}Ni might be formed instead in an exploding white dwarf star, the structure Fowler and Hoyle had suggested for the Type I supernova.

So we decided to focus on that small Type I object for our paper. Type Ia supernovae remain today the brightest prospect.

Stirling Colgate enters the problem

My Rice colleague Dick Wolf returned from an American Astronomical Society Meeting in Virginia and told me that Stirling Colgate had presented an interesting paper there. Jim Truran, who had collaborated with Bodansky, Fowler and me in testing the quasiequilibrium idea, had alerted Stirling to abundant natural ^{56}Fe being a radiogenic daughter from ^{56}Ni decay, and Stirling had produced his own take on it. His interest in how the optical light from supernovae declined with time, the so-called *supernova light curve*, caused him to calculate that the roughly 60-day decline in the light output observed after many supernovae might be explained by the radioactive power from the 77-day halflife of ^{56}Ni daughter decays. This was the idea that Stirling had presented at that meeting. A radioactive cause for the optical light was an old idea that offered an escape from the physical requirement that supernovae should dim quickly as they expand. The idea that the declining light power was powered by radioactivity had been first introduced[2] utilizing ^{7}Be as the radioactive nucleus and later taken over by the equally incorrect but more celebrated californium hypothesis. Stirling now was advocating ^{56}Ni as the correct source.

Now why didn't I think of that? I was familiar with the light-curve arguments and had published with Craddock a paper based on the californium suggestion. I had led the science that identified ^{56}Ni nucleosynthesis. Despite this readiness, the light-curve connection had not resurfaced within me owing to my focus on the idea of testing explosive nucleosynthesis and hopefully confirming with gamma-ray lines that the explosive nucleosynthesis theory was correct. Identifying the radioactivity by its gamma-ray lines seemed to be the big payoff to me, so that is where I invested my attention, and probably why I overlooked the supernova light curve. When and why scientists think of a new idea is a mystery of psychology that cannot be explained as a simple product of preparation.

I came to regret the next incident. I telephoned Stirling Colgate to ask if he would visit Rice as our weekly colloquium speaker and describe his work on the supernova-light-curve idea. After discussing his ideas for a while, I did not restrain my enthusiasm, nor did I hide it, instead saying to Stirling,

"One reason we are so interested is that we are working on a related idea, that the gamma rays might be detectable from outside and might identify ^{56}Ni by the 847 keV energy of the gamma ray. We have a program in gamma-ray-line astronomy here, and I am excited to move that field forward."

"That's very interesting", Stirling replied, and agreed to come.

Stirling arrived in his inimitable style. He flew in his own private plane and brought a friend along with him. I somewhat envied his bravado. Mary Lou and I took them out to dinner at a good restaurant, a standard courtesy toward department guests. On the next day Stirling sat for a couple of hours in my office while Jerry Fishman and I explained our results to him. If Stirling had also been working on these gamma rays, he made no mention of it. Then we awaited his 4:00 pm colloquium talk, which was delivered with Stirling's unique style. Near its end he stated that another consequence of his work is that gamma-ray telescopes might record the gamma-ray lines that would also be expected to be detectable as the remnant thinned. I never forget the abrupt manner in which I turned to Jerry Fishman and he simultaneously to me, our eyes locking in a puzzled shock. Why had Stirling added that remark while saying nothing of our work? Why did he seem to be presenting it as his own idea? Had he in fact also been working on gamma-ray consequences? If so why had he not mentioned that fact, neither in the initial telephone call, nor in his AAS Abstract from Virginia, nor in the conversation with Jerry and me in my office? Why had he surprised us with this seemingly territory-claiming remark to our department faculty and students? This was very awkward. Not knowing what to do, and not wanting to risk a battle over attribution of the gamma-ray astronomy, I suggested to him that we coauthor a paper on the gamma-ray lines. He accepted that invitation at once. I gave him a draft of our paper with a title page bearing the authors Clayton, Colgate and Fishman. A few weeks later we received his additions to the text and his correction of the title page to read Colgate, Clayton and Fishman. I sent him the final draft with the original author sequence restored, stating to him, "The authors shall remain alphabetical." We submitted it for publication on May 20, 1968, just before my family and I departed again for Cambridge.

Our *Astrophysical Journal* paper[3], "Gamma Ray Lines from Young Supernova Remnants", was selected three decades later as one of the most significant papers of the 20th century and reprinted as such in the *American Astronomical Society's Centennial Issue*, compiled to celebrate a century of publications in *The Astrophysical Journal* and in *The Astronomical Journal*. I was asked to include a "Commentary" on that paper, which I did. But my relationship with Stirling Colgate was never a comfortable one after this 1968 visit to Rice University[4]. Confirmation of the predicted gamma-ray lines would have to await a momentous astronomical event in 1987.

1968 Return to Cambridge

In 1968 I and my family flew to Heathrow for our return to Cambridge. Arriving was irrationally exciting to us, and much easier because *White Cottage* and Kathleen were awaiting us. Being able to move back into our previous home was intensely sweet to me, as if it were my own beloved home. All of us did love it, but especially me. It allowed me a delicious self deception, that of feeling that I was coming home, of feeling myself a Cantabrigian. So I gave in to that self deception and enjoyed it. Kathleen and her dog Turie had been frequent topics of conversation, becoming extended family in our hearts. And it all looked just the same, matching our every memory. We held frequent garden parties on the short lawn in the *White Cottage* garden. Donald huddled close to me as we set out the glass frames for the strawberries and spread anti-bird nets over the raspberries. Devon, after ceremoniously burying the head of a turkey that Kathleen had prepared for a dinner, wrote a small epitaph, "In memory of a turkey's head", which he mounted on a cross above the burial site.

I slipped easily into a fantasy that I already was sharing with

Willy Fowler, a dream that we would make of Cambridge a second home. To that end the Fowlers bought a home on Oxford Street, a home that Willy hoped to make a retirement home. He encouraged me to seek ways to make Cambridge my ongoing summer home as well, so that we could continue our close relationship in research, not only with one another but also with Fred and his *Institute of Theoretical Astronomy*. Such talk fueled the emotions that I was already feeling for Cambridge, so much so that I would be locked in its grip for five more years. Willy and I shared one of the large *IOTA* rooms as our joint office and prepared at once to help with this summer's *IOTA* conference, *The Anglo-Australian Telescope*. Fred was diving headlong into the leadership for British astronomy for this telescope. It matched his hope that *IOTA* could become a theoretical planning center for the modernizing of British astronomy. The conference attracted distinguished astronomers, still in 1968 all dressed in coat and tie. Excellent buffet lunches were served daily in *IOTA*. Hoyle's history of the *Anglo-Australian Telescope* can be read in his autobiography.

After the conference ended and distinguished visitors drifted away, I delivered one of the regular weekly research colloquia of *IOTA*, which I entitled "Confirming Nucleosynthesis with Gamma-Ray Detectors". My goal, that gamma-ray detection of live ^{56}Ni outside young supernovae could provide positive proof of the correctness of supernova nucleosynthesis, was very enthusiastically received by that audience. One staff member told me that it was the most exciting new thing that they had heard since *IOTA*'s founding. I also mentioned the light-curve implications, but labeled that interesting aspect "Stirling Colgate's work." This made *IOTA* the first research institution to hear of my work with Fishman.

Fred was in very high spirits over these events and suggested that we take a well deserved break to climb in the Highlands. We spent two or three nights at a series of three hotels, Loch Loggan Hotel in Newtonmore, Aultnagar Lodge, Andrew Carnegie's highlands home in Sutherland, and Inchnadampf Hotel in western Sutherland. The weather held good, making for an exceptional week on the hills and in the dining rooms. One hilarious incident occurred at dinner in Aultnagar Lodge, when John Bolton, famous Australian radio astronomer, said loudly two times, without turning or even lifting his head from his plate, "Would the person banging his fork on his plate please stop!" He stopped.

During the remainder of this fast-passing summer, we could not resist several few-day excursions about England. These were romantic forays to places either ancient or famous. We learned a great deal of historic lore. For subsequent visits to Cambridge I was appointed Fellow of Clare Hall, a refreshingly different sort of college for graduate students and distinguished visitors. There I met Brian Pippard, who resided there with his family as first president of Clare Hall and who would later as Cavendish Professor be in a key position when Hoyle's institute came to an untimely end. I greatly enjoyed the astronomers' lunches in Clare Hall and on several occasions took Stephen Hawking along with me in my car. He still spoke without synthesizer at that time. It was all thrillingly *English*. But it all too quickly would come to an end.

Back once again in our Tangley Street home in Houston, Mary Lou found a job with The Alley Theater as a technical editor. She was always good at sniffing out interesting opportunity. The Alley Theater is one of the three oldest resident theaters in the United States. Its founder and director, Nina Vance, became a celebrity within Houston and within the world of theater. Mary Lou wrote or edited copy about the theater that was useful in dealing with foundation or government grants. Such work was familiar to her from jobs she held in Pasadena. The job was not full-time, which she could not have accepted, and accordingly not full-pay, which we did not need. I felt emotional conflicts over this job, as I had with every job she had held since the birth of our sons. My negative feelings sprang from my wish that she could be happier being a wife and mother, a job with unending tasks for the good of the family. This was a selfish wish. With my busy academic and explosively growing research career I wanted my life to be simpler. Probably I was slightly threatened by her independence from our family. And I always had a gnawing at my stomach at my wishing for more emotional nurturing of our boys and nurturing of our marriage. Perhaps I simply longed to recreate the family of my childhood. My positive feelings were pride in her and in the Alley Theater. We attended almost all productions during these years at the old Alley Theater on the south end of downtown Houston. I had fun with its social life, which Mary Lou sought and adored. This job definitely added much culturally to our lives, just as Mary Lou had done in many ways throughout our marriage. I was always more inward looking, wanting simplification, avoiding intrusions, seeking a clear mind in order to think more deeply and more uninterruptedly about the origins of our material world. In 1968, the Alley Theater moved into its present home, adjacent to the Oscar F. Holcombe Civic Center, at the corner of Texas and Louisiana in downtown Houston.

Stan Woosley and Dave Arnett

Another graduate student from my nucleosynthesis class was to become very significant for my career. His name was Stanford E. Woosley. Stan was one of the many Rice University physics seniors who were attracted to the modern thrust of our new department. The great achievements that Stan would make arose not only from my guidance but from that of W. David Arnett as well. Dave had arrived at Rice for his first faculty position in 1969. I had argued loudly and clearly for creating the faculty position for Dave after working with him in Cambridge, where he was a visiting scientist. We had agreed that joint study of the several aspects of explosive nucleosynthesis in supernovae should be an enlightening undertaking, and within a year we would publish[5] in *Nature* the article "Explosive Nucleosynthesis in Stars", in which we laid out the methodology of that approach and including some early results. For me personally that 1970 paper marked my return to Hoyle's approach as laid out in his pioneering 1954 paper. Our paper announced my refocus on *Hoyle's equation*, as I later would name our approach. Many of the tools for that study constituted the subjects of my graduate course, and it did not take long for me to notice the problem solutions that Stan handed in. They were neat, quantitatively accurate with the mathematics, and often included extensions to the problems that Stan had seen as relevant to it. I do not mind admitting that I made Xerox copies of many of his solutions for my course notebook. One day Stan came to me after class and confided that although he was engaged with another faculty member on research into the atomic structure of lithium atoms, he found that to be unexciting, much less so than my course. He would much rather begin research on my problems, if that were possible. I explained to him that no student was captive to their first exploratory research project, that we encouraged students to continuously examine their goals in space physics.

"To have enough energy for a research career depends mostly on the motivation that you feel for the topic," I explained. "Unless you really want to work on your research more than anything else, you will not have the staying power that success will demand."

I also shared with Stan a personal belief that became a credo for me. The most important start to an outstanding career is the guidance of a knowing faculty advisor into a problem whose answer the world awaits. I had a list of topics that were ones that the world would notice. I explained a few of these, and Stan immediately said that he wanted to work on them. So we agreed that henceforth he would be my advisee, at least after finishing up his obligation to the other faculty member. For all of my students I carefully selected initial research problems that would, when answered, become outstanding publications, even pioneering ones. Being given a good problem is more important than the time and frequency of hands-on advising, because in the end the student must himself learn what he needs to know. I felt so strongly about this because of my awareness that my having been given the *s*-process problem by Willy Fowler was the cornerstone to my growing career. Only world leaders are able to effectively choose the right problem. They understand best what the burning issues of science are. This truth can explain why the greatest research careers emerge from the greatest research institutions. Equally talented students in universities having less distinguished faculty do not become pointed as effectively from the beginning. Reflecting on this made me increasingly aware that Frank McDonald's recommendation that I apply to Caltech from SMU had made everything else possible.

The problem that I posed to start Stan's research was one

that would be decisive in my mind for proving that ^{56}Fe had been synthesized as radioactive ^{56}Ni. The reason that the world would care is that the entire new astronomy of radioactivity depends on the answer being radioactive ^{56}Ni. Nucleosynthesis had shown two different ways to create ^{56}Fe, the most abundant heavy metal. One way produces ^{56}Fe as itself, stable from the outset. The other process produces instead radioactive ^{56}Ni, which later decays to ^{56}Fe and emits gamma rays in the process. The natural data for Stan's study lay in the shape of the so-called iron abundance peak, which is a run of high abundances that, plotted against atomic weight between A= 52 and A=60, showed a huge peak in that region. It is a mountainous peak, one at its highest in the isotopes of iron and nickel. Imagine that one could scoop up an average sample of our solar system, mostly sun but partly earth, that contains 1,000 atoms of silicon. That sample would also contain the following numbers of isotopes, ordered downward from the most abundant of the isotopes: ^{56}Fe = 825; ^{54}Fe = 52; ^{58}Ni = 33.7; ^{57}Fe = 20; ^{60}Ni = 12.9; ^{52}Cr = 11.3; ^{55}Mn = 9.6; ^{58}Fe =2.5. Other isotopes of chromium and nickel quickly fall to values less than unity, and to values much smaller yet below A=50 and above A=62. Clearly nature showed great favoritism for the production of nuclei at masses A= 56, 54, 58, and 57.

Some type of restricted equilibrium was certainly at work in setting this abundance pattern. Both Hoyle's pioneering *e* process and my silicon-burning quasiequilibrium were such processes. The first, Hoyle's *e* process, produced the group ^{56}Fe; ^{57}Fe; ^{55}Mn, and ^{58}Fe, each in its own stable nuclear form, but failed on the others. By contrast, silicon quasiequilibrium produced ^{56}Fe, ^{54}Fe, ^{58}Ni; ^{57}Fe, and ^{55}Mn, with radioactive parent nuclei ^{56}Ni and ^{57}Ni. These last two were the main hopes of gamma-ray-line astronomy. That endowed this study with great significance for the future of nucleosynthesis. I argued that nature's ^{58}Fe, which is bypassed by explosive nucleosynthesis, was in fact an *s*-process product, having resulted simply from neutron irradiation of ^{56}Fe and ^{57}Fe in other stars.

I suggested to Stan that the high natural abundance of ^{58}Ni might be the decisive evidence in favor of the silicon quasiequilibrium, with the consequent hope for detectable radioactivity. He searched with his computer calculations through the abundances of all silicon quasiequilibria, including those having a wider range of excess of neutrons over protons to determine if it was in fact necessary to synthesize the radioactive ^{56}Ni and ^{57}Ni nuclei in order to achieve the observed large abundance of stable ^{58}Ni. Stan's results were convincing. Gamma-ray astronomy could go optimistically forward because of the impossibility of synthesizing abundant ^{58}Ni without also synthesizing the radioactive ^{56}Ni and ^{57}Ni. We submitted our paper[6] to *Astrophysical Journal* in January 1969. This paper, coupled with that of Clayton, Colgate and Fishman, established Rice University as the center of the expectations for nuclear gamma-ray lines for astronomy. It also indicated that the lead in nucleosynthesis had moved from Caltech to Rice University.

Armed by Stan's now expert understanding of quasiequilibrium, Dave Arnett and I suggested that for Stan's PhD thesis he examine the spectrum of quasiequilibria that could be generated when a strong outward moving shock wave slams into the oxygen shell of the presupernova star. What we sought was the nuclear physics of oxygen thermonuclear explosions. We also sought the superposition of shocked oxygen that could reproduce the solar abundances. No implications for weapons exist because it is not possible to reach the requisite high temperatures for oxygen explosions on earth. But the rebound shock from the stellar core could do it. This shock wave is launched by the bounce of matter falling onto the newly born neutron star and strengthened by absorption of neutrino power from the new neutron star. We reckoned that passage of this shock wave might cause the temperature to jump to values in the range 3-5 billion degrees, at which time nuclear reactions stemming from the heated oxygen would burn furiously for only about one-hundredth of a second before the expansion lowered the temperature enough to quench the fusion. Stan's job was to examine those nuclear reactions and how they couple together to cause a rapid evolution of the abundances. This required taking the nuclear-reaction-network approach rather than that of assumed quasiequilibria. Dave Arnett provided his network solver for this, a technique that I had not yet used. Stan rewrote the code with several useful new features. This immense project would last over the next four years, into Stan's half year Postdoctoral Fellowship at Rice.

Our *Astrophysical Journal* paper[7] resulting from Stan's Ph. D. thesis was exhaustive in its delineation of the possible paths for overheated oxygen. It showed that the abundances produced by lowest initial temperature, explosive oxygen burning, matched the natural abundances quite well up to atomic weight 40, including hose of the heavy isotopes of silicon, sulfur, argon and calcium that had been notably absent in the Caltech work with Bodansky and Fowler. Our paper also showed how explosive oxygen burning merged smoothly into silicon-burning quasiequilibria if the initial temperature set by the strength of the shock wave was sufficiently high, in excess of 4.5 billion degrees. Other portions of the abundance bridge were filled in by burning at that peak temperature. And our paper delineated an altogether new type of quasiequilibrium, one that occurs when the oxygen is so strongly heated that it decomposes into a large number of helium nuclei that have insufficient time to reassemble into ^{28}Si nuclei. Because unassembled alpha particles remain after such expansion and cooling, this was called "the alpha-rich freezeout". This process yielded a large ^{44}Ti abundance, sufficient after its radioactive decay to account for the ^{44}Ca natural abundance. Radioactive ^{44}Ti was of great importance for gamma-ray astronomy, for which purpose the Clayton, Colgate and Fishman paper had called out ^{44}Ti as a prime prospect for gamma ray emission, but which was synthesized in insufficient abundance in the gentler two outcomes of explosive oxygen burning. The alpha-rich freezeout is also essential for the synthesis of nuclei more massive than A=58 in the iron abundance peak, a major problem for other nucleosynthesis conditions. Four natural stable isotopes of nickel, ^{58}Ni, ^{60}Ni, ^{61}Ni, and ^{62}Ni, appear with abundances almost exactly matching their natural, puzzling "fishhook" pattern.

If we considered that all three degrees of burning would plausibly occur in differing shells of a shocked supernova, we found that virtually all of the computed abundances between atomic masses A=28 and A=64 were in good agreement with the abundances measured in our solar system. Those abundances resembled a bridge hanging between A=28 and A=64, where the mass A is a proxy for distance along the bridge and the abundance is the height of the bridge, which is high but not equally high at two points, A=28 and A=56, and sags greatly to a lowest level halfway in between. Nor was the bridge smooth, but possessed jagged ups and downs along its general trend, and then drops downward again at masses greater than A=56. What was stunning was the remarkable degree to which the shape of that bridge resembled the shape of the natural abundances. Stan's PhD thesis, which was absorbed into this published paper, was a model of scholarship that has hardly been exceeded in subsequent decades. Every new researcher now must study that paper to catch up to the level of human understanding.

Owing to this incredibly rich fabric of nuclear and supernova detail, such as described only briefly above, one must refocus through the haze of its details onto the sublime central problem to feel its pulse. The goal is no less than to explain the creation of the chemical elements. The signs were Old-Testament-like in their drama. The natural abundances contain so much numerical detail, detail that has been painfully won by astronomical and meteoritical observations, and contain such dramatic structures showing neighboring isotopes differing in abundance by factors of one hundred and more, that I became fond of saying, "Any incorrect theory will be ruled out by the overwhelming detail of the nuclear abundances." It can not be emphasized too strongly that that is what science does; namely, it rules out theories that do not work. Science never proves that a theory is true. Science is a refiner's fire, one that achieves a nugget of gold by discarding the dross.

Correct understanding of the origins of our chemical world can be discovered only once in the entire history of the human race. After hundreds of thousands of year of evolution of *homo sapiens*, that moment had arrived during my lifetime. What a thrill it was for Stan, for me, for Dave Arnett to have been there, to have seen the stunning agreements that existed between our computer simulations of exploding supernova interiors and the jagged patterns of natural abundances. As we looked at these comparisons we knew that we were privileged witnesses to a process of creation. The review paper by B^2FH by contrast, had achieved no such successes and had no calculable models of the nucleosynthesis of isotopes between A=28 and A=54. We had been graced by providence, the first human beings to glimpse the truth beyond doubt of explosive nucleosynthesis in supernovae. What pleasure I had during summer 1969 to explain to Fred Hoyle that his theory of creation of the iron-peak elements in supernovae was now demonstrably correct. But it was not the *e* process that provided the proof, for indeed his 1946 *e* process had been incorrect in essential details, but rather the smoking gun had been found in the explosive expansions of oxygen and silicon. What's more, our great success at explaining the natural abundances of the isotopes between A=28 and A=62 used exactly Hoyle's picture as laid out in his pioneering 1954 paper. I now saw that that paper of Hoyle's was not only earlier but also more important than was B^2FH. Still the flood of citations of B^2FH continued, and those of Hoyle (1954) were almost non-existent. I regret that I did not make this point visibly at this time[5].

Haymes's Gamma Ray Telescope in Action

At about this same time Bob Haymes's gamma-ray telescope was being launched hanging from a high-altitude balloon from Palestine, Texas. Research with one of his students, Neil Johnson, was conducting the first measurements in gamma rays from the center of our galaxy, the Milky Way. They were astonished to find a peak in the gamma-ray flux near 476 keV[8]. I published a short paper[8] in *Nature* to show that the annihilation of positrons with electrons could perhaps account for this peak and to argue specifically that positrons emitted following the decays of radioactive ^{56}Ni and ^{44}Ti were sufficiently abundant for the task. The subsequent history of this would leave unresolved doubts when later versions of Haymes' study showed that the annihilation energy was near to 511 keV and that the gamma rays were rather strongly oriented toward the galactic center, perhaps too central to reflect the greater spread of supernovae. It is generally felt that Johnson, Harnden and Haymes' result[8] was a pioneering discovery of a phenomenon from the galactic center but that the calibration of the energy was inexplicably off by about ten percent. But for me personally, this episode drew me squarely into the world of gamma-ray astronomy, where I would remain.

A phone call from NASA Headquarters invited me to submit a research proposal, and I was subsequently awarded a research grant entitled "Prospects for Nuclear Gamma Ray Astronomy". During the next five years I published several works clarifying the nuclear expectations. The first[8], in 1971 in *Nature*, argued that nucleosynthesis of ^{60}Fe, with a radioactive lifetime of 1.5 million years, would be produced by intense s process and be ejected from supernovae. And ^{60}Fe decay was accompanied by two strong gamma-ray lines. Those gamma rays from individual supernovae would not be detectable, but I emphasized that tens of thousands of supernovae would occur during its decay lifetime of ^{60}Fe, so that the interstellar medium would shine in gamma rays from these accumulated events. They could collectively produce a measurable gamma-ray signal. This nuclear case was the prototype monitor for measuring the average rate of nucleosynthesis during the last million years of our galaxy, although ^{26}Al, which has very similar properties, would be detected first by orbital gamma-ray telescopes capable of measuring the gamma-ray energies.

Bob Haymes had been able to hire a postdoc for his program, J. D. Kurfess. Jim did many good things for that program. In fact, Jim Kurfess and Neil Johnson became such respected leaders of gamma-ray astronomy that they later became Principal Investigator and Chief Scientist respectively on the *Oriented Scintillation Spectrometer Experiment (OSSE)* on NASA's *Compton Gamma Ray Observatory*. What also interested me about Jim at the time was that he was an excellent basketball player. He was a good dribbler with a deadly jump shot. He had played in college for Case Institute as it was then called. It was our habit for years in the Space Science Department to stage halfcourt noontime basketball games. We had a lot of good players. For some years this maintained my physical conditioning as well as running the three-mile perimeter of the campus did, and it was a lot more fun. The high point occurred during one of Jim's years there when our department intramural team won the intramural basketball championship. I still have my championship certificate.

Another Research Student from my Course

I took a third research student into my general program of explosive nucleosynthesis research with Dave Arnett, also a student from my Rice University class in 1968-69. William Michael (Mike) Howard was keen to join the studies of explosive nucleosynthesis. Mike came from Baylor University in nearby Waco, and although he was not initially as well trained as Rice graduates, he quickly proved himself. I liked him very much, not only for his delightful personality but also because his lighter academic background reminded me of myself arriving at Caltech from SMU a little behind some entering students. Small in stature, weak of eyesight, and unathletic, Mike was unimpressive physically; but years later we would often jog together on the country roads just north and west of his Livermore home during my consultancy there. He learned to love fitness. Mike's rapid move forward intellectually after entering graduate school stirred memories of how it had been for me. In that way we had much in common. He was well liked by everyone for his love of physics and his unusually good temperament. When he asked me about research opportunities I was very glad to take him on. The Air Force Office of Scientific Research was supporting my research program with adequate funds for another graduate stipend, an important consideration that is not commonly understood by the public. Research universities expect that graduate students working full time on research will have their stipends paid by external grants that support that research.

Dave Arnett and I had plenty of problems to explore. Mike Howard liked the one described to occur when the outward moving supernova shock wave hits the shell of helium that exists far outside the oxygen zones that Stan Woosley was studying. The passage of the shock wave through the helium left it hotter than helium normally becomes during the normal evolution of stars. That heated helium shell was a spherical layer that had been established in a massive star by the fusion of hydrogen into helium earlier in the life of the star. The so-called hydrogen shell source used the CN cycle[9] of reactions to effect this fusion. Hans Bethe had discovered this sequence of reactions, which are capable of fusing H into He. Those reactions have the side effect of converting the initial carbon and oxygen atoms into nitrogen atoms. As I described this to Mike I remembered fondly that my own graduate work had included a study of the proton reactions with the ^{13}C nucleus during the operation of that CN cycle. I had described the CN cycle[9] in my textbook, which had become indispensable for initial training of new students. So Mike first studied the CN cycle, and then joined with Dave Arnett to compute its operation in the star prior to the supernova shock wave. This was more than just practice, because it was needed to evolve all abundances to their appropriate values at the time of arrival of the supernova shock wave.

Mike's results were eye-opening. They demonstrated that if the ^{14}N nuclei, which are, except for the hydrogen and helium, the most abundant ones to survive the CN cycle in stars, had never become hot enough to be destroyed with helium prior to the arrival of the shock wave, the high temperatures between 500 million degrees and 800 million degrees that would be caused by the shock wave would cause several reactions to rapidly destroy ^{14}N nuclei. Radioactive nuclei ^{15}O, ^{18}F, ^{19}Ne, and ^{22}Na would be created abundantly from the ^{14}N seed nuclei and cast into space. In our resulting *Astrophysical Journal* paper[10] in 1971, once again formulated and published with Dave Arnett, we suggested that the high ejected abundances of those radioactive nuclei and of stable ^{21}Ne might be nature's parents for stable ^{15}N, ^{18}O, ^{19}F and ^{21}Ne. The origins of these rare stable isotopes were then a mystery that had received little attention. Uncertainty concerning their origin remains today, because the circumstances contributing to their abundances are more varied and complex than we then appreciated. But we nonetheless glimpsed the general route by which these rare isotopes, which are totally missed during the main processes of stellar evolution, could be filled in by processes similar to the one we had calculated. Of special interest is the radioactive ^{22}Na created, not in sufficient quantity to account for the abundance of the daughter of its decay, stable ^{22}Ne, but to which I would return with a new idea just a few years hence. We also turned attention to a second related problem; namely, how heavier nuclei initially present in small amounts in the star could be changed during the burning by neutron reactions into rare unexplained species[10]. These details may not have stood the test of time in all particulars, but they were pioneering in their examination of another new aspect of nucleosynthesis.

Robert Rossellini and the intended film about science

In the midst of all these new results in nucleosynthesis theory, a glamorous adventure that was totally extraordinary came out of the blue. During the spring of 1970 the celebrated Italian film maker, Roberto Rossellini, came to Rice University. He was a featured guest of the *Rice Media Center*, and his visit was supported by John and Dominique de Menil, wealthy Houston patrons of the arts. The Menil Collection, the Cy Twombly Gallery, the Rothko Chapel and The Byzantine Fresco Chapel Museum are celebrated inner-city consequences of their largesse and high goals[11]. Rossellini offered evening presentations to the public, which I attended, at the *Rice Media Center*, and he there expressed his ambition to create a feature film about science. He described this general goal and then said,

"I seek two Rice University scientists who would like to be my science advisors on this, and who would like to create a strategy for the filmscript."

It was tantalizing, so after a few days thought I submitted the requested short proposal. Within a week I had been selected, along with my friend Clark Read, Professor of Biology at Rice University. What followed was a two-week whirlwind of working dinners, some at the de Menil home, a Houston modernist landmark which had been commissioned in 1948 from architect Philip Johnson, who was the leading American disciple of Mies van der Rohe, then the exemplar of European modernism, and some dinners at *Maxim's*,

John de Menil's favorite continental restaurant in Houston. The dinner bills were never even presented to John, who paid them later. I then and later discovered that Roberto Rossellini loved fine dining as well as he loved films; but films were his intellectual passion. He was animated and intense when discussing films, similar to my intensity when discussing nucleosynthesis. These dinners were totally stimulating in a way that I had not previously experienced and that was uncommon in 1970; namely, they concerned not science research so much as the dream of presenting science in a compelling way to the public.

My proposal differed from that of Clark Read, and Rossellini and de Menil liked both approaches. Clark advocated taking the great themes of science, making them understandable, and making clear how the future of the human race depends on applying those abstract concepts to human affairs. I loved his idea, which would be strong even today. My proposal had been to film the life of an individual scientist as he struggles from apprentice in graduate school to later leadership of astrophysics. My point was that astrophysics necessarily uses all of those great scientific principles in building its understanding of the universe, an aspect of the charm of astrophysics. I thought that following the development of the protagonist would humanize it, perhaps being more attractive to film audiences. All were attracted to that concept too.

The upshot of all this was that it was agreed that I would come to Rome for three weeks in June before I would again settle in Cambridge at Hoyle's institute. Because I was delivering a paper in Helsinki in May concerning the needs for neutron nuclear data for astrophysics at a conference held under the auspices of the *International Atomic Energy Commission*, it was agreed that I would come to Rome from Helsinki before returning to Cambridge. Perhaps fired by visions of three weeks in Rome, I mostly played during a somewhat unrestrained week in Helsinki, drinking wine and beer while strolling the pubs with the locals and the girls of Helsinki. During long hours when the summer sun crawls tantalizingly toward the horizon, beneath which it dips for but two hours before rising again, the city was alive with summer leisure. I explained the sun's behavior repeatedly over wine to my latest acquaintances, using a rubber ball on which I marked the equator of the earth and the outline of Finland. On a couple of these near all-nighters I left with a girl. The summer streets are unbelievably full of people, just walking and drinking and savoring the long sunlit hours from which they are so long deprived by a lengthy winter. Helsinki in May was a really remarkable scene. I slept until about 11:00 AM each morning.

Dominique de Menil's private assistant, stylish and thoughtful Helen Winkler, was sent to Rome by the de Menils to assist our three week effort there and to serve as my tour guide into matters of importance to the history of art. They booked and paid for two hotel rooms in lovely Hotel Raphael by the Piazza Navona. That hotel is favored by artists. There we lingered frequently beside the Bernini fountain that graced its core. My biggest problem arose from burning my candle at both ends, so that I could not awake before 10:00 AM. Of course this was continuing my schedule from Helsinki. I was relieved to find that this was regarded as normal by artistic Romans. The puritan background of my life made me feel that I should be working early. But work we also did. I outlined science issues and events in the life of my putative astrophysicist, and the assistant tried to type them up during the days. Then Rossellini would come by the hotel to pick us up for dinner, where we would consume more wine and talk about the ideas that were being set down. Roberto drove us to many of his favorite restaurants and always paid the bill. I spent very little during these three weeks. We dined at traditional Italian restaurants, pasta course first, then fish or veal, then salad, then coffee. We would not finish dinner before 10:00 PM. At the last of these privileged evenings, on June 17, the streets of Rome were, quite incredibly, empty! Italy was playing Germany in the semifinal match of the 1970 World Cup in Mexico City. No one was about. Even the taxis sat empty. Later that night it was clear that Italy had won, 4-3 in the only game in World Cup history to have produced five goals in extra time. And through these working dinners we returned repeatedly to how the life of my fictional astrophysicist would imbue his science with a personal touch that the audiences would relate to. Roberto kept drawing me out, questioningly, striving to understand each nuance of the science. He was a remarkable man, sincere, idealistic and deep.

Roberto invited us twice to lunch in his home, somewhere along the Appian Way I believe, where his Indian wife prepared the lunch. The food was excellent. On one such occasion I met two of his children, "Robertino" and one of the twin girls, Isabella, daughter of Ingrid Bergman. Even as a teenager, and indeed I did not know her age, Isabella was definitely gorgeous. Roberto also showed off his Mazda, one with the rotary engine that fascinated him extraordinarily, and peeking deeper into his garage, a Ferrari. Roberto napped after lunch, but I kept writing on my script and ideas.

During the second week Rossellini took me and Helen, Menil's assistant, along with him on a flight to Sardinia. In Alghero he was to be a featured speaker at a conference of film makers. Because Helen and I dressed as glamorously as we knew how, photographers kept photographing us when they could not get Roberto and puzzling over the identities of the two "unknown stars" that had accompanied Roberto. One of these photographs showing us stepping out of the airplane appeared in the Alghero newspaper on the next day. We took our own photographs[12] of Roberto, seated with me at the water's edge in conversation. This seaside resort was well regarded, and I swam in the bay waters, interested by the large population of small octopus that scampered about. We ate some for dinner.

*The author and Roberto Rossellini at the beach
near our hotel on Sardinia, June 1970.*

Where did this Roman episode with Rossellini lead? Not very far in terms of its stated goal; but it did have a long-lasting consequence that I have never disclosed. Roberto was, within weeks, swept away by other projects, pulled unceasingly this way and that, so in the end the leadership at the top was insufficiently decisive. Clark Read and I therefore became unsure how to proceed. By 1974 it was clear that the project had quietly died. It may have drawn inadequate financial support for all I know, as I never asked. But after its demise a new idea sprang from the experience. I was spending a lot of time during 1972-74 explaining my life in science to Annette, my new young wife, and that contributed to the notion that I might write a book based on my concept for the film. I did this rather quickly, and *The Dark Night Sky: A Personal Adventure in Cosmology* was published in 1975 by Quadrangle. I never mentioned the aborted film project in that memoir; but now I can. I think that 1975 was very early in the genre of mid-career memoirs designed to share scientific life and its issues. Although the book was well received in its few newspaper reviews, the publisher's impetus to push it was slight, and it faded away with a single printing.

Back at Rice University

During the 1968-72 period my Rice research was closely related to my Cambridge program of research. My plan for productivity in Cambridge was simple; namely, to export my Rice group to Cambridge! Hoyle accepted my strategy without blinking, and found space for them on his invitee list. So, while pursuing our research in explosive nucleosynthesis, Stan Woosley and Mike Howard accompanied me to Cambridge. I offered to pay their transportation costs if they could live on their continuing Rice graduate student stipends. They were as excited as kids, and did this on two occasions, although they dipped into their savings and into family support to enjoy it more fully. We had very good times together there. We became personally closer than would have been possible without the camaraderie of these relocations. Stan and Mike were both so much fun, and the experience meant to them much of what it meant me. Much of our research was conducted at IOTA, on their IBM 360. We flaunted this affiliation on our journal publications by using both institutional bylines, Rice and IOTA. My Rice group in Cambridge grew large, so large that it came to be called *The Texas Mafia* around IOTA. Its members included Stan and Mike, Dave Arnett, Ray Talbot, Mike Newman and Kem Hainebach[10]. Realize that this large Rice contingent was not simply able to come on my whim. There was no room at IOTA for scientists to come on a whim. Rather, it reflected my decision with Hoyle that doing so would create an especially intense identity for IOTA. I take pride that our research in nucleosynthesis was so productive at IOTA. Our contribution was later described by Martin Rees, the future director of the institute, as one of the distinguished episodes of IOTA history. I was achieving what Fred Hoyle had hoped from me, fulfilling his initial plan for nucleosynthesis at IOTA.

The Rice Mafia and Fowler at a pub lunch in Grantchester. From left, Kem Hainebach, Stan Woosley, Raymond J. Talbot, my postdoc at Rice, W. David Arnett, Assistant Professor at Rice; Fowler in shirt and tie, David Morris from Rice, and the author, Professor at Rice.

When this Stan-and-Mike-and-Dave epoch was coming to an end I was finally able to get back to the gamma-ray lines from supernovae radioactivity. It seemed to me that this possible test of explosive nucleosynthesis was perhaps the most exciting aspect of my career to date. It continuously generated interest wherever I traveled. In a small luncheon group in Cohen House on the Rice campus one day Curt Michel suggested that it could even win a Nobel Prize if the lines could be detected. I was determined to not dream such dreams, but the excitement was palpable. NASA Headquarters informed Rice of its grant award to support my ongoing investigations of gamma-ray astronomy. For all of these reasons it was time to return to that problem. In the interim since the Clayton, Colgate and Fishman paper I had realized that we had overlooked radioactive ^{57}Co as a gamma ray emitter as it decays to ^{57}Fe. The ^{57}Co was itself the daughter of the synthesized ^{57}Ni. The community should be alerted[13] to the importance of ^{57}Co as a gamma-ray source, because its decay was accompanied by two low-energy gamma rays, 136 keV and 122 keV, and its 270-day halflife was sufficiently long that those gamma rays might be detectable for years, yet not so long that their rate of emission from a fixed number of atoms would be too small to detect. This fortuitous halflife would prove itself fourteen years later[13]. I also wanted to again write a research paper without coauthors, excellent though those coauthors were. I wanted to explain simply the significance of my ideas within the format of a short, single-author paper. That style suited me best, and I would use it numerous times during the remainder of my career.

In that ^{57}Co publication I was also motivated to present a clearer discussion of the differing times of escape of the ^{56}Co gamma rays as the expanding supernova thins to become more transparent to the gamma rays of differing energy. The amount of overlying matter required to halve the gamma intensity by absorption was about 10 grams per square centimeter[13] for the 0.84 MeV line from ^{56}Co decay but twice as great for the 2.60 MeV line. That difference in absorption causes the 2.60 MeV gamma line to reach its maximum flux almost three weeks earlier than the 0.84 MeV line. The magnitude of that time lag reveals the supernova structure overburden combined with its speed of expansion. In my euphoria I did not realize that currently planned gamma-ray telescopes would not have the collecting area and sensitivity to measure this time lag; but I saw my job to be explaining the expected effects and letting Haymes and the other gamma astronomers worry about the sensitivity. In a similar vein I showed in this paper that the lines from the ^{57}Co decay became stronger than those from the ^{56}Co decay after two years despite the fact that the initial ^{57}Co abundance was by about forty times the smaller of the two. Moreover, both could be detected from any supernova within our Milky Way galaxy! We needed only to have an observing telescope above the atmosphere and, of course, we needed such a dramatic astronomical event to actually occur. Mankind had last seen a supernova in our galaxy about 320 years ago, in the direction of the Cassiopeia constellation.

In one final inspiration I observed that if the supernova does not maintain its spherical layering during the explosion, if instead some fingers of newly created radioactivity are hydrodynamically propelled outward through the overburden to a position closer to the surface, the radioactivity might become detectable much earlier than one might otherwise expect. Such an early appearance would be evident thirteen years later from a celebrated 1987 supernova[13]. I was able to do all of this with hand calculations and one figure in a three-page paper. That was my preferred style.

Despite this timely foray into explosive nucleosynthesis and gamma-ray astronomy, I was by no means finished with my first scientific love, the *s* process. It is said that one loves the topic of his PhD thesis for his entire life. It kept me returning time and again.

With Rice graduate students Mike Newman and Richard Ward I continued to explore the mathematical properties of that chain of successive neutron captures in stars. We published a sequence of papers[14] having that aim in *Astrophysical Journal* bearing the general title "*s*-Process Studies" along with a subtitle clarifying the specific goal of each study. These studies revealed some amazing and wonderful aspects of the mathematics that kept my name at the forefront of the process that I had first evaluated in my PhD thesis. Mike and Richard played a big role expanding mathematical knowledge of the abundance relationships generated by the *s* process. I would remain in *s*-process studies until the 1990s when a group of brilliant Italians in stellar evolution, led by Roberto Gallino, built revealing computer models of where in stars it occurs. They devised a pulsed-reactor model, with intermittent mixing between pulses in certain red giant stars. Their models revealed in detail how the *s* process occurs naturally within the evolution of red-giant stars.

The elephant in the room

My glance forward to the conclusion of this epoch has gotten me ahead of my story. In 1969 a fresh wind, swirling and magical, was in the scientific air. This was especially true in my department at Rice. Everyone was captivated by the *Apollo* program at NASA. Scientists were somewhat divided over it. Critics felt that it was money wasted, that so much more science could be achieved with less money. That position was heard from those that believed that science was the main goal of a lunar landing. President John F. Kennedy had called for the lunar landing in a speech delivered in the Rice University football stadium without, however, advancing it as a method of doing science. Something much bigger was at stake. It was the spirit of human adventure and exploration and the competition for world supremacy. Every time I left my Rice office I saw that hulking football stadium dominating the west end of the campus, with only a spacious field of wild grass between it and the Space Science Laboratory. Maybe it was my living in Houston that made me more receptive than some scientists to the thrill of this adventure. A photograph showing me and the Rice stadium from the outer corridor of my office is <1982 Clayton and Space Physics Building >. Now, if all goes well, it was about to happen.

As the summer of *Apollo* approached, I was proudly planning for my family's third consecutive summer in Cambridge. I passionately wanted to return in summer 1969, as my agreement with Fred Hoyle had called for. Cambridge was very much at the core of my dreams.

CHAPTER 12. IRISH TROUBLES

The year 1969 brought my life to a pinnacle and to a bottom. Mankind scaled a pinnacle in the history of the human race. We traveled to the Moon in 1969, and walked its surface, and brought pieces of it home to earth for study. Although the emotional impact of that unique event was immediate, its scientific fallout in my life continued to grow for three decades. In sober retrospect today, I judge our exploration of the Moon to have been the defining scientific happening of my lifetime. Many significant colleagues studied those first samples in their laboratories. Those bags of rock and soil had been carried down like the stone tablets of Moses. They brought truth and changed us. Man's launch into space ranks with the discovery of quantum mechanics in the 1920s, pace setters for their respective semicenturies.

Men are carried away by the events of our own lives, those trivia that seem so meaningful to each at the time. This was the year of publication of my paper defining the hope and future of a new astronomy, gamma rays from radioactive nickel nuclei outside exploding stars. It was the first year for widespread use of my textbook *Principles of Stellar Evolution and Nucleosynthesis* (McGraw-Hill, 1968). Owing to its publication and to the escalating competitive search for top faculty in US universities at this time, I was promoted to Full Professor of Space Science in April 1969. My letter from Frank Vandiver, distinguished Professor of

History and then acting president for Rice University for a two-year period between the presidencies of Kenneth Pitzer and of Norman Hackerman, carried a nice raise to $17,700 per nine-month academic year, which is how university faculty are normally contracted. The roughly $23,000 per year when summer pay is included constituted a fine salary. It equaled the amount we had paid two years earlier for our nice house on Tangley Street, just north of Rice University. In May 1969 my *PHYSICS TODAY* paper," Origin of the Elements", would be published, and in the same month our family of four again relocated, poetically and unforgettably, for the third consecutive summer to England and to Hoyle's Institute of Theoretical Astronomy. Summer 1969--summer of *Apollo 11*, of breathtaking "Munros", of astrophysics, of life in a country village, of my wings melted by the heat—that summer haunts my memories. It stands a pinnacle among smaller hills in my life. It was to be our final summer as a family.

To my mind this momentous year really began with Christmas 1968. We had just celebrated Mary Lou's 33rd birthday, and drove to Dallas to spend Christmas holidays with my large family there and with Mary Lou's parents, who also lived there. The Rice University fall semester was at an end. My parents had roughly half of their thirteen grandchildren at their home, with all the usual trimmings. But the wind of history was blowing. *Apollo 8* had been launched on December 20. That mission achieved translunar injection[1] while we were driving the 240 miles from Houston to Dallas. On Christmas Eve, astronauts Borman, Lovell and Anders became the first men to see with their own eyes the far side of the Moon. *Apollo 8* transported the first of the human race to orbit the Moon. Television from space carried their Christmas Eve message. They read the first ten verses of the book of *Genesis* to us on Earth, concluding their broadcast with "Merry Christmas and God bless all of you--all of you on the good Earth." It was simple. It was profound.

Because White Cottage was being offered for sale, we had to plan for our summer without prearranged residence. When we reached Cambridge we would simply stay in a *Bed and Breakfast* until we found something. My only attempt to book lodgings in advance had failed. I had written to the Vice Master at Churchill College, Kenneth McQuillan, telling him of our needs and reminding him of my two meetings with the Master, Sir John Cockroft, who was not only a great person in physics history but also very interested in Fowler's use of electrostatic accelerators to measure the nuclear reaction rates in stars. Having dined with Cockroft in the Caltech Athenaeum during his Caltech visit, and having walked the tow paths along river Cam with him during the May Bumps, cheering on the Churchill crew, I had hoped for some wedge into summer lodgings in Churchill. How much it was worth to know someone important in Cambridge was always a mystery to me, but in my youthful imagination I had enlarged its possibilities. Furthermore, I knew that Geoff and Margaret Burbidge stayed in nice rooms in Churchill, and would do so again. Cockroft had died in 1967 shortly after my departure from Cambridge, however, and Vice Master McQuillan regretted that there existed no available rooms for summer 1969.

Many advised that the students would have lodgings booked until mid June, so we decided to spend May and early June visiting Ireland and Scotland. With that in mind I wrote Fred Hoyle to suggest that we meet in the highlands for some climbing during early June. In the 1960s touring was still an inexpensive thing to do, despite the high value of the Pound Sterling. It traded near $2.40; but it was a time when one pound bought a great deal in England. For example, we had found that we could buy more bread for a pound in England than we could buy for $2.40 in the United States. Exchange rates are tied to matters more important than bread.

For summer departure we decided to close up our house near Rice University, to drive to my parents' house in Dallas, and to fly from there. I always kept close contact with my parents, and this schedule allowed us to visit and share goodbyes. It also meant we would see them again first thing upon our September return. I booked four seats on Aer Lingus from New York Kennedy Airport to Shannon, Ireland. I was reimbursed for my fare from my Sloan Foundation Fellowship, which still had money left from its initial $20,000 awarded three years previously. Because the Sloan Foundation is private, I was not required to fly a US flag carrier, which is a requirement for government supported travel. I believe that Donald and Devon were sold half-fare seats as a family promotional. We would use the 450 pounds promised by Frank Westwater, IOTA Managing Secretary, for local subsistence in Cambridge. On the evening of May 18 we watched with my parents in Dallas the news reruns of the launch of *Apollo 10*, which was approaching the Moon by the time we departed for New York on May 21. On the morning of May 22 we were in Shannon.

Quickly through customs we emerged into a misty, cool, overcast Irish morning. It was this comparison with Texas bright sun and summer heat that seemed to whisper enticingly to us as the damp, cool, dimly illuminated morning air revived our love for all that is English. The terminal and tower of Shannon airport were tiny, and directly outside its exit lay a massive circular bed of tulips surrounding the statue of a saint. I photographed Donald across that sea of tulips, his head bowed and hands together in prayer, standing next to that statue, imitating its pose. Donald loved to ham it up. But the wondrous excitement of the U.K. is what I remember. The first sights and smells of Shannon recalled all of the magic of our two previous summers in White Cottage on Storey's Way in Cambridge. There my Englishness seems to have been born as if by magic from some forgotten tomb of ancestral memory, perhaps from my grandfather Kembery, whose own father had emigrated from the family farm in Somerset. The Isles, their weather, all that lore of childhood, all felt to me as if from an earlier life. Many Americans share this same reaction, this same feeling of somehow coming home. Certainly my father and mother felt it. It is a powerful feeling.

Now we were eager to explore Ireland. I had found that I could rent an auto at the airport, drive it around Ireland, and drop it off in Belfast. Then after a ferry across to Scotland I could pick up another booked auto and leave it in Cambridge. I booked all of this in advance. Car in hand, we started our explorations with nearby Bunratty Castle, and finished the day at a hotel in Killarney. Donald and Devon fell asleep in the car during the afternoon drive.

On this same date, May 22, 1969, two of the *Apollo 10* crew, Colonel Tom Stafford and Commander Eugene Cernan, undertook and completed a daring maneuver. They blasted apart from the Command Module and lowered the Lunar Module onto a descent path toward lower orbits, paused in a lunar orbit nine miles above the surface of the moon to take two passes over the intended landing site, then reburned to return to and dock with the Command Module piloted by John Young. We heard at the Hotel in Killarney that it had all occurred splendidly. Many in the hotel settled in front of the TV set in the lounge after dinner. We were the only Americans in the hotel. I was excited of course, but no more so than the English and Irish staying there. This required a second pint of Irish bitter. It was the last trial run before the lunar landing on the next mission, *Apollo 11.* Those in the TV lounge were very interested to learn that I was a space scientist from Houston; but I suspect they would have been skeptical but for the presence of Mary Lou and the two boys. Donald had blurted out this news in pride, as if his father had something to do with that daring mission. During that descent to nine miles, Tom Stafford took a photograph for the ages. "Earthrise from the Lunar Module" shows the half-Earth, blue and misty white, hovering on the lunar horizon from which it was rising. This photograph became a symbol of lunar exploration, of the Earth in space, and of a deep new awareness for the human race.

Next day we savored Killarney, our first taste of Ireland. At Kate Kearney Cottage in the Gap of Dunloe, Mary Lou lost her gold charm bracelet, one to which she affixed gold souvenirs of our worldly visits. She did not notice it gone until we had driven on many miles. She was crestfallen. I turned around and went back, not very hopeful of recovering a lost bracelet of value. But we did. A keeper of the Kate Kearney Cottage said, "Oh, yes. I have it right here". Mary Lou beamed. I made the keeper accept a five-pound note, which he seemed genuinely unwilling to do. The recovery of that gold bracelet was very sweet, like astronauts plucked from the sea after flaming reentry, reaffirming our sense of something very special about this summer. I photographed Mary Lou with the keeper outside the cottage. Looking at the photo it seems like yesterday, though she and the bracelet are now long gone. On this late-May day of 1969 we were on both occasions alone at Kate

Kearney's Cottage, only the cottage keeper and ourselves, a remote quietness that I am told is seldom found today at that place. Our visits to these isles in the 1960s were not the commonplace travel events that they have become today. Their seeming uniqueness in my life paralleled the uniqueness of travel to the Moon.

We drove in almost no traffic to Muckross House, Abbey and Gardens. It was beautiful, and we lingered for hours exploring the abbey ruins. Banks of azaleas in bloom lined a walkway to the gardens. We read in wonder of a 15th century Franciscan abbey that had been ruined by Cromwell's troops in 1652. Still, despite the forlorn sense of stones cast down, the Abbey's beauty remains intact. A very old yew tree spread its branches above the cloister walls. Exactly the same unmistakable glimmerings of something ancient but now lost are found by me in Tolkien's trilogy, *Lord of the Rings*. Photographing Donald and Devon climbing through the Abbey ruins, I strove to identify mystical impulses within myself and to separate them from historical truths that I did not know.

Continuing around the Ring of Kerry we paused on the stony beach at Waterville to cast stones into the sea. We strolled its old village streets. Then to Blarney Castle, where I and my sons hung upside down to kiss the bottom side of the Blarney stone. Donald laughed with glee at this. Devon lost his second tooth of the month while doing so. In the 1960s all Americans knew of the Blarney stone, and of the gift of gab that kissing it was said to bestow. As a child in Iowa I had heard my relatives exclaim of me, "Why he must've kissed the Blarney Stone!" So to me it seemed another meaningful moment no matter how trivial it really was. That sense of significance remains difficult to ascribe to valid causes, but many may identify with my feelings. I cannot say with assurance but I doubt that young people today have the same experience. How many know of the Blarney stone? Television probably changed our cultural formulae, replacing old oral traditions with new visual ones. Then we took an old hotel in nearby Cork, where we spent the next halfday in walking the old town, admiring its churches and pubs.

Next day we drove to Waterford, where Mary Lou made the first of her large number of purchases to be made during summer

1969. It was crystal, of course, of renowned beauty. We ordered six sherry glasses and a sherry decanter, which we were delighted to discover they could ship to her parents in Dallas. In 1969 the Waterford Crystal Works showroom was located in a large rustic wood-frame building, perhaps part of an old mill, west of the town. I loved those glasses and decanter, which I still own, not only for beauty but also because drinking sherry is *so British*. We ended the day in a tranquil *Bed & Breakfast* called Assagart Farm, along the lane to which we had paused to witness a simple hay wagon pulled by two aging horses. We watched until the hay cart was out of sight, leaving me with images from Iowa childhood. Assagart Farm was a small working dairy, with its milking shed behind the farmhouse, which was itself white plaster over stone, in the fashion old Ireland and of the English Lake District. Together our family watched the cows return for milking at 5:00 PM, as I knew they would, just as they had done in Iowa decades before. We were invited into the milking barn to watch the primitive milking machines do the work. It was earthy, suggesting sexual function, as indeed most things did in my 34th year. I told Donald and Devon that they then had to read their books quietly in their room while Mary Lou and I "napped" until dinner at 7:00 dinner in the farmhouse dining room.

Sex was a challenge for us, and this was no exception. We both approached it with inward looking anxiety. Our emotional baggage was sufficient that poor function plagued us. It did so for fifteen years of marriage. The romance of this day could not overcome deadening well-worn tracks. We bathed after, quietly, in our tub, and joined our sons for a somewhat subdued dinner. Nor could we find the resilience to rekindle passion later by giving ourselves up to the needs of the other. We never really discovered the needs of the other. How sad. We remained blind to true passion. This episode, but one out of many, grieves me. It sticks in my mind, haunting my memories. It presaged our divorce, which lay just around the corner, although nothing could have been further from my mind during May 1969.

By next morning we were ready for new adventure. We departed Assagart Farm after the breakfast consisting of tea in a large pot covered by its cozy, fried eggs and fried tomato, fried bread, bacon, and toast with orange marmalade. I wanted to linger over it. Thus fortified it was easy to again feel the good in our family of four. We were bound for Dublin! Mary Lou was again a font of research information on The Abbey Theatre, James Joyce, William Butler Yeats, Charles Stewart Parnell, and assorted "troubles". *En route* we had a fascinating luncheon stop in Wexford, where we saw the first signs of the new troubles. Walking the docks on Wexford Bay we saw graffitti--several signs painted in bold letters, "JOIN THE IRA". The Irish Republican Army was stepping up the pace of its attempt to eject the British from the North. Late afternoon we reached our hotel of two nights on O'Connell Street, just north of the River Liffey. We tired ourselves with two days of headlong exploration of Dublin.

In order to not miss our ferry in Belfast, we departed on the afternoon of the second day for Armagh, a historic town of Northern Ireland about 35 miles southwest of Belfast. In Armagh one of the world's most inventive astronomers, Ernst Opik, an Estonian, had continued his career safe from the communist Russia troubles after World War II, although he would have then found himself in the cauldron of the Irish troubles. I knew of Opik primarily through his early work on helium burning in stars. His calculations of that rate were later corrected by Salpeter, who introduced the ^{8}Be resonance between two helium nuclei as the accelerating first step. Opik also calculated in 1938 that stars became red giants by building up a core composed of helium ashes of the hydrogen fusion. Because I had never met Opik I did not presume to meet him on this occasion. I believe he would have welcomed it, but we had not enough time programmed into our plans. And fate grabbed our attention in Armagh with a historic look at the troubles. Arriving in Armagh we drove straight to the top of the Cathedral hill overlooking the town. Armagh has two cathedrals, one catholic and one protestant. We visited them both, enjoying the way in which their presence dominated and gave meaning two the town and its people. It was in the protestant cathedral that we saw the posted announcement: a revival tonight on the fairgrounds; the Reverend Ian Paisley speaking. This name rang bells in our memories. It had been in October 1964, less than a year after the assassination of President John F. Kennedy, that the Rev. Ian Paisely had led a crowd, perhaps a mob, of irate royalists into a catholic area of Belfast to tear down the illegally displayed Irish tricolor. In the ensuing riot in Belfast on that October 1964 day, the police moved in with pickaxe handles and wounded more than 50. It had made our TV news and our newspapers, and for some reason Mary Lou and I had been interested in it at that time. Witness to that riot was the 16 year old Gerry Adams, today's leader of Sinn Fein, the political wing of the Irish republican Army. Of that riot, Gerry Adams later told an interview for a June 10, 1996 feature in *New York Times*: "That really did it. I don't even know then if I knew what I was getting into, to be honest. I think we had a sort of innocent naivete. We were activists in a civil rights struggle. I think we actually felt what we were looking for was so reasonable. It was all part of the 60s--the anti-Vietnam protest, the student uprising in Europe, the American civil rights movement". Gerry Adams may today be a hope for peace in Ireland as he leads Sinn Fein without the bitter recall of the past that afflict so many populist leaders, of which Ian Paisley is a notable example.

We decided to attend the revival of the Reverend Ian Paisley. Out of curiosity we first drove by Ian Paisley's Headquarters, which was but a one-story metal building erected at the end of an alleyway in the town. It was closed in the late afternoon when we walked up to it; but we read a confirming announcement of the revival meeting, spoke with a lay helper, and studied the map to the location. We finally caught up with him in Dungannon, a description of which I defer to Mary Lou's words, below, written in summer 1969. There we took a hotel room for the night, had dinner, and dressed for church. Good that we did too. Arriving at the fairgrounds, which were simply an open green outside the town of Armagh, we were waived into a small carpark and took our seats in the building. It was a large tent-like structure, having perhaps twenty rows of ten wooden folding seats arranged in aisles, and a simple pulpit at the front. I was fascinated with the people as they walked in, mostly older men or older couples. Youth was not evident. Every man, without exception, wore a black wool suit, white shirt, and tie. They reminded me of my grandfather Clayton when he dressed for church. Simple men, hair closely cut and parted, and humble but purposeful demeanor. They greeted one another with simple words, or just silent nods, without any trace of gaiety or feigned sociability. "These are", I remember thinking, "salt of the earth".

This serendipitous event was the highlight of this trip to Ireland. Its dark implications presaged the stormy upheaval of our lives to come. I find it fitting to share Mary Lou's gifted essay on this experience, written that evening and given to me.

IN SEARCH OF IAN PAISLEY
Mary Lou Clayton (deceased; written in 1969)

Finding Ian Paisley's church, Ravenhill, on an arbitrary summer Sunday in Armagh, however, is not so easy for a stranger in the town. Graceful tree-lined streets are empty. Shut shops and silent homes give no indication of discord. The imagination is hard-pressed to see confrontations here between righteously angry, stone-throwing mobs. Protestant majority and Catholic minority evidently agree at least in their observance of the Lord's Day. After a while one is tempted to chuck it and say "I went to Armagh, but it was closed." Only a few people are around on a late Sunday afternoon.

A "regular" Presbyterian clergyman unlocking the gates of his church is able to point out the general direction of Paisley's church but adds that it is a bit complicated to find. Three men talking on a street corner in the "general direction" are more specific; but after circling the narrow streets of the area several times in vain, we pull up in front of two women talking across a hedge that separates their semidetached houses. "Where", I ask, "might the Rev. Paisleys' church be?" We have been up and down that same street, yet they point out an alleyway some few yards from where we have stopped and stand staring at us as we cross to the other side of the road. At the end of the alleyway stands a small frame one-storied building painted baby blue. I cannot resist the temptation to take a picture of it even though the two women are still staring at us over their hedge.

The door is open, and the interior of the building proves as stark as the exterior but less colorful. Wooden benches line either side of a longish aisle leading to a platform with a plain wooden pulpit. Voices are coming from a room in the back, and a gentle tap on the door produces the first smiling face I have seen in the whole town. The smiling face on a shiny bald head and a slightly rotund figure whispers me back up the aisle to the vestibule. Although I have hiked my skirt as far down as possible, I am conscious that a good deal of my legs are still sticking out, but the man smiles on. No. he is not the Rev. Paisley, he gently laughs, but only one of his lay workers. Picking up a framed picture he says, "Here's the good man himself, taken right here at Ravenhill." The color photograph shows a tall man in a clerical collar smiling next to a banner, "For God and Ulster." You can buy one for 12s 6d.

"Will the Rev. Paisley be preaching here tonight?" I ask. "Ay, no," the man replies. "Tuhnight he leads a meeting at Dungannon." "Is that far from here?" I ask. "Ay, no," he replies, "about twelve miles north", and still smiling he begins to draw a map. "Yuh're from America"--half question, half -statement. "We have many friends in America." "Yuh'll come tonight to hear himself?" He doesn't question why I would be interested in hearing Paisley. Handing me the map and a half dozen copies of the Protestant Telegraph, he repeats again, "Yuh will come to hear himself?" and waves to me as I leave.

As we approach the intersection of the M1, which runs westward from Belfast to just south of Dungannon, County Tyrone, we see the gray frame building right on the highway next to the petrol station, just where the smiling man said it would be. Since the meeting won't begin for a while yet, we go on the mile or so to Dungannon to find a Bed and Breakfast for the night. It doesn't take too long to ascertain that either there are none or else they close on Sundays too. Dungannon does, however, have one hotel, a reasonable one at that; so we take the room available, change into something suitable for Sunday evening service in Paisley-land, and retrace our steps to the "gray frame building right on the highway next to the petrol station"--the Dungannon Free Presbyterian Church.

The scene has changed in a few hours' time. Cars fill both sides of the road for yards and yards in either direction. The service has begun, so we quietly work our way around through the crowd outside to a window on the far side of the low flat-roofed building. The people inside are listening attentively to a "warm up" speech being given by an earnest young man describing his road to salvation. "Duh yuh know," he says, "that the hardest thing tuh do is tuh tell yur luved ones yuh've been saved--especially yur drunken father." Murmurs of encouragement from the crowd counterpoint the ebb and flow of his story. His particular road, it seems, began with the Salvation Army. But he was privileged to have been saved twice-- first from sin by the Salvation Army and secondly from the Salvation Army by Ian Paisley.

When the personal testimony is through, the smiling man from Armagh touches my arm and asks us to follow him. Firmly pushing through the crowd, he commandeers two chairs, then wedges them into an aisle in the back of the building, and insists we must sit down. A rustle of paper follows as the earnest young man starts to lead the crowd in hymn #1 from the printed sheets we'd been handed at the door. There is a distinct odor of mothballs in the tightly packed room, of Sunday-best clothes and feather hats allowed out of their storage chests for the occasion. During the singing of the hymn, several children openly stare at us, nudged back to attention from time to time by more discreet adults who take those occasions to get in a good look themselves.

A large man in the middle of a seated row of collared clergy facing the crowd rises as the last verse is finished, laughingly commenting that he might as well carry on since he can't move in any other direction anyway. He asks us to stand for a prayer, and I pray no one will yell "Fire!" Every inch of space in the whole building, aisle and all, is occupied by several hundred people who have somehow managed to squeeze into the room.

As offering plates are started from the back after the prayer, the clergyman gently pokes the pockets of the crowd, commenting that although the offering is twice as large when it is taken for Dungannon, he hopes the good people of Dungannon know how much .Ravenhill appreciates their help. Every penny of tonight's offering will go for "the new church", he adds, cajoling ever more of those pennies out of their hiding places. The new church in Ravenhill will have its official opening Sunday in October, thanks to someone in Armagh whom he mentions as "generously vacating the land for us." "We already have raised over 80,000 of the 150,000 pounds needed for the church--and we didn't have to hold a single jumble sale or raffle to do it!" The news and the not-so-subtle Catholic joke draw appreciative chuckles and a ripple of applause from the people. "Just this morning," he says, "I received an anonymous gift of 200 pounds." Suddenly it clicks--the easy joking, the rapture of the crowd. This is not another part of the warmup. This is "himself" speaking. I feel slightly annoyed. I came in search of Ian Paisley, but he has slipped up on me.

Paisley's form covers the pulpit. He is a big man with big features, a large smooth untanned face with slightly boyish fullness and brown hair graying at the temples slicked up and back in pompadour fashion. The new church, he continues, will have closed circuit television; but since there will be room for only 3000 people inside and 1,300 seats have already been committed, he urges those who want to be there on opening day to fill out request-for-seating slips, available after the service. The irony of the building that he is describing to a predominantly rural crowd most probably not accustomed to closed-circuit TV slips blissfully by the Northern Irish farmers and wives decked out in their mothproofed best tonight. They are with him.

Abruptly he changes his subject to appeal for funds to help those who, along with "himself", will briefly leave the continuing battle in Ulster to protest the invasion of "Popery" in Geneva--"John Calvin's home." In mercurial fashion he oscillates, now jocular about "Popery" and the absence of "Miss Bernadette Devil" from Parliament, now reaching a peak of outrage as he shouts: "Duh yuh know that thuh Presbyterian Assembly is going tuh welcome and recognize with a special seat Mr. de Valera himself?--that scoundrel who in 1916 bore thuh blood of out kith and kin on his hands!" "No!" gasps the crowd in disbelief. "If there are any Presbyterians here tuhnight, I say tuh yuh tuh tell yur minister what yuh think when he gets home." Leaning further over the makeshift pulpit, he adds, "Someone asked me what I would do. I would speak up, get up, get out, and stay out!" Heads nodding agreement bob about the room, accompanying a rumbling "Yes!"

During this "News-of-the-Protestant-World" monologue, Paisley has been talking in an intimate fashion to the people, gesturing, leaning back and forth, reaching out in the manner of a one-on-one conversation, his deep resonant voice rising and falling dramatically with each anecdote, playing his audience casually at ease. Just as swiftly, however, his posture, expression, and voice shift yet again as he announces that his topic for the evening will be "Ian Paisley and the Priest". "Himself" is "on"--the minister ready to address his flock. Slowly, emphatically he repeats his topic, word by word, and the crowd as if on cue sits a little taller as he intones his text from "Sa-ams" in a high-pitched voice that he maintains throughout.

His sermon is a classical presentation of the evangelical style: text read, text reread slowly, thesis extracted, thesis explored, word by word, examined, explained, involuted, turned out, and emptied out for "meaning." The message distilled in this manner for Paisley's flock tonight is crystal clear. He booms it out: "Jesus Christ is thuh only priest who can forgive sin. All others who claim to do so are false priests!" Question: And where, dear brethren, are there more false priests than anywhere else? Answer: In "Popery", of course. Paisley's performance is a credit to the honorary doctorate he holds from Bob Jones University of Greenville, South Carolina. Close your eyes and Ulster fades away, meshes with a hundred like scenes as Paisley intones, "Thuh only priest for me is Jee-sus."

As evangelical sermons go, Paisley's has produced no great surprises, no shockers. He even makes no reference tonight to Communism, a word that the <u>Protestant Telegraph</u>, which he edits, fondly compounds with Catholicism. He is a bit cut and dried tonight, but then this is his third or fourth sermon today. The crowd doesn't seem to mind, though. There are predictable responses to each of several references he does make to the time he spent in the Crumlin Road jail "for God and Ulster" (i.e. for holding a parade without a permit). Conjuring up how it feels "tuh hear thuh jail cell door bang shut behind yuh", he recreates the sights and sounds of jail for the awestruck audience. Since Paisley actually served less than *one-third of his six-month sentence, it seems likely he will spend a great deal more time referring to his "jail cell" than he spent in it.*

In the characteristic beat of these meetings, the time eventually comes when the text and its variations have been exhausted and the seat has absorbed more than the mind. At approximately that moment, the evangelist leans forward again on the pulpit, lowers his voice theatrically, and plucks a "personal story" from his life, prefereably his childhood, the coup de grace *calculated to leave no eye dry when he calls for the final hymn--with eyes closed and head bowed:*

Just as I am, without one plea,
But that thy blood was shed for me,
And that thou bidst me come to Thee,
O, Lamb of God, I come. I come.

"Will yuh raise a hand if yuh know the Saviour?" Paisley asks the crowd. A rustle of movement. "Fine," he says reassuringly."Now if thur's one here tonight who would like tuh say 'Yes' tuh Jesus, will yuh raise yur hand. I won't ask yuh to do more." Not many other options are possible, under the circumstances.

Three choruses and one prayer later it is done. People shake hands all round, drift slowly out to their cars, and leave.

CHAPTER 13. APOLLO
AND ICARUS

Daedalus warned his son not to fly too close to the sun, as it would melt his wings. They saw still other coasts recede and fade, when Icarus, emboldened by the ease of flight, darted out of his father's track and steered to higher zones with boyish daring. But the threatened punishment came swift and sure. The powerful rays of the sun melted the wax which held the feathers in place, and before Icarus was even aware of it, his wings dissolved and fell from his shoulders. The unhappy boy tried to fly with his bare arms, but these could not hold the air, and suddenly he plunged headlong through the sky.

-Gustav Schwab
Gods and Heroes, p. 84, Pantheon Books, 1946

The ferry in summer 1969 from the Belfast docks to Scotland departed before noon. It was a magnificent ship, taking us and our auto to Ardrossan, on the Firth of Clyde west of Glasgow. The duration was about eight hours, and the sun was low in the west when we drove away on Scottish soil, a bit the worse for wear and ready for the *B&B* that we quickly found.

Anticipation for the highlands escalated while we drove north past Loch Lomond. After just two previous visits, I loved Scotland; but the priceless opportunity of it was being there with Fred Hoyle and Willy Fowler. They stood on the pinnacle of nuclear astrophysics in 1969, and their output throughout the 1960s had been prodigious. I was, along with Al Cameron, the "new boy", a fourth voice making a science from their ideas. I was Willy's protégé and becoming one of Fred's. Anyone who has ever been in a privileged relationship with the world's most important thinkers in the topic that dominates his life will know what I felt. I was impatient to rejoin them; and in the Highlands I had them almost to myself.

A highlight of the drive became an endearing image in my mind of the highlands in the 1960s. Approaching Fort William from the south one faced the problem of getting around Loch Leven, a spot of great beauty famous for Glencoe, site of an historical massacre by the hosting Clan Cameron of the visiting Clan McDonald. Loch Leven sticks its long finger ten miles eastward between the mountain ranges, and in those days one had either to ride its length eastward on the south side and return that same length on the northern side before continuing north or find the ferry to be operating across its mouth at Ballachulish. Luckily the ferry was operating on this day, which was fine and clear. I watched Donald and Devon throw stones into the water as the ferry returned across the narrow neck of Loch Leven to its southern side. This ferry was a small open barge, capable of parking six cars at most, but seldom full. What a romance for this American, awaiting a ferry to transport an automobile across a quarter mile of water so that a grand adventure could be consummated. These were still the days when the British Isles possessed adventurous scope, when distances seemed large and cultural pockets historic and understandable. I found this to be a precious aspect of many experiences, of which the Ballachulish ferry is but one. So many of them are now gone, destroyed by the capabilities of human engineering. Today a modern bridge reduces Loch Leven to no obstacle at all. Allowed by the ferry to resume our travel, we slept one night in Fort William, on the western foot of Ben Nevis, the highest point in the U. K., before continuing along glorious Glen Shiel, past *The Saddle* and the *Five Sisters of Kintail* and other famous Munros[2] to The Loch Duich Hotel. This same hotel had been used during our first visit to the Highlands in 1967. I arrived in euphoria, again to be with the two most important friends of professional my life.

In addition to Fred Hoyle and Willy and Ardie Fowler, who were already settled in when we arrived, the party contained Wal and Anneila Sargent, Stewart Harrison, and Vahe Petrosian. All but the last two had been on our first visit two years earlier. All but Ardie Fowler and Stewart Harrison were astrophysicists of the highest order. Ardie, Anneila and Mary Lou amused themselves in other ways while we climbed a sequence of memorable peaks above 3000 ft. We climbed *Aonach Meadhoin* (3284 ft), rising above The Cluanie Inn on the western end of Loch Cluanie, and *Ciste Dubh* (3218 ft) in the Five Sisters, *The Saddle* (3317 ft) on the opposite, south side of Glen Shiel, and *Ben Sgritheall* (commonly *Sgriol*, 3196 ft)), each climb begun from loch level. This area provides some of the finest mountains in Scotland.

Conversation during these outings took on its own peculiar rhythm, returning without preamble to a topic from hours, or even years, earlier. Firstly, it must be understood that Fred Hoyle was our alpha. Soft spoken, modest, always contemplative, we all silenced our jabbering when Fred decided to speak. He was our leader, whether on Munros or in astrophysics. For this reason we spoke to Fred

The author atop Liathach, photo taken by Fred Hoyle. We had climbed this summit alone.

and to Fred's interests most of the time. Willy and I made several excursions into the synthesis of radioactive ^{56}Ni and its subsequent effects in supernovae, because our paper had just appeared in print at the year's beginning and because all of us shared a special fondness for the lengthy struggle toward a correct explanation of the high abundance of ^{56}Fe, the topic of Fred's 1946 initiating paper on nucleosynthesis in stars. But excursions into our favorite topics were usually short and terminated whenever Fred spoke what was on his mind. Vahe Petrosian had written an important paper with Salpeter about neutrino emission processes in stars, and it got some topical excursions because neutrino emission influenced the ^{56}Ni story. Fred had pointed out in his magnificent 1954 paper that neutrino emission would dominate the penultimate epoch of the advanced evolution of the massive stars. But Vahe's interest had lately turned to cosmology and the information from the observed numbers of radio sources, where to some extent it was to remain throughout his distinguished career. And that topic happened also to be Fred's topic in 1969. Therefore the conversations during this week analyzed in various ways the convictions that Hoyle had reached in 1969: that radio sources and quasars were fed by a massive (ten billion solar masses) central object confined to no more than 10% of a light year from the galactic centers; that the observed red shifts in their light indicated causes that existed in them in addition to the Hubble expansion of the universe; that these central masses cause giant outbursts with very short durations for such massive astronomical objects; that we should follow Eddington's example by believing the astronomy and forcing the physics to adapt; that the creation of matter led to some version of a universal steady state by regulating that very creation of matter. It was a Hoylean vision, much of which is believed true and much of which is taken by most astronomers as heresy. We discussed these topics a good deal, and it was always exciting to do so.

We always took dinner in the hotel dining room, not only during this trip but also during all of the many Scotland trips with Hoyle. The hotel provided the best dinner around, and after a day on the tops one does not want to travel in search of an alternative. Donald and Devon ate at the earlier sitting, and Mary Lou would sit with them to be sure they did it all properly. Willy always ordered the wine for dinner. This was because he cared much more than anyone else. His great love was the white Burgundies, Chablis, Chardonnay, Montrachet, Puilly Fusse, so one of those usually got the call. Willy's other job was orchestrating things so that dinner was slowed down, lengthened, dwelled upon. This involved much discussion of the relative merits of Scottish fare, the merits of the wine list and the menu, the right timing for coffee and for desert, and a proper after-dinner drink. Fowler was a born master of ceremonies, and could usually stretch this out to almost two hours. This allowed time for the most wonderful experience of all, conversation among seekers of truth. Then, next day, we did it all again.

Donald's first Munro

On the eighth of June my son Donald achieved his first Munro. He was enthralled with our talk of the glamorous mountains and of our obvious enthusiasm for the strenuous hikes. He said to me," Dad, I want to go along with you on the last day."

"Do you think you could keep going all day long?" I asked. He assured me that he could. "How have your boots been?" I asked him, although I had done so several times already. Good fit in boots is the most important requisite for Munros.

"They are great!" he replied. I knew they were up to the climb, and that Donald had experienced no blisters from hiking about with his mother. So it was settled. Donald went with us on the last climb of this memorable visit. He was 9 1/2 years old, beside himself with pride over my decision, not to mention the agreement of Fred Hoyle that Fred also thought he could do it. Our target was to be Ben Sgritheall near the mouth of Loch Hourn, where it empties into the Sound of Sleat opposite to the area known as Sleat at the southern tip of the Isle of Skye. Donald's going posed no envy problem for 8-year old Devon. Devon was too young for such a stress, and he did not evince the enthusiasm for it that would suggest that he could do it. Indeed, Devon loved to play about the shores of Loch Duich with two girls of similar age.

Fred's Lotus was full, but before I could offer to drive along in my car Willy, bless him for enthusiasm, said, "Donny will ride up here on my lap!" And he did. We had to drive southeast, around the entire northeastern perimeter (11 miles) of Loch Duich, where, just a way over the Shiel Bridge, a small one-lane road branched off to the west, through Glen More. I watched Donald looking up at every passing peak, wondering how it would be. After another dozen miles the road reached the sound at Glenelg, from which another five mile drive south along the sound reached the dry upward northern flank of this beautiful but gentle 3196 foot climb. The day was fine beyond all reasonable expectation, and the midday sun reflected glittering fireworks from the waters of the sound. After two hours of steady upward work our car looked a toy, as in the ascent of an airplane. Donald shot ahead, so full of energy and enthusiasm that he was willing to squander it on speedy maneuvers. I shot a role of color negatives with my 120 box camera in order to prove the heavenly wonders. Sweat poured down us all on the steep upward pulls. Donald began to flag; but the appearance of the summit, just 3/4 hour away, revived him to his glory. I was, and am, very proud. I jotted some lines of on a piece of paper, lines to become my poetic tribute to my son in later months when I could polish it. But first came lunch on the top. Then came communal gaiety. Then came more talk about massive objects in astrophysics. Finally, reluctantly, we began the fast walk down the long incline to the south ending in the small village of Corran. Afternoon tea was in a small guest house, after which Fred hitchhiked the ten miles back to his auto at our starting point. Even after three decades this day remains one of the bright moments of life, one by which others must be measured. Share my small verse to my son, Donald, on his first Munro:

DONALD'S FIRST MUNRO

That look of pride at being asked
To his first climb with the men,
Garrulous exhultations claiming
It really was a common thing.
Riding there on Willy's lap,

Studying each passing peak and
Hearing every comment, he
Made judgements of his chance.

Up the first one thousand feet
Deer-like, soaring, swift,
Launching rocks and shouting far
To hear the faint return.
Somewhat spent he straggled
Toward the rear so near the top,
But my disguised encouragement
Was superfluous to him.

That triumphant smile deserved
The cloudless sky and countless summits
Lying round our feet.
Oh bright sun! Brilliant son,
Ben Sgritheall could not hold down
The nine years of your waiting.

The Grange, in Great Wilbraham

We reached Cambridge on the 12th of June, 1969, and took a pair of *B&B* rooms in a house on Chesterton Road, roughly opposite the pedestrian bridge leading over the River Cam onto Jesus Green. This was so altogether pleasant that we could have spent the summer there but for the high price--3 pounds and 6 shillings daily for mine and Mary Lou's bay-windowed room overlooking the river and the green, but a cheaper 1 pounds 9 shillings for Donald's and Devon's modest loft at the rear. Devon observed wryly, "We always get the low-rent district". Devon was always aware that things would not work out very well. Nonetheless, the 4 3/4 pounds stirling per day was more than we could long indulge. Our family of four had returned with 119 pounds remaining in our account at Barclay's Bank on Sidney Street. On 13 May I had instructed our Houston Bank to transfer $1000 more to that account. This tidy sum, almost a month's take-home pay at Rice University in 1969, was to augment the 150 pounds monthly that Frank Westwater's promissory letter had committed to our monthly maintenance. So 4 3/4 pounds daily was quite a pleasant extravagance. Mary Lou loved the extravagance. To her, life presented a fascinating string of extravagant opportunities. I was more inclined to worry over dwindling funds.

From that Chesterton Road B&B we began our search for a suitable place to live. In June finding a nice three-month residence in Cambridge is not easy. The students are studying until about the end of June, and they occupy almost all of the transient space. Cambridge in the 1960s was very much a "town-and-gown" city. Although always favored by discerning tourists, especially intellectual ones, it was not then the tourist Mecca that one finds today. Ordinary people worked there and lived there, and students almost exclusively filled the rentals. To find anything we were advised to check with the university lodgings office. At that time their office was in a grey brick house on Maid's Causeway, the beginning of the Newmarket Road as it emerges from the traffic circle at the lower end of Jesus Lane. This was but a nice walk across Jesus Green to Victoria Avenue and, across it, another short walk along Midsummer Common. The two ladies in the lodging office were very helpful, but their short list of possibilities looked discouraging. After examining two student apartments that had been vacated early in some student realignments and finding them unappealing, we decided to perhaps waste a little time driving out to see something called *The Grange* in the village of Great Wilbraham.

We had again booked a car for the summer from Marshall's on Jesus Lane. They sold and serviced new Austins, but also rented out late models for longer hire. We again took an almost new Austin Mini from them and kept it for the entire summer. We surely cut a humorous sight, excited sardines in a very English can, Mary Lou and I seated in front and Donald and Devon peering out the two large rear windows at all passing things. Rental of this car was ridiculously inexpensive by today's travel expectations. I cannot remember the exact sum, but it was near forty pounds monthly, an amount that does not get one week today. With this Mini we examined every lodging opportunity that seemed at all appropriate. *The Grange* seemed that it should at least be examined.

We drove out Mill Road through Cherry Hinton toward the southeast in high, adventurous spirits, through the small town of Fulbourn, with Fulbourn Fen farmlands spreading flatly northward from the road, and a few miles more through pleasant farming land to the town of Great Wilbraham. Great Wilbraham was larger than its Little Wilbraham counterpart, just 1/4 mile beyond it. These were small historic farming towns, arranged traditionally with dwellings in town and farm lands surrounding. They appealed to this Iowa boy, but Mary Lou reminded me that the quality of the house was what counted. We proceeded through the tiny town centre past its last house, *The Grange*, where a creek named Little Wilbraham Waters divided Great Wilbraham land from that of Little Wilbraham. Just 100 yards before that creek stood *The Grange*. One look told us that it was for us.

The Grange was a painted-brick three-story white farmhouse, c. 1910. It apparently functioned historically as a northern gate to the large house and estate that lay sufficiently far behind its gardens to be out of sight. Only grazing cows adorned that field, separated from the gardens behind the house by a split-rail fence and a wooden gate having a wire loop for a lock. The house, grounds and context leaped into my heart. Large fields of barley stood along the northern side of the road. Fortunately *The Grange* also was fitted with stove and refrigerator, making it a feasible option. Suddenly my wish that we should live within walking distance of Hoyle's *Institute of Theoretical Astronomy* seemed not so important after all. The potential for a romantic family life was asserting precedence over astrophysics within me. Instead of the short walk of my previous two summers, I would simply have to drive. We took it immediately.

It was clear to Donald and to Devon, then nine and eight years old respectively, that this was the place to live. They ran about the garden, comparing it with the more formal garden architecture that we had known on Storey's Way. They ran up the stairs to the second floor, looked out the windows shouting "Look. Cows!", then ran up the narrower stairs to the third floor. Then we heard more shouts of "I'll take this one", and so it was settled. Donald took the small bedroom on the front of the second floor, where our bedroom also would be and, curiously, where his bedroom had been for two summers in *White Cottage*, and Devon took a third floor bedroom having a spectacular view, but muttering "Why do I get put up in the keep?" His allusion to castles showed that, despite his sardonic pessimism, Devon saw the romantic side of the architecture, especially after Donald later came to envy the superior easterly view from Devon;s window, from which *Wilbraham Temple*, the manor house of the estate could be seen. "It's pink!" Devon shouted, "Behind the cows."

Writing now I think of Marcel Proust's fictitious town of Combray. As with the aunt's house, there were two ways from *The Grange*. The first lay along the road in front with the barley field on its northern side, to the left along Fulbourn Road to the center of Great Wilbraham, such as it was, and to the right along that same road across Little Wilbraham Waters and into the smaller village of Little Wilbraham; and the second way emerged from a gate at the back of the garden to become Temple Way, a footpath through the pasture of cows to the pink manor house in the distance. In an emotional coincidence, I too would pass the bucolic summer in this evocative setting in search of lost time, in remembrances of things past, reflecting on haunting memories of it long after we departed. Of all residences in my life *The Grange* with its context was the nearest to the Iowa farmhouses, and that great attractor tugged within me, calling me back. Thoughts of that lost time would be initiated suddenly by views reminding of a remembered scene—the farmer cutting the barley, or a cow trying to reach through the rail fence for a tuft of tall grass behind one of its posts. At the same time a precious thing from my present life was slipping away.

The first thing was get the boys enrolled in Great Wilbraham School. There was no problem in this, and no cost. Children are required to be in school, even though they would attend for only four weeks until early July. We again purchased uniforms endorsed by the school, to which the boys offered no objection and even evinced some English pride of their own. They uniforms consisted of blue shorts and a white shirt and white stockings. The daily walk became that along the Fulbourn Road to their school instead of the walk to Fred Hoyle's Institute. That five minutes afforded the advantage of crossing the village common and cricket grounds adjacent to the Great Wilbraham parish church. On some Sunday afternoons we watched the cricket match on those grounds, seated on the grass just beyond the boundary. We attended several services in the parish church.

Separated as it was by wooden fence and gate from the cow pasture behind, the garden invited savoring life. Summer afternoons and evenings seemed wonderfully cool to Americans. The abundant roses were a splendor. Dozens of standing roses outlined an inner perimeter of the short fine grass favored in English gardens. A couple of large arbors were covered with red climbing roses. The boys loved to climb upon a utility shed that was covered by one such arbor and sit surrounded by roses. And the roses all bloomed, in various colors, but especially yellow, throughout the summer. We sat in canvas reclining lawn chairs and watched the birds. Tiny robins hopped about near our feet, as they had in the more formal garden of *White Cottage*. House martins built their wattle-like homes under the overhanging eaves and plummeted out their tiny doorways purchasing air speed by the dive. One nest was quite low, having been built above the rear double doorway, and from it the martins would not swoop upward until no more than a foot from the ground following their initial free fall. On return they managed to target the small entry holes with an accuracy that the military would have envied. We never tired of watching them. There were no squirrels for reasons unknown to me. The insects seemed few and benign in comparison with the dreaded pests of Texas summers. No wonder England is a land of gardeners.

Of the inner lawn I managed to create a cricket pitch, purchasing sets of stumps and a tennis-like ball and bat (designed for children) at a sporting goods store on Sidney Street. Donald and Devon would alternate as batters, at opposing wickets, and I would bowl to each, allowing each to total his runs scored. Mary Lou would only occasionally participate in this. Playing at cricket allowed me to share my expanding comprehension of this English sport with my boys. I learned the real stuff from Fred Hoyle, who routinely invited Willy Fowler and Geoff Burbidge and me to his home on Clarkson Close to watch whatsoever Test Match was in progress. There Willy and I learned the true intricacies of cricket, which I shored up by reading the daily analysis by John Woodcock, cricket correspondent for *The Times*. Slowly the sport began to make sense, and when it did I strove to pass understanding on to my nine and eight-year old sons. I thought this important rather than trivial because it was so English. Our best matches occurred on Sunday afternoons, however, when Willy and Ardie would come over to visit. Willy was very eager to participate, even with a tennis ball and a toy bat! Fowler was a very good sport, sensing immediately what the boys were capable of and playing to their strengths. And just as at Lords', we took a tea break. At the end of the designated over, Willy was allowed to announce, "That's it. Tea", echoing his knee-slapping hilarity over the same call by the television announcers at the test matches. We were all playing at being Englishmen, a genuine retirement option being considered by Willy and a growing attraction to me. If only the joy between Mary Lou and myself could blossom as well.

In this summer of 1969 I began to run. I ran just for the joy of running. I count my discovery of running as one of the really important ones of my life, steering me on a course of lifetime fitness. Then 34 years old, I was already becoming soft. The unrelenting fascination of scientific research had gradually separated me from my physical side. Science is a sedentary profession, one marked by long periods of sitting at a desk and within which aerobic activity often amounts to pacing the floor deep in thought. But the roads around Great Wilbraham were so inviting that it occurred to me that I could enjoy just going out and running along them! *Occurred to me* is an intentional word choice rather than a cliché of speech, because jogging was then so rare. The 1960s preceded the jogging mania that would soon appear in America. Today every runner, surveying his new home, will immediately think, "Where will I do my jogging?" But I knew no other runners in Cambridge; nor did one then see them jogging about the town in 1969; nor was I already a jogger or exposed to its merits. To my recollection, I never encountered another runner while running during the summer of 1969 in Cambridgeshire. As a result, I have felt all of my life that I made a big discovery. In order to spend the day at the *Institute of Theoretical Astronomy*, I arose and jogged before breakfast. At the sporting goods store on Sidney Street where I had obtained the cricket equipment, I purchased what were actually soft track shoes, not shoes designed for road running. I knew no better. Without delay I donned them and my shorts and went out the door. My favorite route turned north out of town, through the hamlet of Little Wilbraham, then southeast along a little used road to the village of Six-Mile Bottom, whose parish church tower inspired me along my way. I loved that name, Six-Mile Bottom. The run to Six-Mile Bottom required, door to door, about twenty minutes at good speed. I was especially fast over the mile on the return from Six-Mile Bottom to Little Wilbraham. The ritual became that on getting home I would collapse in the kitchen, uttering a plaintive cry, "Tea! Tea!" in a high falsetto because I had become known to Donald and Devon as "The Tea Bird". I would cool down drinking water while I prepared a large piping hot pot of English breakfast tea, usually Twinings. I persuaded Mary Lou that my run required English bacon or pork sausages, so incomparably better than ours in the USA, and grilled tomato, which I learned of in countless *B&B* overnight stays. I was taking on everything English! The un-English spectacle was my running along the roads in the mornings, a strange act indeed to the locals, who eyed me suspiciously.

I may speak lightly of my transformation to Englishman, but I do not make light of the transformation of my physical conditioning. I have essentially run three miles on every other day during the subsequent 40 years since that summer. My weight has not wavered from 170-175 pounds during that entire time. However, even while I run today I never forget my discovering the joy of it along the road from *The Grange* to Little Wilbraham to Six-Mile Bottom and back again.

After breakfast and, until school ended in July, after walking Donald and Devon to the local school, I would either drive the Austin mini to IOTA or Mary Lou would drop me there. The latter was more common, because she was off on adventures of her own. She rubbed at least a dozen testimonial brasses in Cambridge and environs. These mostly lay in the flagstoned aisles of churches, gravemarkers for some cleric or knight of significance to parish history. These brasses were much more open to public than today, when access to them has had to be more restricted.

Mary Lou was not one to stay happily at home. Her discovered hobby was antique shopping, often with Ardie Fowler. Hardly a week passed that I did not make a return drive with her to view her latest find and make a lightning decision about its purchase. These were seldom in antique shops, I might add. Most were at estate sales, either at the estate home or at a local auction house in a town near the estate. Two such sales occurred in Cambridgeshire vicarages that were to pass into private hands after they were no longer to house a priest. The 1960s were a memorable time for antique hunting. Many more pieces of quality appeared upon local markets as estates were sold off than one finds today. And of course, they were incomparably cheaper. We acquired enviable pieces for 25-50 pounds sterling, and spent only about 500 pounds in acquiring a goodly number. And we adored the inside look at English life that attended such auctions or sales. I shared · enthusiasm for such opportunities, seldom voting against a find. During that summer we slowly filled *The Grange* with antiques. The Fowlers also acquired several pieces for their Oxford Street home. When finally we left at summer's end, we hired a shipping agent, who constructed at The Grange two sturdy wooden shipping crates roughly 6 ft x 6 ft x 4 ft, 144 cubic feet of antiques. Incredibly today, this cost too was modest. We could actually afford to do this on the salary of an Associate Professor. The crates made their way to Houston undamaged, although one Grandfather clock and one bracket clock were damaged, as was one of the two Chesterfields that we brought home. Regrettably, I would live with them for only another six months, the last of my 14-year marriage. My marriage did not survive the trip home as well as the antiques had.

A discovery of a very different kind

It was during our July pub lunches at *The Green Man* in Grantchester[2] when my eye fell on a lovely young woman, and, I noted, hers on me. For just that second our pupils locked onto those of the other, giving that moment considerably more psychic time than clock time. She was laughing and having a good time with two other young women. Our eyes met again, and later again, and we shared smiles. With 3600 seconds in an hour, our eyes had been drawn to the other simultaneously. Then one of her companions turned to look my way and said something that caused them all to laugh. Willy noticed all this, more than others engaged in heated astrophysics. When Willy and Gary Steigman got up to suggest leaving, I excused myself, saying that I needed to stay and go to town for something and could Willy please ride back with Gary. Willy eyed me suspiciously with his standard parting benediction, "Don't do anything I wouldn't do." These words were always delivered with jolly sarcasm. Such was Willy's style.

"What is that you're eating to have such a good time?" I asked after sauntering to their table. "I want to order that next time."

"It's a ploughman's lunch", she answered.

"Stay and have an ale with me," I invited, because they were rising to leave. She did and we did. Now our eyes could linger more openly on the other. "Your dark eyes are the most beautiful I have ever seen", I said with feeling that expressed more than its triteness could convey. We never finished that second pint, feeling happy and hot inside.

She was a graduate student from Spain and resided with one of the other two in a flat on Chesterton Road, near the River Cam. "I always wanted to see one of those flats. Show it to me," I said. She did. I never really saw it on that occasion as our hot mouths met in fierce kisses just inside her closed entrance door. Quickly we fumbled with each other's clothing, my right hand lowering an undergarment. She began to help as I removed things from her and from me, kissing throughout. The sex was the most gratifying of my life, a pleasing surprise at key moments when I often felt self doubt. This derived from its raw intensity rather than from any meaning, from a hot ferocity leaving no doubt of mutual desire. I had not known that in daily life. It liberated me in some new way from my anxieties, some known and some unknown.

Our passion produced erotic feelings of loving this dark-eyed girl, no doubt from loving a new experience of myself. My self esteem skyrocketed. We shared phone numbers to facilitate the setting of times for best contact, and we met repeatedly for the remainder of this summer. We met at differing times of day. These indulgences were a therapy for me, becoming even a kind of love with which I had not been familiar. I did suffer guilt each day, for I was a Clayton and Claytons are an ascetic lot; but I was also liberated and empowered every day. I arose with more happiness each morning, loved my morning jog and pot of tea more than ever, and was more energetic during every hour of each day. Willy, by the way, teased me knowingly after my return at 4:00 on that first occasion. Willy was a lifelong womanizer, with radar for what had occurred.

For the remainder of my life I would puzzle that Mary Lou and I had not been able to love in the same way. Young and naïve, I could not formulate my questions clearly.

Research at IOTA

My days in IOTA were consumed by nuclear astrophysics. Its questions were my muse. Intense conversations with Willy and Dave Arnett were painting new directions for nucleosynthesis. But we saw little of Fred Hoyle, other than when watching cricket on television at his home on Clarkson Close, his respite from a busy state of decision making. He was busied with high matters of state, scientifically speaking. Two years previously, after my first summer at IOTA, he had been appointed chairman of the Science Research Council's Division of Astronomy and Space. He was also member of the Anglo-Australian Telescope board, whose goal was to build a world leading optical telescope in Australia. I sensed without ever discussing it that Fred was done with nuclear astrophysics and nucleosynthesis in stars, the subject to which he had given birth just after WWII. The combination of his overriding interest in the cosmological problems and the challenges of science politics allowed him little time to keep up with new challenges of nucleosynthesis. It was during one hour watching the Test Match from Lords Cricket Ground that Fred said pointedly to me in response to my question, "Don, I have always found that it is necessary every five-to-ten years for me to shift my concentration to a new astrophysical problem." It was both his way to acknowledge that his nuclear astrophysics period was at an end and his way to respond to my question about choosing my own scientific future. Should I stay with nucleosynthesis or branch toward new related science questions? I would never forget this remark. It lived vividly within me, evolving into a personal credo for myself. His remark, so carefully composed, endowed me with power to move into new directions during the decade to follow. Perhaps my sexual affair added its own kind of power. I doubt that I would have achieved what lay ahead for me but for Fred's remark, because I was if anything reluctant to give up a great investment. The easy road for me was the road retracing where I had been and what I had achieved.

Willy Fowler and I again shared an office and also shared all day long the problems that lay before us. This included daily lunch at one of the nearby pubs set for lunches, in Madingley, in Grantchester, in Trumpington, or in Coton. We always had a pint of bitter, and had to restrain ourselves from more, not always succeeding. The lunch itself was typically a "Ploughman's" special of bread, cheese, and chutney or one of the many variants of meat wrapped in pastry of which the English seemed so fond.

Joseph Silk approached me in July 1969 with a fascinating idea. He had read our prediction in the January 1969 *Astrophysical Journal* of gamma rays emitted by the radioactivity produced by the supernova explosions and of the chance of confirming nucleosynthesis with gamma ray detectors. Joe suggested that if it is correct that supernovae eject radioactive ^{56}Ni and that the gamma rays lines given off when it decays escape into space, then the universe should be full of those gamma rays. Our January publication had focused on detecting them from the direction of an individual recent supernova, of seeking the telltale gamma-ray-lines from the direction of that object. Joe offered the thought that he could sum over all of the supernovae expected in the universe if I could estimate the number of gamma rays emitted by each supernova. Gamma rays from ancient supernovae would have been redshifted toward lower energy, but still traveling freely within the universe. We were fired up by this good idea, finding excitement in the thought that the history of the universe might be measurable from its background of photons. This idea had been made familiar with the 2.7-degree thermal residue of the early hot universe; but the gamma rays are incredibly more energetic, and they are spread out in their creation over time rather than from a single moment of cosmic time. This made them quite different in their potential significance. Within three days our excitement mounted when I found a new way of looking at the idea. Namely, if all of the iron in the universe was associated with the emission of a gamma ray into the universe, then their average densities must be equal. There exists approximately one gamma ray for each iron nucleus. Using this concept and the estimates of the density of iron in the universe we found that the background could be as large as the counting rate of gammas near 1 MeV that had already been observed by detectors aboard space missions.

What poetry I sensed in this idea! When each supernova ejects radioactive ^{56}Ni, those nuclei and their daughters decay with a halflife just in excess of two months, emitting the gamma rays as a byproduct of the decay. From supernovae more than a decade old they will have all decayed already. Viewing supernovae in nearby galaxies, where they are close enough for their gamma rays to be identified as coming from that single object, one must therefore look at those that have exploded during the past decade. Older ones are already extinguished. But in the vastness of the universe the situation changes. Distant galaxies, say 300 million light years away, just to pick an example, emitted radioactive gamma rays that are just reaching Earth today from supernovae that exploded 300 million years ago! The intervening 300 million years have been spent in gamma ray travel from that galaxy to ours. When one examines the numbers one finds that the numbers of galaxies that are 300 million light years distant is very great. This causes the number of supernovae that exploded within them 300 million years ago to be very great as well. Their gamma rays are reaching us in detectable numbers today even though we can no longer discern from which supernova or galaxy they have come! And this robustness remains true for increasing distances, so that their supernovae bestow gamma rays upon us today from ones that exploded ever earlier in the universe's history. This opportunity of reading the history of the universe in the gamma rays from ancient and distant radioactive decay called to my mind the words of the poet John Greenleaf Whittier, which I used as epigram for one publication:

> *And ever upon old decay*
> *The greenest mosses cling.*

For me it was extraordinarily enjoyable meeting Joe Silk during this 1969 summer, and of the chance to get to know him and to work out this idea with him. Joe is smart and witty in his understated English manner. He has since gone forward to a dazzlingly prolific career in astrophysics. We submitted our paper on the gamma ray background of the universe to *Astrophysical Journal Letters* on August 18 and it was published in the October 1969 issue[4]. Three decades later its topic remains a goal of gamma ray astronomers, although it is now not clear that the radioactivity will be the dominant source of gamma rays near 1 MeV in energy. Because the central idea fascinated me, in future years I would twice return to it in order to embellish it with technical improvements, one of its treatment of cosmological models and one of its supernova models.

My mother saved postcards and letters that I wrote to her and my father from England in the 1960s. One picture postcard written on July 2, 1969, carries a black-and-white photograph of thatched-roof cottages along a narrow lane leading up to the great estate of Great Wilbraham, the Wilbraham Temple. The black and white photo has the caption, "5 Temple End, Gt. Wilbraham", printed in white (negative) along the bottom of the photo by a mask used during the printing of the photocard--very old-fashioned and lovely in its lack of pretension. "Temple End" refers to *the end of the short lane* leading to the Temple, as the great estate house behind The Grange was known. The cottages along Temple End, of which number 5 is a superior example, were built in an earlier time and lived in by craftsmen who supplied the little village of people who worked the lands of the estate. The lane is very quiet, with only a rare automobile.

Mary Lou and I had been invited to The Temple. The invitation was for tea, at 4:00 p.m. on a Sunday. The lord of the manor wondered what we were doing in Cambridge, where we had come from, when we would return, and if going to the Moon was a sound idea. They thought so but wanted to hear it from me. We also discussed briefly the investiture of Prince Charles at Caernarvon Castle. Indeed, on the other side of this postcard I wrote, "We watched the investiture of Prince Charles today, live and in color at Hoyle's house. We loved the ceremony and have much admiration for Charles himself". My admiration, which continues to this day despite the social perils of his life, was seconded by others-- by the Hoyles, by the family of Wilbraham Temple, by Robert Brown, Captain in the Queen's Navy, who lived in one of the very old thatched-roof cottages adjacent to the pub and who captained *The Cutty Sark* on her annual ceremonial voyages to America, and by many other Englishmen. Nowhere in 1969 did I hear, "The crown is a crock". I find it hard to believe that this harsher attitude about the crown, which one can hear today, represents Englishmen at their cores. The handsomely large airmail postage stamp for this postcard cost 9d (about 22 cents), showing a stone Celtic cross with the words "Tywysog Cymru 1969 Prince of Wales".

Dave Arnett was also a visitor to IOTA during summer of 1969, and we talked nuclear astrophysics constantly. Arnett was one of a group of distinguished students trained by A. G. W. Cameron at Yale Univesrity in the mid 1960s. Cameron started there a school of nucleosynthesis similar to mine at Rice, with the goal of evaluating Hoyle's equation for the nucleosynthesis of the elements from carbon to nickel. I was very impressed by the computer program Dave wrote for the IBM 360s of a calculation in which carbon in a supernova shell would be ignited by the heat of a passing shock wave launched by the rebound of the collapsed supernova core. Dave's first calculations were tuned to quickly burn all of the carbon, which was achieved within seconds after the heating of the shock wave and the cooling resulting from its rapid expansion. Moreover, its ashes ejected at high speed gave good resemblance to the abundances of the isotopes of neon, magnesium, aluminum, and silicon. Dave called it "explosive carbon burning", and his paper appeared later that year[5]. I was reminded of my discussion with Al Cameron in Kellogg in 1963, in which he advocated this view of the origin of the elements, and Arnett's calculation showed me how well Cameron's vision had been implemented at Yale University.

As for me, I decided that Dave Arnett was to be the new leader of explosive nucleosynthesis in stars. I wanted him to move to Rice University, where I was confident that we could lead the world in its study. I was able to persuade Dave to move to Rice University, and also to persuade Alex Dessler, our Chair, to offer him an Assistant Professorship[5] to make it reasonable. The 1960s were a golden time for obtaining good faculty positions, which today are very hard to win. It was also to be a golden time for Rice University. This was Dave Arnett's first faculty position, and we became close friends and colleagues, starting with this 1969 summer. He remained at Rice for only two years and we wrote together a number of significant exploratory papers elaborating that idea of explosive nucleosynthesis.

The Eagle Lands

On July 16, 9:32 AM EST, *Apollo 11* lifted off from Cape Kennedy. At 12:16 it commenced translunar injection, the term used to describe the burn[1] moving its velocity to an orbit that would carry it toward the Moon. I remembered that Ed White was to have commanded this mission. It carried three men unknown outside NASA astronautic circles, *Apollo 11* Commander Neil Armstrong, Command Module Pilot Michael Collins, Lt. Col. USAF, and Lunar Module Pilot Edwin (Buzz) Aldrin, Colonel USAF. The astronauts were test pilots for the most part. That was deemed to be "the right stuff" at this time. Armstrong and Collins had gained their first near-Earth experience piloting the X15 rocket plane. Aldrin held the record for most time spent outside the capsule in extravehicular activity at that time. I watched this on BBC news on a TV that we rented from an appliance store, and as I did the excitement mounted within me. Our astronauts were coasting toward the Moon at a speed near 10,000 miles per hour.

At about the same time The Institute of Theoretical Astronomy was hosting their annual summer conference on astronomy. The topic was infrared astronomy, and many famous scientists were there. Their offhand remarks reaffirmed the love/hate relationship between astronomers and NASA. Many astronomers had criticized NASA for spending so much money putting men into space when they could have done so much more science for less money. This attitude was wrong, although its bare premise was correct; but the purpose of NASA's mission was not to do science. It was to extend the human race into space, and to put a foot down on another planet. And, oh yes, NASA also wanted to plant a flag; and this was one of many NASA theatricals that scientists made fun of. There were always cynical jokes mixed in with the scientific interest in this grand adventure, which grabbed scientists by the throat just as surely as it did the man on the street. I was little different than the others. But watching from *The Grange*, set in the barley fields of Great Wilbraham, the TV news of the departure of that trio for the Moon, I sensed that it was something more. Indeed it was. This was the defining moment of the 20th century.

On July 20 the *Eagle* landed. I do not remember the exact time, but it England it was in the very early morning, and I had set an alarm for what seemed to be the middle of the night in order to turn on the TV. We were part of the largest assembly of viewers in history. We watched Aldrin focus the camera on the ladder by which Armstrong would descend. Then, a blurred shadow of a man, Neil Armstrong placed one foot after another down the ladder. "That's a small step for a man, a giant leap for mankind" he said as he became the first human to plant his foot on another planet. After he was down, tears came to my eyes as Armstrong walked slowly about, showing that the surface would hold and was navigable. Aldrin came out and literally *bounced* over the surface, his exhultation apparent. That early morning space walk lasted 2 1/2 hours. They placed a plaque with the words "Here men from the planet Earth first set foot upon the Moon, July 1969, AD. We came in peace for all mankind". They deployed experiments, including some designed in my department at Rice University. And they examined and picked up rocks, precious keys that would unlock the history of the Moon's origin so near the earth. As the sun rose in England, they reentered the Lunar Module. The visit was accomplished. On the afternoon of the next day, July 21, the ascent stage was reignited. It blasted off the descent stage, which had no more use to them and would remain on the Moon, and rose, a tiny gnat of a spacecraft upward going in the weak gravity of the Moon, to dock with the *Columbia*, still orbiting the Moon with Collins at its controls. Columbia reignited to move from lunar orbit into an earth orbit, and on July 24 it splashed down in the Pacific Ocean, about 80 miles from Hawaii. The entire world exulted.

I find some irony in having watched perhaps the two most exciting events of man-in-space from vantage points in the United Kingdom. Although I lived in Houston and its space glamour, although I was now Professor in Rice University's Department of Space Science, a department having my friend and department colleague, F. Curtis Michel, appointed Scientist Astronaut, although I had lectured to astronauts at Johnson Space Center, despite all of this I had watched *Apollo 10*'s swoop past the lunar surface from a hotel lounge in Ireland and had seen *Apollo 11*'s "one small step for a man" from an old farmhouse in Cambridgeshire.

At the same time I was seeing United States life and attitudes in an altogether new light, one in sharp contrast with the England of the 1960s. One cannot know how sweet life can be for a person like me except by living in a privileged manner in England during the 1960s. We, on the other hand, were locked in a hopeless and mistake plagued war in Vietnam, divided over the blacks' march toward their civil rights, and suddenly preoccupied with personal guns and our "protection". And yet Americans had nonetheless extended the human race to another planetary body, even if only for a few steps and for moon rocks of inestimable scientific value. My view of my country was challenged by viewing us from outside, and the subsequent three decades have not made me more comfortable although they have allowed me to see our national virtues with more pride. It was Zeitgeist too that my concept of marriage was suffering the debilitation that would sweep the USA during this decade. The essentially puritanical intellectual disguised with bravado, namely myself, and the frustrated, fun-loving, compulsive, perhaps manic-depressive, Mary Lou, would not be able to hold together for even another half year. Our marriage required restrictions to function honorably and safely. Seductive freedoms and the sirens of self satisfaction had arrived too suddenly for us to deal with.

The year 1969 is to me all of those thoughts tangled in webs of pride and failure. At this very time I was offered informally the position Science Director of the Johnson Space Center. Despite the glamorous success of *Apollo 11*, and notwithstanding the fact that I was thrilled by the accomplishment, I declined this opportunity. Not only were my own views of the world and of life changing too rapidly, but I knew that the glamour of that position did not really carry much weight. Science was not what *Apollo* was about. Science was performed only as a spinoff from manned spaceflight and from growth of space technologies. The latter were the main goals of *Apollo*. That was OK, but the real power lay with the manned space program, not with science to be performed by that program. I was also aware that the previous Director had resigned over disagreements at his attempt to inject more science into Apollo. I knew that my Scientist Astronaut colleague, F. Curtis Michel, had resigned as Scientist Astronaut when it appeared to him that the test pilots had too firm a lock on the planned trips. So when I wrote to my parents again from *The Grange* on 17 August, I said:

> *This will probably be my last letter. We take Austrian Air Lines Friday 27 Aug. for Vienna. Then after three days the train to Budapest for XIII Intl Cosmic Ray Conference. I will deliver a talk there. Boys stay in* The Grange *with Miss Waterson, a middle age schoolteacher on a modest summer job. Confident they will be fine in Great Wilbraham. On Sept 2 we take train back to Vienna. Then Lufthansa to London via Munich. We have three days to be sure the antique furniture is picked up. Four pieces are having new fabric before we return on Sept 6 London-Chicago-Houston, a bit late for the start of classes. I have had super summer writing two papers of which I am very proud. I wrote to decline the job at MSC (science director). Last director quit, as I read in papers that did Curt Michel did also. I just feel I have found the life I like. When I change from university professor it will be to more directly help society. Too much NASA revolt over science at MSC.*

End of a marriage

Not surprisingly in retrospect, Mary Lou and I continued to drift apart during this summer. We lived together, and our family undertook countless activities together, but an invisible gulf between us widened. Our marriage fell into coded patterns of safe role playing. We certainly did some of both. We shared enthusiasm for the village life in Great Wilbraham, and that would normally have been glue enough, but our psyches were fatigued. Our resilience for overcoming our unspoken failings flagged. Leisurely strolls to the village pub raised my sexual aggressiveness, not through ale alone but also through the human sweetness of the lovely village pub and, yes, through frequent passion with the Spanish senorita. Surprisingly to me at the time, the latter raised my fire for Mary Lou as well, but my self-doubting performance reasserted itself when confronted by Mary Lou's habitual lack of enthusiasm. That stalled my reborn vigor. Had she reacted differently than our accustomed roles, I think there may have been a chance for us at that moment. The senorita might have been just the right therapy for me; but we were unable to build on my sexual maturing. That saddens me as a possible chance lost. Mary Lou was an always interesting and entertaining woman. I learned a great deal from this, but not soon enough to stem an unspoken tide of dissatisfaction. As the saying has it, "Life lets us flunk the course. Then it teaches us the lesson."

I found the most memorable aspect of Budapest for *XIII International Cosmic Ray Conference* to be not the conference itself nor the lovely old iron-curtain city but an August 31 outing to Lake Balaton, which is the largest landlocked lake in Europe. I spent the day with a friend and cosmic-ray physicist, Rochus Vogt, from Caltech. We separated from others because we both wanted a hike rather than the souvenir shops and cafes. Robbie, as I knew him, suggested immediately a wooded trail above the lake. Surprised that he knew the terrain I asked him when he had been there. His story was horrifying to me. He had been there as a 13 year old in 1943 as a member of the elite "Hitler Jugend", dedicated youths in Hitler's service. He had been here with his 14 year old brother, also a Jugend member. As we walked he pointed occasionally to special points of interest to his personal trauma. As resistance became desperate, the Jugend were ordered to attack allied troops, even though they were not actually fighting outfits. Many resisted and fled despite knowing that flight was punishable by death, many were shot down by Germans for fleeing, many fought and died, including Robbie's brother. When he went down, Robbie, in grief and fear, realized that it was no good. Hiding now, he took the uniform from a fallen allied soldier, put in on, and walked to freedom. Because this relies on Robbie's memory from 1944 and my memory of what he told me in 1969, it may have a fact or too misstated; but there is no doubting the basic story and its numbing impact upon both of us. Nor is there doubt of Robbie's emotions as he saw the site of his nightmare. I could not imagine such a trauma, and I fumbled for some appropriate words of sympathy when Robbie showed me where his brother fell. Awkwardly, I embraced him. There were no more festivities for me on that day.

Despite a flood of work activities at Rice in the fall 1969 semester, worries about the future of my marriage were never far away. I was so ambivalent about it. My experiences in Cambridge had made me long more openly for better sexual love; but I had no idea how to achieve that with Mary Lou. That sounds strange, but it seemed the truth. I confess that I blamed Mary Lou for our lack of intimacy; but her difficulty also provided my alibi for my neurotic reactions to our relationship. I was fantasizing over new experiences with women; but I loved my sons very strongly and I cared about Mary Lou and our rich mutual involvement with those boys and with the goals and shared experiences of life. I was frozen in a state of passive inactivity, of avoiding the facing of reality. Only decades later did I believe I saw what the way would have been to have righted our ship. This was so much harder than physics and astrophysics. I preferred to stick to those intellectual efforts, ones for which I had a gift.

For Mary Lou it was also a time of crisis. She was smoking more than ever at the same time when I was trying to stop. She was drinking too often and too heavily, and in one embarrassing moment at the Dean of Science's home threw up on their carpet because of it, and seemed dazed over it, near tears. An outpouring of sympathy from me was sincere, reminding me of wanting to comfort young Devon when two years earlier he spilled his Horlicks on his new suit on our first evening in White Cottage. My heart ached similarly for them. But that moment of my compassion, of wanting to shelter her from pain, did not show me the way to genuinely help her. Returning home one afternoon to our Tangley home, just three blocks north of Rice Stadium, she met me at the door.

"Don, we need to talk," she said.

"Yes, I agree," I replied, and followed her to sit on one of our Cambridgeshire Chesterfields.

"I want a divorce."

I was not sure I had heard properly. My head immediately swam in denial. Surely that was not what she had said.

I who always suffered from a peculiar sense of loss, both amorphous and over specific things lost, now sank into my greatest loss. It might be argued that a sense of loss lodged within my very character, a residue of early unremembered or imagined losses that had left only that character trait. Perhaps that trait was actually more significant than divorce. Millions of men might agree with me that losing one's sons to divorce is the greatest trauma of our lives. I speak specifically of that psychological bond that a man has to his sons and that no doubt has its roots in the evolution of animal species. For better or worse I relived my own development through that my sons, so that I taught them to bicycle and play baseball because those skills had been important to me. In 1969 it was divorcing fathers who relinquished living daily with their sons, as I now experienced moving from my home on Tangley Street. Despite the trauma, I had not the will to prevent this divorce. I longed too strongly for the seeking of new sexual love to be able to find the strength to oppose this divorce. I no longer knew how to seek that with Mary Lou. So to Mary Lou and my sons I left the Tangley home, the antiques that filled it, and the equally beautiful red Volkswagen Beetle convertible that we had obtained in London, and I agreed to a liberal $400 monthly in child support. In 1969 that was about three times the amount of our monthly mortgage payment and twice our monthly grocery bill. It was generous at the time and proved adequate with Mary Lou's salary from the Alley Theater. I took a small furnished apartment, concerned as always with economy, but I would live there only four months until departing again for Europe in May 1970. During that time I accepted from Lee Rudee, Master of Wiess College at Rice, the invitation to reside in Wiess College as a Residential Faculty Associate on my return from England. That proved to be wonderful for me, because becoming acquainted with and advising and residing among all those young Rice men helped to fill a personal vacuum.

For me the huge year of *Apollo* comes to its end on January 6, 1970 with Fred Hoyle's keynote lecture at the old Rice Hotel in downtown Houston during the first *Lunar Science Conference* after *Apollo 11*. First of all, this conference was a glittering affair. All those planning to be key players in the lunar sample science to follow were there. And if it was too soon to have many science results to report, it was not too soon to get in line for the grants program from what would become NASA's Cosmochemistry Program. There is always a lot of public posturing for NASA administrators at these annual meetings. I would attend this annual event for the next 35 years, doing some posturing of my own, and count the experience among the highlights of my scientific life. From the outset there was a palpable sense of the history of mankind being written, all of us sharing the "one small step for a man" feeling. Most of the important delegates had become known to me when I participated at the *Physics and Chemistry of Space* Gordon Research Conferences held in New Hampshire's *Tilton School* in 1963 and 1964. These two had been historic meetings, forerunners to the sample-science conferences that were now commencing. I had been excited to learn that Fred Hoyle would be coming to my home turf for this event, invited to deliver a theme address in his inimitable fashion to the banquet.

In his banquet talk Fred did not get into the controversies about the nature of the lunar surface or the origin of our very large Moon so close to the earth. Instead, Fred turned toward the cultural and evolutionary effects of the Apollo landing. He reminded the participants of a prediction that he had made in 1948, and which he quoted:

> *Once a photograph of the earth, taken from outside, is available—once the sheer isolation of the earth becomes plain, a new idea as powerful as any in history will be let loose.*

Then returning to the moment, Hoyle continued:

> *Well, we now have such a photograph, and I've been wondering how this old prediction stands up. Has any new idea in fact been let loose? It certainly has.*
>
> *You will have noticed how quite suddenly everyone has become quite concerned to protect the natural environment. Where has this idea come from? You could say from biologists, conservationists and ecologists. But they have been saying the same things as they're saying now for many years. Previously they never got on base. Something new has happened to create a world-wide awareness of our planet as a unique and precious place. It seems to me more than a coincidence that this awareness should have happened at exactly the moment man took his first step into space.*

The famous photograph, *Earthrise behind the Moon*, is duly celebrated. It may be the most beautiful, the most frightening, and the most important photograph ever taken.

In my own way I felt isolated in space. It had been hard to tell Fred that we were not inviting him to our home because I had moved out. Admitting the failure of my marriage, admitting my failure as a husband, lowered my self esteem to its nadir.

CHAPTER 14. FALLING STARDUST

> *Some strong pressure is certainly at work. ..this is a critical time for him, a lesser climacteric--a time that will settle him in that particular course he will never leave again, but will persevere in for the rest of his life. It has often seemed to me that towards this period men strike out their permanent characters; or have those characters struck into them...then some chance concatenation, or some hidden predilection (or rather inherent bias) working through, and the man is in the road he cannot leave but must go on, making it deeper and deeper (a groove, or channel)...*

> Stephen Maturin--
> in *Master and Commander,* p. 181
> --by Patrick O'Brian

At forty years of age my personality was reshaped by events almost volcanic in their eruptions. Subsequent events flowed down the channel that was formed, and, like water through a volcanic caldera, carved that groove deeper. During this 1975 eruption I was similar to Captain Jack Aubrey, Maturin's fictional colleague, also no longer young but not yet old. Much naiveté and insecurity had been dealt with, and without recognizing the urge I was seeking new identity and new scientific mission--some goal to take me beyond being the formulator of nucleosynthesis systems. Stumbling upon this channel would give me a new character and a new career as a meteoriticist. My new scientific adventurousness was reinforced by the directions of subsequent work. It may be that it can not have given me a character for which I had no hidden predilection.

I can not judge how my hasty remarriage entered among the forces in my impending transformation, but I think it may have been a factor. Near the end of summer 1970, just one year after the breakup of my marriage, I had met a beautiful young woman on Jesus Green in Cambridge. From one vantage point Annette Hildebrand was a simple German girl from the Schwarzwald (the Black Forest in southwestern Würtemburg), but from a different vantage point she was a bundle of adventurous uncontrolled emotion. She was attending an English language school in Cambridge, one among the dozens of European girls that flock to the summer language schools that are a cottage industry in Cambridge. Unknown to me because I did not proceed slowly enough to find out, she was also looking for a new exotic life away from the Black Forest. I on the other hand was looking for sexual love, looking as only one who has felt himself long starved does, giving that pursuit top priority. It was my weakness, to dive right in.

After some later romantic visits to the Schwarzwald, I invited Annette in spring 1971 to join me in Cambridge during the summer and fall of 1971. This was to be my first sabbatical leave from Rice University. Annette was enthusiastic about living together and obtained a student admission into an English school for foreigners. I had retained for a second consecutive year the townhome at 8 St. Clements Gardens, a row of eight brick terrace homes on the eastern side of Thompson's Lane, a short two blocks between St. Johns College and Park Parade on Jesus Green. A photo can be seen at <1970 Clayton residence>. It was then owned by the Bursar of Emmanuel College, who lived on the ground floor and rented out its top two floors. Except for unanticipated stress that my hasty invitation to Annette injected into my Cambridge life, this was a charming residence for those seven months. This townhome is now out of private hands because the entire row of eight houses was purchased by Trinity Hall College as a hostel for their students. Cambridge University owns a great deal of Cambridge. The reader may sense the risks that I was taking.

Despite seven months of evidences that Annette harbors irrational thoughts and emotions, I was smitten by her loving sweetness toward me. We had become lovers at once during our meeting on the previous summer, and thereafter I found myself in that constant tension between clinging to her devotion to me, which was stronger sexually than I had experienced with Mary Lou, and silent warnings to extricate myself, not only for my benefit but also for hers. It was not a period of decisiveness in me. That story, its evidences of inappropriate behavior, its passion, its irrational reactions, do not bear retelling. Some resolution one way or another about the continuity of our love had to be made, and, as I returned to Rice University in December, I chose to follow my heart rather than my best judgment. Indeed, my best judgment was deactivated by Annette's presence. I longed for a passionate loving marriage, so I chose to believe that I would find it here, where the clinging of love was so strong. In an act of faith I believed that this young woman would bloom in her confidence in a setting of committed love, not only toward me but also toward the intellectual life that she had never had. After another winter visit to her home in the Black Forest, which was enchanting, we were married in summer 1972 before both sets of parents and her large cohort of brothers and sisters in a ceremony in the protestant church in St. Blasien, high in the Black Forest. In fall 1972 I brought her home with me to Rice University, where I had already taken up a residential appointment in an apartment in Wiess College, one of seven residential colleges on the Rice campus. The Master welcomed Annette into that position with me, so that is where we set up home.

Scientific Transformation

This huge change in my life set the stage for my scientific transfiguration, which would become more fundamental and long lasting than the reader might suppose. An inner eruption began with a radical scientific idea and festered on my indignation at the resistance of others to that idea. A new scientific persona was established by the scrimmage. The idea itself, that interstellar grains of solid material, dust grains older than the Earth, had survived the formation of our solar system 4.5 billion years ago, was not solely mine. One finds brief casual mentions of that idea; but I became its formulator and developer scientifically for the first 25 years of what would become a scientific revolution. I put scientific flesh on those bones. I presented calculations for a scientific hypothesis having specific testable consequences. Doing so differs greatly from casual mention of a possibility.

The first clues that inspired my new picture were discovered by nuclear chemists making measurements of the abundances of isotopes within meteoritic rocks. Let me explain. Meteorites are large rocks that fall to earth from the sky. But they are solid collections of smaller rocks of differing characteristics. For specific chemical elements, nuclear chemists measured the relative numbers of the distinct isotopes of that element and compared those relative abundances with those on earth. Tiny differences could be seen, but they did not dissuade the notion that all components of the meteorites had been chemically assembled from a common pool of solar gas during the formation of our planetary system. What I laid out was a theoretical map describing specific differing types of presolar solids, remnants of astrophysical presolar history, that one might look for and what their telltale isotopic signs would be. Presolar solids were not made in our solar system but instead already existed in the interstellar gas. This bold idea was radically original.

I constructed a new theoretical description of the contents of interstellar matter. I focused especially on the idea that small solid grains would condense from hot gaseous vapor within the inner shells of expanding supernova explosions, and I emphasized the isotopic signatures that these grains would carry in the interstellar medium. Incredibly, that seemingly fanciful tale later proved true, right down to numerous details. But my early predictions were resisted, sometimes irrationally and unprofessionally, because of their revolutionary nature. I was castigated openly by a few scientists at the peak of their own scientific influence and covertly by anonymous scientific referees trying to reject the papers that I wrote advancing the new idea.

It was the nature of that resistance that enraged me, stiffening my spine.

I was not so naive as to expect that others would simply accept my ideas. Science is seldom like that. But my blood boiled at unscientific prejudices that sought to prevent their fair hearing. Scientists are not "believers", no matter how attractive our personal wishes might seem to us to be. Popular press often wrongly suggests that one scientist believes one thing, but another believes something different. The word belief is not appropriate. Their difference is not one of belief but rather of judgment of the relative plausibility of differing proffered explanations of the facts. Scientists are pragmatists who seek the most internally consistent explanation of known facts. But prior to establishment by facts, we seldom accept a revolutionary vision. So I did not expect acceptance of my vision-- just a scientific hearing. My foray into the world of meteoritics was, by the way, given courage by Fred Hoyle's advice to me that he found it best to change his scientific focus every ten years or so. My time had come.

The concept of this revolution interested scientists everywhere, because it lay in the natural history of matter, in much the same sense that cosmology, geology, evolutionary biology and anthropology busy themselves with natural history of matter and with the scientific knowing of it. The historic origin of the atoms themselves had always been my focus. What I introduced in 1974 was different--the description of solid presolar chunks of matter--atoms bound together chemically into structures like tiny rocks, or perhaps I should say "gems". Today we hold them in our hands, in laboratories on Earth, gemlike solids that existed before there was an Earth. Today we study objects that predate our entire solar system--grains of sand prior to any terrestrial beach. But at that time presolar, solid grains were unknown. Finding them has been revolutionary because it had long been thought to be impossible—to physically hold an object older than the earth. Although astronomers had long studied, albeit at great distances and with telescopes, objects that are believed to predate the sun and Earth, those astronomical objects can not be handled, are not amenable to laboratory measurements of properties and structure. All solids on Earth and all planetary bodies and solids within them were believed to have been assembled in the early events of the formation of our solar system, or in some cases even by later processes occurring on the planets. But the time of formation of Earth, planets and sun, that time when our solar system first graced the heavens, was regarded a veil through which we could not see. Just as the writer of Genesis could declare, "In the beginning, God created the heavens and the earth", scientists could believe, on apparently good grounds, that every solid object that we can handle was also created at or after solar birth.

I was declaring that it is not so. The veil was pierced in the mid 1970s. And the piercing of it, with the associated expansion of natural philosophy and of astronomical and geophysical sciences that accompanied this revolution, so engaged me with struggle, struggle with facts and with strong conflicting personalities, that my new persona was inexorably forged. And it was at just this point that Stephen Maturin's words could also be applied to my transfiguration from steady worker and thinker to young Turk and eventually to sage, from nuclear empiricist to radical thinker, thence to scholar and finally, perhaps, to prophet. My character accreted new characteristics. If this sounds strange, indeed it is. But the strangeness lay not so much in me as in an essentially human process. I began to think outside the box. But instead of robbing me of my humanity, the accretion of reactionary characteristics was, I believe, accompanied by increased humanity within me.

A Question of Isotopes

Scientists stand on the shoulders of other scientists. I can pinpoint those that boosted me to my breakthrough. On 2 November, 1973 three chemists at the University of Chicago published, in the weekly journal *Science*, a remarkable finding. Small rocks having exotic oxygen isotopic abundances could be extracted from meteorites. Oxygen is composed of three stable isotopes of that element, those having atomic masses 16, 17 and 18. All three isotopes coexist in all oxygen compounds on earth. The oxygen in that meteoritic rock was exotic by having more of the mass-16 isotope within its three-isotope constitution than did every other known sample of solid material in the solar system. All three isotopes, ^{16}O, ^{17}O and ^{18}O, each behaving chemically as oxygen but differing in the numbers of neutrons within their nuclei, had previously always been found to occur in the same ratio to one another, at least within a few parts per thousand. But these meteoritic rocks measured by Robert N. Clayton, Lawrence Grossman, and Toshiko Mayeda in their Chicago laboratory, were demonstrated to be 5% richer in the mass-16 isotope, written ^{16}O. Their Abstract[1] stated:

The effect is the result of nuclear rather than chemical processes and probably results from the admixture (to the solar system) of a component of almost pure ^{16}O. This component may predate the solar system and may represent interstellar dust with a separate history of nucleosynthesis.

Here is the nuclear situation. Each oxygen atom has eight positive charge units in its nucleus. These are provided by the eight protons that inhabit each oxygen nucleus. Each oxygen atom has eight negative electrons in orbit about its nucleus, making the entire atom neutral. But three different stable isotopes of oxygen exist; namely, those nuclei containing eight neutrons yielding an atomic mass A=16 from combined number of protons and neutrons, those having nine neutrons yielding an atomic mass A=17, and those having ten neutrons yielding an atomic mass A=18. It is the relative numbers of these isotopes that concerned the Chicago finding. In all previously known samples of solar system material, were one to extract a portion containing 200 atoms of the mass-18 isotope, ^{18}O, it would contain 38 atoms of the mass-17 isotope, ^{17}O, and a whopping 99,760 atoms of the lightest isotope, the mass-16 isotope, ^{16}O. Whether air, ocean, rock or Moon, the relative numbers were always the same. The Chicago group had shown that if one takes 200 ^{18}O atoms from one of these special rocks that they found within the meteorites, it also would contain 38 atoms of ^{17}O, just as did the earth sample, but it would contain an extra 4,500 atoms of ^{16}O. The relative numbers of ^{17}O and ^{18}O isotopes remains the same, but the extra ^{16}O is a large 4.5% increase of its usual abundance.

This surprise was both puzzling and stimulating. Where had the extra ^{16}O atoms come from? Searching for some parallel, some precedent, the discoverers pointed out that "galactic cosmic rays are the only certain example of matter within the solar system with a separate history of nucleosynthesis." They had reported a second example. This was stunning. It was the first solid sample that in bulk differed from solar system isotopic ratios. But what did their cryptic words, "may represent interstellar dust", mean? Represent? If the rocks were not themselves presolar, in what sense did they represent interstellar dust? The authors did not answer that question.

I noticed the article when it first appeared. I subscribed to *Science*, so weekly it crossed my desk by mail. Entitled "A Component of Primitive Nuclear Composition in Carbonaceous Meteorites", it caught my eye and has puzzled me ever since. But at the time I was busy with many other things, especially an impending visit from Fred Hoyle and some personal issues related to my remarriage and to concern over my two sons from my first marriage. I did not concentrate on science very well through those big changes in my personal life. Perhaps my personal fomentation enhanced scientific thoughts lying outside the box. In any case the thought of rocks having 5% too much of the mass-16 isotope had registered indelibly. In reading their article I had noticed that they speculated that the suspected ^{16}O-rich component might be "the result of carbon burning in a supernova explosion", because supernovae were indeed known to be huge ^{16}O producing machines. They speculated that gaseous ejecta from that supernova might have been mixed here and there into the otherwise homogeneous solar system. But in that case they would have to expect large correlated isotopic effects in magnesium and silicon isotopes in the same samples. I immediately doubted the correctness of their idea. These associated effects, and more that they did not consider, would never be found. No. That idea did not provide the correct explanation. Alternatively, they also suggested, some ^{16}O-rich dust might have survived the origin of the solar system and been incorporated into these rocks, which were themselves called "calcium-aluminum-rich inclusions", or CAI, because their constituent minerals were so very rich in oxides of Ca and Al. Surviving presolar dust might have been widely sprinkled into the early solar system, a salting and peppering of the solar gases. These were bold ideas that flew in the face of accepted belief that the solar system began as a hot gas. Presolar dust was discounted because it could not survive the incorrectly imagined hot beginnings of our solar system.

I could not fathom how interstellar dust, even if some portion of it did condense *near a supernova*, could be made sufficiently ^{16}O-rich that a small smattering of that dust could account for a 5% excess in rocks to be formed later inside our solar system. The Chicago discoverers simply dropped off that comment without any recognition of the scope of that problem. Five percent is a lot of excess ^{16}O. On top of that, those authors had not realized that the supernovae were the galactic producers not only of ^{16}O but also of ^{18}O, but were deficient as contributors to the interstellar ^{17}O abundance. Thus the total ejected gas from a supernova would be characterized by a deficiency of ^{17}O rather than an excess of ^{16}O. I estimated that interstellar dust could not be made 5% richer in ^{16}O from that process while maintaining the same ratio of the numbers of ^{17}O atoms to those of ^{18}O atoms; and even if it were possible, the authors could not seriously suggest that these rocky inclusions called CAI had been assembled totally of interstellar dust—preposterous. My quick calculation that made that so implausible to me concerns the very nature of supernova explosions.

Supernovae are explosions of large stars in which gaseous masses comparable to tens of millions of Earth masses of hot atoms are ejected into interstellar space at incredible speeds, near

10,000 miles per second. An imaginary object moving at that speed could circumnavigate the earth in two seconds rather than the 5400 seconds required by earth satellites. That is fast! How shall gas moving that fast slow down? It will have to collide with tens of thousands times more mass of interstellar atoms to tamp that great rush, to slow it to speeds normal for interstellar matter, to absorb its great momentum and to allow for cooling of the shock-heated gases. By that time that much interstellar matter has been added to the ejecta the bulk material can hardly have had more than 0.1 % excess of ^{16}O, fifty times less than the final ^{16}O excess found in meteorite rocks. To make things worse, those rocks still remained to be forged somehow within the solar system, surely introducing further dilution of the excess ^{16}O. Even if the authors' suspicion that supernovae were involved proved correct, how can the final excess be a huge 5% in solar system rocks? These puzzling questions had not been asked in 1974 when I seized upon them. The Chicago group had been fishing for an explanation to a very knotty problem.

I was not so worried about the other problem that puzzled meteoriticists so, which was the requirement for presolar dust survive until the time of the formation of the planetary bodies. I thought that could easily happen. That question totally mystified solar system chemists, however, because their argument, then approaching institutional dogma, had been heretofore that the early solar system consisted initially of such hot gases that dust could not survive[2]. Indeed, one of the discovery authors of the ^{16}O excess, Lawrence Grossman, had won considerable early fame by his chemical argument that such a hot-gas model, slowly cooling, could account for the chemical composition of the very calcium-aluminum-rich inclusions (the CAI) that the meteorites now revealed, the same inclusions within which the ^{16}O excess had been found. These stones were almost a centimeter in size. Those CAI, Grossman argued, might be assembled from the first gas atoms that had become able to condense as the solar gases cooled. That means that the system is so hot that all dust would evaporate save for this CAI composition. This might roughly be analogous to the condensation of snowflakes from hot steam when it has cooled sufficiently. He showed that those hottest possible solids would be composed of oxides of calcium and of aluminum, which would become stable when the temperature fell below 2000 C. This physical picture of the initial hot gaseous solar system had already been advanced by others and had become an accepted dogma within this community, but it was Grossman who calculated[3] this seemingly convincing relationship to the newly discovered calcium-aluminum-rich inclusions (CAI). The problem was that in Grossman's picture the CAI could not be enriched in ^{16}O at all, because solar gas atoms have no excess ^{16}O. Grossman seemed to have ruled out his own model with the discovery of excess ^{16}O within the CAI. For the next several years I made a thorn of myself at meetings by standing to publicly object each time some chemist made an argument derived from assuming a hot, totally gaseous initial solar system[4]. The only model consistent with both aspects seemed to some to depend on excess ^{16}O-rich gas blasted into the solar system by a neighbor supernova. It did not address how the CAI could contain so much extra ^{16}O, however.

Al Cameron introduced me to his arguments that the solar gas might not be sufficiently hot to evaporate all interstellar dust during the birth of our solar system. That reinforced my own thought that the solar disk was not very hot. The more I thought about it the more I found that his arguments, and my own, made more sense to me than the current picture of a solar system beginning from the cooling of a hot gas. I could not see what could make it so hot when had initially been a cold dusty cloud. Robert Clayton *et al.*. [1]explicitly cited Cameron's suggestion[5] that presolar dust might survive in meteorites, but could then not understand the existence of the CAI rocks. I felt the irresistible tug of an issue related to the origin of the elements in stars.

I was, incidentally, intrigued that Robert Clayton bears the same name as my great grandfather. After we quickly became friends, Bob, a Canadian, and I would publicly refer to each other at NASA conferences as "cousin Bob" and "cousin Don". This was all in good fun. Since scientists often make written statements like "As Clayton (1973) has shown…," I thanked Bob for increasing my fame, and he was generous enough to allege the converse.

I have recounted these issues in some detail because they were huge in my life and in the scientific study of the meteorites. These were the issues that led me to STARDUST in 1973-75. They provide my entry cue to a new scientific life.

Fred Hoyle's first trip to Rice University

Other important things were about to happen in my life, however, so before postponing more thought on this curious new discovery I called my Rice friend and colleague, Dieter Heymann in the Geology Department. I asked him to present a discussion of the excess ^{16}O in the minerals from the Allende meteorite at one of the weekly discussion groups on astrophysics that I organized, a group we called "The Astronomical Unit" as a pun on the solar system distance scale. Dieter agreed to discuss this at our scheduled meeting on Friday November 9, 1973, the date I chose because Fred Hoyle would then be in residence at Rice University. It relieved me to count on Dieter to reopen these issues as seen from his eyes, considerably more knowledgeable than my own where meteorites and chemistry were concerned, and even more importantly in my intensely busy state, to prevent me from forgetting this new finding entirely. My mental state was flying, so that I could not be confident of remembering things in the rush. Buttressed with that reassurance, I banished excess ^{16}O from my thoughts.

Other goals beckoned. Fred Hoyle visited Rice in order that we might work together on ideas about nova explosions and also enable him to speak to the public in The President's Lecture Series, an annual lecture at Rice University sponsored by its president, whose goal was to present a famous person to the public and to Rice students. It was a big local honor. Many nationally famous people have appeared in this Rice University series. All expenses were paid by the President's Office, and an honorarium of $1500, a nice bonus in 1973 and an amount that I hoped would be useful to Fred in the wake of his resignation from the Plumian Professorship at Cambridge during 1972. I proposed Hoyle and the committee had accepted my nomination as the 1973 President's Lecturer. Fred's memorable keynote speech at the first Lunar Science Conference in 1969 had swayed any undecided members. Surrounding Fred's lecture we hoped to finish some publishable work, a goal that I had proposed to him in letters suggesting these few weeks at Rice. I described a specific idea for a study that might be quite important and which I thought we could finish up during his visit.

I was constantly aware of the strangeness of me attempting to help one of the world's greatest scientists. After all that Fred had done for me during the previous six years in Cambridge, I found myself now trying to do something for Fred. It seemed presumptuous of me to imagine that I should or could. I did worry that his elopement to private life above Ullswater in the English Lake District might become cash poor, and it did worry me that despite his enormous intellectual gifts he might in so remote a place fall from nuclear astrophysics altogether (which did in fact happen after our four research visits with one another during the next three years). So whether appropriate or not, I attempted to help Fred Hoyle retain his preeminent position among nuclear-astrophysical theorists. In a telephone call with Willy Fowler we coordinated plans so that Fred would eventually depart from Rice to an even better honorarium at Caltech.

In my letter I described to Fred a new observable astrophysical opportunity that might exist in the nova explosion on the surface of a white dwarf star. So armed with ambitious hope for a modest repayment of my debt to him, I picked him up from Houston airport on October 31, 1973. As a Resident Faculty Associate of Wiess College of Rice University, I had arranged guest rooms for Fred in the College in return for an evening discussion with its students on November 14. It was hardly St. Johns College, Fred's college, but it was good enough for Fred, a man without pretension.

The new scientific idea that I had proposed to Fred was that the positrons emitted by radioactive nuclei created by the thermonuclear outburst on the skin of a white-dwarf stars, positrons that deposited the radioactive power that made nova ejection of matter possible, might be observable by gamma-ray telescopes. The novae are much smaller explosions than are the supernovae, but they are believed to be thermonuclear explosions.

The radioactive nuclei that are produced during their outbursts then decay by emission of the antiparticle of the electron, the positron. On this issue I was conscious of standing in a first-person stream of human knowledge.

What do I mean by that? Carl Anderson had discovered the positron and had become the youngest Nobel Prize winner, the same kindly Carl Anderson who as my adviser had enrolled me into my first graduate courses at Caltech and had spoken to me of their accepting me for that opportunity. Willy Fowler had discovered the previously undocumented nuclear reaction, called "radiative capture", by bombarding carbon with the proton beam in Kellogg Laboratory, and by showing that radioactive nitrogen was created by the collision; and he had confirmed the identity of the positrons emitted by the radioactive nitrogen using Carl Anderson's original apparatus to show that the radioactive ^{13}N emitted positrons, the same Willy Fowler that had given to me an experiment chance concerning carbon in the CN cycle of the stars. Willy's pioneering laboratory experiment had studied the same nuclear reaction that provided most of the positrons in the nova. Fred Hoyle had created the theory of nucleosynthesis in stars using the assembled nuclear knowledge, the same Fred Hoyle who had with Willy annointed me high priest of nucleosynthesis, and now I was proposing to Fred that we employ gamma-ray telescopes to confirm that all of this history was central to human understanding of the nova explosions, which we would suggest to be a laboratory for confirming explosive fusion in the cosmos. That is the continuous stream of knowledge to which I refer. I had waded shoulder-deep into that stream of primary discovery by great pioneers who had influenced me deeply. I was baptized in its waters by those who had baptized me.

This intellectual heredity bears reemphasis. The idea of positrons emitted by radioactive ^{13}N decay in a nova atmosphere creating detectable gamma rays by their annihilation had not sprung from the world's assembled books, nor from a flash of insight by someone far removed from the historical path of the discoveries, but rather from first-hand efforts by flesh and blood beings from whom I received first-hand inspiration in the involved issues. They had personally prepared me. Now the tables turned, and I was in turn suggesting to Fred that we examine the consequences in joint work.

Forewarned, Fred had also begun thinking about the dynamics of the nova atmosphere as it expands from its overheated state. So from the moment he arrived we were ready. Two days after his arrival, on November 2, 1973, I presented our analysis of the visibility of the annihilation gamma rays at our Friday afternoon Astronomical Unit. Good questions and ideas arose, as they will with a hot idea. Exactly one week later we submitted our paper "Gamma Rays from Novae" to the Astrophysical Journal Letters. Being clearly a new idea of importance for astronomy, it was quickly accepted and published[6] in 1974. The entire chain of exciting publishable research had been covered in ten days, maybe a record speed from conception to submission. We thought at the time, and I still do, that the nova is an outstanding laboratory for doing explosive physics and hydrodynamics. The Milky Way Galaxy graces us with dozens of visible novae annually, so they are frequent enough and varied enough to be regarded as an experiment, one over which mankind exerts no experimental control but one in which nature provides such a sufficiently varied spectrum of conditions that full understanding might be achievable. Not only could a measurement of the 511 keV gamma-ray flux created by the annihilation of the positrons with the electrons in the white-dwarf atmosphere provide a new unique diagnostic, it could simultaneously prove the thermonuclear nature of this explosive event. It was a great dream then, and still is-- but nothing more. No space experiment has yet achieved the sensitivity to detect that gamma radiation from the events that have occurred. They have been too distant. We would await a closer nova with NASA's Compton Gamma Ray Observatory following its launch seventeen years later; but to this day the right opportunity has not presented itself. The theory has grown more secure than it was in 1973 when we set it forth, but it remains unconfirmed.

The author and Fred Hoyle at work in Texas.

On Saturday November 10 Annette and I took Fred to his first and only Southwest Conference football game. It was Annette's second American football game. This too had a sense of recompense, in this case for the careful introduction to cricket that I had received from Hoyle in Cambridge. Surprisingly Fred is quite keen on American sports, and watched them with studied appreciation. On this particular day he commented on the triumph of the kicking game. Time after time the booming punts from the Rice punter had pinned the Arkansas Razorbacks to near their own goal; and when it was all over the Owls had won 17-7.

"It was the punter who won it," Fred commented afterwards. From huge Rice Stadium we strolled to our rooms in Wiess College, where Annette cooked liver and onions for our celebration of an excellent week. Annette's sweetness shone, and her eagerness to please was winning. But it was a little disconcerting how far removed she necessarily was from the spectrum of interests that Fred and I shared during this three week visit. Fred's presence with us forced me to think about this, slightly embarrassed by the disparity but constantly loving and optimistic. Fred and I must have seemed to her esoteric intellectuals pursuing scientific understanding of the universe, and she remained the Schwarzwald bride to us. Although that particular gulf between us was admittedly large, her sweetness and enthusiasm made up for the understandable disparities of age and opportunity. My romantic dream was at stake here. I was confident that our love and the many opportunities of the mind that we would share could open her intellectual life and minimize our differences. It made no difference that she would not share nuclear astrophysics, but it was my hope that she would nurture her own intellectual interests. On several occasions I proposed that she might audit a Rice course of interest to her, and I presented the course lists and looked through them with her; but she always rejected the offer despite, or maybe because of, her sense of inadequacy over her education, which had not completed her Abitur, the top diploma for university-bound Gymnasium graduates. She had withdrawn from the Waldshut Gymnasium and obtained instead a general diploma from the Realschule before pursuing language schools. A sense of failure hovered like a black cloud over her lack of an Abitur; so I tried to find avenues for making that imagined deficit whole; but I would not succeed in this.

Hoyle's Presidential Lecture on November 13 demonstrated his public appeal. Only Carl Sagan could pull the public into a scientific lecture the way Hoyle's name did. He titled it simply "Stonehenge", the megalithic monument on England's Salisbury plain. Its puzzling and uncertain purposes had long fascinated Fred, as it did the public and the mystical fringe. Rice University President Norman Hackerman and his wife Jean hosted an elegant pre-lecture dinner in a private dining room in Cohen Hose, the Rice Faculty Club, inviting Annette and myself as his local hosts, Dean of Science W. E. Gordon and his wife, and others. Strolling afterwards from Cohen House toward the student-center auditorium, we saw the problem. The audience had overflowed the auditorium. Audio technicians quickly hooked up loudspeakers in the adjacent live-oak-canopied courtyard, allowing the overflow to sit outside on a pleasant November evening to hear Fred speak, even if unable to see his slides. At first embarrassed, I and the others soon realized what excitement this represented. This was Apollo-era Houston, home of the astronauts and of Johnson Space Center. The high tribute to Fred Hoyle symbolically linked common impulses of those builders of an astronomical site 4000 years ago to the current explorers of the Moon. Stanley Kubrick encapsulated the relationship in his amazing movie 2001. Rice officials quickly saw with relief that what seemed at first to be an embarrassing lack of adequate planning actually affected the audience in a very positive way. Excitement was palpable. It was an unplanned happening! Suddenly I seemed a sage to the lecture invitation committee, although I had not envisioned this.

During that same two weeks Fred delivered the weekly colloquium talk to the Department of Space Physics and Astronomy, a student seminar in Wiess College, and an evening lecture "Copernicus" sponsored by our department. Each engaging topic was presented with straight talk and humility. I never heard a better lecturer than Fred Hoyle.

But now the Houston phase of Fred's visit came to an end. At 5:30 AM on the morning after his "Copernicus" lecture, we arose to my already loaded 1962 Buick Sport LeSabre Wildcat and drove west in the darkness—witnessing sunrise behind us near Austin, breakfasting in San Antonio, lunching in Del Rio near the Mexican border, flying past the Amistad reservoir, over the great bridge over the Pecos River after a stop for the breathtaking view, past the Judge Roy Bean ("the law west of the Pecos") Saloon and Museum, on west toward Marathon, a small hamlet between the Glass Mountains and the Wood Hollow Mountains, and finally the last 80 miles south to our cabin in the Chisos Mountain Basin. Scanning the approaching wilderness with Fred I decided to place my photograph of a vulture on that road in the mid-career memoir I was then writing[7]. Exactly twelve hours after setting out we were in our cabins, stone cottages 106 and 107 in the Big Bend National Park. We sat on the rustic porch with a drink, and watched one of the abundant translucent small scorpions crawl past our feet. The sun set, the air chilled fast, and we were all in bed by 8:00 PM.

The Chisos basin is the eroded center within the rim of an ancient volcano. The rim walls rise to about 8000 feet and fall precipitously on the outside into the surrounding deserts. Next morning after a friendly breakfast we hiked to the top of the Lost Mine Trail, and from there across rough country to the peak of Lost Mine itself. I would later use this setting in my scientific novel, *The Joshua Factor* [8] (Texas Monthly Press 1985). Throughout this drive and hike we talked alternately of nova explosions on white dwarfs and of the geological forces whose results lays all around us. We were turning to the possibility of rapid sequences of neutron captures occurring in explosions on the skins of white dwarfs-- perhaps even a site for the r process. Within two months we would submit those ideas to *Astrophysical Journal* for publication[9]. Next morning we took the more strenuous hike, up to and along the south and east rims. We packed lunches from our cooler and were careful to carry many bottles of water. I must say that Fred did that hike very well, though almost sixty years old, and finished as strongly as did Annette and I. Next day we hiked into the Boquillas canyon. Here and in the Santa Elena Canyon on the eastern side of the gap one stares in wonder at canyon walls carved by millions of years of water flowing along the Rio Grande. Immense grooves eroded ever deeper by time make a suitable image for the personality groove of which Stephen Maturin soliloquized and which was deepening within me. Of course, all of this hospitality was a further wish to repay Fred in kind for his gifts to me, in this case for the introduction to Scottish mountaineering during the 1960s. Happy at the favorable outcome, we drove him to the airport at Midland, Texas and dropped him there on November 20 for his flight to Los Angeles.

During this road trip not a further word was spoken about the University of Chicago discovery of excess ^{16}O in meteorites. Our conversations were too busy! So many good ideas lay on our plate already that we were sated. It was a shame, in a way; because Fred Hoyle had already published an idea for the origin of interstellar dust that I would soon rediscover, forgetting in excitement where I had learned it, as I rushed to apply that idea to a possible physical reason for the existence of isotopically anomalous interstellar grains. But that formation mechanism had not yet surfaced as we drove. The ^{16}O mystery lay dormant within me for six months, and by the time I would create a possible explanation for it, it would become subverted to related issues involving extinct radioactivity, thereby placing the ^{16}O excess secondary to the issues of extinct radioactivity, issues that would embroil me in criticism and controversial stances. I had so many good ideas for cosmochemical applications of supernova dust that I used poor judgment over which application to publish first. My public stances over the subsequent criticism, my stiffening spine, and the explosive expansion of my imagination together created a new persona, and I accepted the need of that persona in order to force scientific examination of my new contributions. The criticisms grooved that persona to such depths that it become the new me. I was determined to win by stuffing my ideas down their throats until they could no longer be neglected.

Shadows on Snow

A winter holiday in Germany's Schwarzwald then took me light years from west Texas. Annette and I had been married at this time for one and one half years. Still only 21 years old in a new land, she had homesickness. This caused me no problem; after all, I had asked her to leave the only world she had known and move into my unusual world. I understood her homesickness and enthusiastically undertook several returns to her land, which I easily loved. On December 14, 1973, we departed Houston on Swissair flight 101 to Zurich. It traced the same approach to Zurich that I had made in my first winter visit in 1970. Snow covered the hills and mountains of the Schwarzwald and of Switzerland. Annette's mother, Ira Thoss Hildebrand, met us at the airport and brought us in an hour along the Hochrhein, that east-west leg of the River Rhine that separates Switzerland from Germany, crossing the Rhine at a small border bridge at Zurzach , just east of Waldshut, and thence through farming villages to their home in Wutöschingen. All about were the sensual loves of my childly heart--pine wreaths and ribbons, Christmas trees, foggy breaths, dim winter lights augmented by seasonal colors, snow clumps falling from trees, fruit schnaps, chocolates and marzipan, women in fur coats, old villages, the constant clanging of church bells, tractors towing wagons along the narrow lanes, a big family gathering-- and most of all-- the farms. From the balcony of their house I gazed in contentment across the Wutachstrasse in front of their house, through an apple orchard on its other side, to one of the major farmhouses of the village, lying at the base of the great hill, the Bolhof, at the village's eastern boundary. All were covered with snow. *I could live here*, I thought to myself.

For some years I indulged in that mental exercise--should I say fantasy?--of imagining a gradual escalation of my life in Germany. That thought began just as simply as I have described, with affection for beauty and for the dearness to me of Germanic traditions known in my Iowa childhood. Only three years later my fantasy would also become a scientific plan with the beginnings of my affiliation with a Max Planck Institute in Heidelberg. My hoped for allegiance to England had been dashed by Hoyle's resignation. *Why not Germany?* The very question reflected my reluctance to embrace US society as I then perceived it. Perhaps I rejected the US in which I lived daily in favor of a remembered sweeter childhood on the Iowa farms. I can not say. I had then, and still have, misgivings over the American tendency to destroy tradition and beauty in our determined path toward corporate profits and in our temptation to try to control the world. We had just escaped from Vietnam and a war that I had publicly protested; and as I write today we attempt to escape from misguided entanglement in Iraq. This 1973 Christmas visit was already my fifth visit in a list that would grow long.

A dark shadow emerged during this visit when Annette requested that I not acknowledge the existence of my two sons. I acquiesced in that when I should not have done so, thinking wrongly that my silence was not so much a deception or a denial as an effort to not upset parents who were presumably extremely sensitive to my having two sons. I hoped that this lack of disclosure would be but a temporary thing. As these visits progressed, I began to look more and more at the architecture and settings of homes and of life, with an eye toward becoming more German. After all, had I not embraced a huge romantic dream in marrying a young

Schwarzwald girl. I was continuously deepening my affection for her village as well as for herself and her family. In so doing I did not give sufficient awareness to Annette's wanting little to do with the Schwarzwald, other than visiting her parents, which for some contradictory reason she wanted to do often. For the first time I perceived her love-hate feelings toward her father. I did not give sufficient weight to her denial of Donald and Devon. The meaning of such contradictions went largely unquestioned, as happens when one fears the answers.

It was during this Christmas that my eyes first fell on the old castle in the neighboring village of Schwerzen, just one mile's pleasant stroll from Wutöschingen. I learned from the Mayor of Schwerzen that it could be purchased for a song by anyone willing to commit to its upkeep. Could I manage that? In retrospect the answer was "No!', but I admit that for a few years I entertained the question. I took long strolls on the snowy hills. We took a bus to Konstanz to see "Egmont" in the theatre there, with Annette's brother Horst Hildebrand, an actor, in the lead roll. We went with the family and friends to dine, often at a rustic restaurant in Witznau up in the high Schwarzwald, where we bowled (" zu kegeln") in the small attached Kegelbahn. I mailed one chapter of *The Dark Night Sky,* namely "The great reactor in the sky", to Rice for typing following lengthy caressing of its words at a writing table having a view of the snow-flecked garden. Frau Hildebrand invited local friends, which I then mistakenly assumed to be numerous, to dinner at home, in part to show us off. We walked and skied near Rothaus, where the brewery and Gasthof featured its good Pilsener, Rothausbräu, amid the rarer chilly winter air at 1000 meters. I practiced speaking German daily with Annette's teenage brother, Volker Hildebrand. Friendship grew with yet another brother, Herbert Hildebrand, and his family, who also lived on the banks of the river Wutach but on the northern edge of the village. We traveled again to Konstanz to the theatre having eldest brother Horst starring in "Revenge". We celebrated Weinachten with beauty and grace, many old ornaments dangling from the family Christmas tree. A third time to Konstanz saw Moliere's "Floh im Ohr". These three weeks did enchant me. So did sequels.

It took much longer to see things that were terribly wrong in the Hildebrand family.

Unexpectedly one day a deep anger arose in Annette, amorphous, unfocussed, puzzling, until it threatened to break out, and then did break out. On January 4 she was in a mad fury most of the day, shouting at me and at Papi, as we called her father. First she did not want to go back to Houston but rather to remain in Wutöschingen. Later she hated Wutöschingen, "smelly farmer pit that it is". She struck me in anger when I tried to calm her. Later she started to strike her father when he talked to her. Some blackness had fallen over her that could not be dispelled. It took time and everyone's concerns. This black mood was deeper and more violent than the jealous rages over nothing that occasionally dotted our first two years together. I was yanked from my winter idyll, whose soft beauty had enveloped me swiftly and sweetly. It takes longer to see the somber question marks. By nightfall she was, as the euphemism goes, "back to normal". Now, more than ever, I questioned what normal was. I was very, very troubled by this, conscious of having mistakenly married this pretty girl out of my own needs rather than hers. Would it not be better, I wondered, almost wished, if she did just stay home and start anew after her adventure with me? But she then reversed and became emphatic in wishing to return with me. I did love her, and our romance had been sweet in the absence of that dark cloud that occasionally appeared. So in hope we embraced and packed our bags. Swissair 100 returned us on January 7, 1974, to Houston. Mike Newman, one of my gifted graduate students, met us at the airport in my yellow Buick.

Issues in Houston

Hans Bethe, 1968 Nobel Prize winner in physics and discoverer of the two nuclear burning cycles by which hydrogen is fused into helium at the centers of stars, thereby providing their immense supply of heat, visited Rice University during January 8-12 as 1974 Houston Lecturer. This physics lecture was not named for the host city. The Houston Lecture is an annual honorary lecture endowed by the physics department in honor of William V. Houston, late Professor of Physics and past president of Rice University. Bethe and Houston had studied together in Munich. His Wednesday department colloquium was entitled "Pions", and the university-wide Houston Lecture on Thursday presented a more humanistic view of physics and its relationship to world stability. On Friday morning at 10:00 AM I had a personal appointment with Bethe. After a lively discussion of the world of nuclear astrophysics, I brought up the subject of the Nobel Prize.

"I would like to encourage the support of Willy Fowler for a Nobel Prize in Physics," I said.

"I totally agree", replied Bethe with disarming candor. "But first I am supporting one prior choice," whom he did not name and I did not ask. We then discussed Fowler's work rather thoroughly, and agreed that its lasting greatness lay in his measurements of the rates of thermonuclear reactions in the stars rather than in his forays, mostly with Hoyle or with myself, into the theory of nucleosynthesis. We discussed Hoyle's priority with the theory of nucleosynthesis in stars, and Bethe stated that he thought that a prize shared by Fowler and Hoyle would be very appropriate. I agreed enthusiastically. He was inquisitive about Hoyle's recent visit and of our current work on thermonuclear explosions on the surfaces of white dwarf stars. This previously undisclosed conversation relates to a controversy that would erupt after the announcement of Fowler's 1983 Nobel Prize.

This was a critical time for me, a "lesser climacteric". My life was teetering. Only its scientific center held. Worry over my young wife and my chancy remarriage only two years after a sudden divorce from my first wife, worry over the two sons of that first marriage, worry over my adequacy as husband and as father, worries about moments of poor sexual performance, and other worries invaded my days. These are heavy issues for men. In my view I failed as a father during this time, trying to balance the need to reassure Annette of her importance to me with the needs of my sons for a father. My time with them was jealously guarded by Annette as being time taken from her. I had some sympathy for Annette being just eight years older than my older son. But nonetheless I was dismayed. Only an impressionistic sketch of a few personal events is seemly here. My 13-year-old son, Donald, came by for his birthday present, stole my Fiat keys, later returning them via a Wiess College student. I held long phone calls with Devon, then 12 years old, over his gloominess and reluctance to reach out for things, including visits with me and Annette. After infrequent talks with Mary Lou I unfairly painted her as an unsympathetic person to Annette in order that Annette would not be angry or suspicious about my talking with her. Mary Lou sometimes hung up the phone when I expressed dissatisfactions with the amount of time I was having with the boys. I could not blame her for that. I knew this was more my fault than hers. I began driving lessons and tennis lessons with Donald. We used my yellow Fiat Sport Spider for short maneuvers in the Rice stadium parking lot, laughing as Donald repeatedly killed the engine trying to start from rest in first gear. Of course, he adored that, being almost 14; but this healthy camaraderie made Annette jealous, so I played tennis twice as long with her. Time for work amounted to no more than a few hours daily, barely enough to keep things going but not enough for creative thought. I needed professional help but did not seek it. I bought a birthday bicycle for Devon three months in advance of his birthday, trying so hard to be a father. I played tennis hours daily with Annette, to be with her doing what she wanted, and to try to help her become a competitive player. She possessed tennis talent, real athletic ability, but repeatedly suffered temper tantrums on the court. I had to put up with Annette's restless sister, Annelie, recently divorced by a doctor in Norway who had had enough of her hysteria. She settled in for her three-week visit with us in our new college rooms in Wiess College, remodeled by the college for our use. I was simultaneously conscious of not being a good Resident Associate for the college students, of being drawn out too thin to have substance for anyone. With so much pulling, like taffy, my weak points stretched toward the breaking point. On April Fools Day I was served court papers to increase my level of child support, even though $400 monthly, paid on time through the courts, was honorable support for two adolescent boys in 1974.

April was the cruelest month. My nightmare festered to a head when Donald, figuratively crying out, stole the Fiat, and was seen by campus security driving it around. Security reported it next day to me. He loved that car, as he knew his father did, and he identified with me and wanted to love me. He may have also wanted to punish me, because he then stole it again and wrecked it, not seriously for him, but enough to damage its front suspension. He nonetheless had been able to nurse it home, totally unwilling to abandon a car that we loved, and left it wounded in my parking place at Wiess College. I had to do something. I mounted, with Annette's support, of which I was always so careful but uncertain, a child-custody countersuit to the child-support suit. Increased child support was no answer, I thought, whereas adolescent boys do need their father. And I had misgivings about Mary Lou as a parenting force ever since her tragic paralysis on June 2, 1973. The red Volkswagen had overturned with its convertible top down in a sloping highway median as she drove, breaking her spine and leaving her paralyzed from the waist down. This was an unspeakably huge catastrophe for her, for the boys, and also for me. I had fetched the two boys from the Conroe hospital, from which they were discharged owing to having been miraculously thrown out with only cuts and bruises.

Not only was my grief great over Mary Lou's paralysis, but it complicated my custody suit. So time, depositions from the boys and from Mary Lou and from their psychotherapists, and considerable money was spent in a losing cause. Because divorce was not yet the societal norm in 1974, we argued new ground, trying to place the welfare of my two sons above petty squabbles between ex-spouses. We hired an attorney ad litem, for the boys, to argue their welfare, and made sure that this attorney was in court--a three-way action. Such positions are routine today; but we were all uncertain what precedents would dominate. It ended with a sad whimper. Putting the boys on the stand on June 24, the judge forced them to answer this totally unfair but superficially reasonable question; "Which parent would you rather live with?" They could not, certainly, turn their backs on their paralyzed mother, although they were more than ready to say neutrally that they felt well cared for with either parent. When they were ordered to answer I knew that my cause for them was lost. I suffered with them at their answers, haltingly, achingly delivered, and rushed to hug them afterward to say that it was OK, that I understood. They remained with Mary Lou, and child care was raised nominally to $475 per month (from $400). But maybe this defeat won something. My sons were genuinely moved by this effort to bring them directly into my life again. I saw the hope of light from this tunnel as new relationships slowly built from this rubble. Annette was impressively mature and dependable through this stressful conflict. That was certainly another bright spot. She had also been severely injured in an auto accident as a child in which her oldest sister was killed. Within three years my sons would themselves agree that living with us would be best; but not before more tragedy, more heartbreak.

I could see that my life was not suiting my main goal. I longed for time free from worry, time that I could spend in simply thinking about the captivating puzzles of my nuclear universe. I saw that my first marriage had given me that time, and ample time for being a father as well. I reconsidered the possibility that I had blundered by not addressing squarely our sexual problems in that marriage. Now it was too late for that, but still I contemplated the contrasts to my present barrage of concerns. I yearned for trouble-free time to simply think. Thinking time had been the secret to my scientific successes and to my inner joy from boyhood on. I wanted to lie again on that branch of the tree beside the front gate of my grandfather's farm and contemplate the sparrows building of their nest above. But I was also deciding that lament solved no problems. I needed to take the hand I had dealt myself and make the best of it. That would take conscious focus and lots of work. I had to take charge of my life and be better than I had been. And just now, at this turning point, I had matters to take care of.

Annette and I finished the move from the small one-bedroom apartment that I had initially been offered as Resident Associate of Wiess College into the remodeled suite of three large rooms there on 4 July 1974. This was a really new thrust for Wiess College, and for the college system at Rice. They were providing for a married Associate (called *Fellow* in Cambridge) an outstanding apartment rather than simply a room. Only the Masters had previously enjoyed such participation in College life as residents. They had even been willing to have my sons as resident college members, an idea that had titillated the boys. I would have needed only to buy the extra meal plans. I was grateful to the College for this; but it only exacerbated my sense of guilt at not being a more effective Associate. My relationship with the students was too superficial in the face of all of my concerns. I had a difficult time trying to explain to Annette what the college hoped for from a Resident Associate but that we were not doing well. Her energy was for her private needs rather than those of college students that were only a few years younger than herself.

A reprise of Cambridge

I needed the escape from all this when on July 5 Annette and I boarded a flight for Scotland. From Glasgow we rented an auto and drove north to Inverness, the gateway to the western highlands. The highlands were my goal, even without the presence of Willy and Fred, and Annette shared the enthusiasm for it. My decision to return to Cambridge as well, without those two great men and friends, was my excuse for the trip. From an overnight in Inverness I telephoned Mr. Morrison at Inchnadamph Hotel. Yes of course he was happy to see us again, and he did have number 5 vacant for the next week. I was elated, but shortened it to three nights when another telephone call to Mr. Millard at Achnasheen confirmed that he could also take us in from July 10. So partly in search of healing for my heart we retreated to the highland Munros that I had first climbed with Fred Hoyle-- up Conival, Ben Mor Assynt, sacked in with misty rain as always, gentle sunny Quinag, baroque Stac Pollaidh, and finally the incomparable Beinn Eighe. Now alone, Annette and I embraced the days and the nights, for love did remain, raising my hopes that all might yet be well with us.

On July 14, 1974, we drove once again into Cambridge. Its familiarity now had acquired a new strangeness. This was to be my last residence as guest member of the Institute of Theoretical Astronomy, now renamed The Institute of Astronomy. Fred Hoyle and Willy Fowler would not be there. I would emotionally wrap up that glorious phase of my life. My ostensible goal was participation in their NATO sponsored Advanced Study Institute, "Origin and Abundances of the Elements". I had conducted robust planning correspondence throughout spring 1974 with Martin Rees, its director. I was to be the kickoff speaker on the morning of July 23, for a session entitled "Introduction to Abundance Problems and Observational Evidence for Stellar Nucleosynthesis". It would be a week later, in my closing talk, however, that I would present my idea of surviving supernova dust for the first time.

On every occasion on which I had returned to Cambridge, I had tried to seize it and not let go. I had loved Cambridge deeply for seven years. I had wanted its traditions to be my own. One way I tried to soothe that yearning was to pretend to be returning home; but this time I knew it was not true. Nonetheless, I had arranged with Miss Hammond, the owner, for us to return to a previous residence, her home at 6 Sydenham Terrace on the eastern end of Halifax Road, a short spur running eastward just across the Huntingdon Road from the Trinity Hall Sports Ground. Although we could be resident only a short three weeks on this occasion, it was the same house in which Annette and I had resided during the entire summer following our 1972 wedding, the summer of Fred Hoyle's abrupt resignation from the Plumian Professorship and the associated dashing of private dreams for Willy Fowler and myself. Miss Hammond moved to the country to be with her daughter during our residences. Sydenham Terrace consists of six red-brick houses, connected in the British fashion by common walls between individual residences, tiny front gardens behind a two-foot brick wall, and a rear yard. Each was two floors, with parlor, dining room and kitchen on the bottom, bath on the stairway landing, and three small bedrooms up top. It was cozy and felt very English, very Cambridge-town.

No more than 100 yards west ran the Huntingdon Road, just beyond the modest baker and grocer at the end of the block. The north and west borders of the Trinity Hall Sports Ground across the Huntingdon Road were terminated by the gardens of Storeys Way, where seven years earlier I and my family had begun my Cambridge life. How time had flown. I came back to this simple terrace of houses ruminating on such thoughts, for how could I escape remembrances of Cambridge days that at this very moment were ending for me. My private anguish resembled that felt by me ten years earlier upon visiting the Kembery farm in Iowa with Mary Lou. A short walk from our door to parallel Oxford Road came directly to Willy and Ardie Fowler's similar but semi-detached house, remodeled by them, now with a seemingly forlorn "FOR SALE" sign hanging in front. My heart screamed "No!" while I took one last photograph of it. With just such emotions, in just this place, a place suddenly for me haunted by ghosts, suddenly detaching myself from it by working through a type of grief as in the death of a loved one, just here, just now, I wrote the paper that would change my life and my personality.

Annette could hardly have had any notion of my feelings. I shared what I could, but I saw no point in trying to share an emotional frame of mind that would surely have seemed to her to be bizarre, exaggerating my sense of loss, and backward looking to boot. She needed to move forward, not backward with me. On the first morning we breakfasted, then walked through town center shopping. As I always had, I dropped in the newsagent in person to pay in advance for *The Times* delivery through August. We invited Stan Woosley to our house for dinner, which Annette cooked, happy to have company in what had been her first home after our 1972 marriage and that was to be my last home in Cambridge. This May-and-September marriage was out of joint in so many ways. I resisted that thought by wooing her, with my enthusiasms, with my long blonde hair and flashing smile, with seeking in her my expression of love. Sexual passion abounded in support of this goal; but I was pushing it. It would take years before I could admit that I had made a mistake. I was never good at admitting mistakes.

On July 15 I met with Martin Rees, Wallace Sargent, and Stan Woosley at the Institute of Astronomy, Cambridge, to set the final program for the NATO Advanced Study Institute, "Origin and Abundances of the Elements". I had stressed in correspondence that the meteorites were yielding puzzling information about abundances of the isotopes within them. Flung to Earth after a million years in Earth-crossing orbit following some excavating collision a million years ago between two asteroids, these stones recorded matter with a differing history than Earth's matter. Isotopic differences demonstrated that difference in a way that had now been discovered but was still dimly perceived. As keynote speaker I entitled my talk "Analyzing the Abundances of the Elements". I had urged Edward Anders' attendance to lead a discussion on meteorites; and that was scheduled for the second morning. Robert N. Clayton, whose laboratory had uncovered the ^{16}O-rich minerals within the calcium-aluminum-rich inclusions of the meteorite Allende, would be the featured follow-on speaker after Anders' introduction. I was pleased with that tandem. There followed much, much more--two weeks more of all that rich tradition of nuclear astrophysics that had become associated with Fowler and Hoyle and myself--but I was focusing on the stage that had been set for my new meteoritic thrust.

We followed this meeting with lunch in the new Cavendish Laboratory, just across the Madingley Road. I did not like it. The Cavendish to me would remain Ernest Rutherford, nuclear physics, and the old buildings on Free School Lane. The new Cavendish was not only new architecture, but in place of nuclear physics it now featured Anthony Hewish and Martin Ryle, who would in two months time be announced as Nobel Prize winners in physics for 1974 for the radio astronomy discovery of pulsars. That evolution of the Cavendish was good and natural, but-- more to my feelings— the radio astronomers had hastened Fred Hoyle's resignation by their adversarial relationship to Fred and to his institute. In my 1974 diary I wrote of the new Cavendish on that July 17 simply, "Ugh!"

In that negative frame of mind I spent no more than a few daily hours at the Institute of Astronomy. I was preoccupied more with dark humors again emanating from Annette, and with my urge to still them. So we rubbed brasses, walked the colleges and took photographs, played tennis, had friends to dinner beginning with Mike Howard and his Italian girlfriend. Another night it was Icko Iben, Dave Arnett and Mike Newman. We had a small party for her 22nd birthday on July 22; but since the people were my scientific friends she did not think it much of a party--and in this I would agree; and yet we did have fun with it. Despite my efforts,

Annette frequently called home to Wutöschingen, once twice on a single day--homesick again, yet glad to be away? Something was very wrong. While I carried this anxiety I experienced sexual unpredictability, intensifying the worry that caused it. In my hours at home I polished chapters of a book, a memoir, growing out of my urge to explain my life to Annette. It had initially taken root as a potential film with Roberto Rosselini; but that adventure had died on the vine. I wrote several chapters of *The Dark Night Sky* there on Halifax Road. On July 27 I posted five chapters of it to E. P. Dutton, hoping they would want it. One week later I mailed it to Quadrangle, who eventually would publish it.

Maybe a continuous period in Germany would release Annette from her homesickness. I spoke with Klaus Fricke[9], who was again in Cambridge for that summer, about the chance of coming to Germany on leave from Rice. That query, which did not fall on fertile ground, was my first attempt at the idea of taking Annette to Germany to reside for a time. I soon found a better place in Germany for the new me. Annette and I would walk to town for kebab at The Gardenia on Green Street; or go to the Pickerel where I could drink a pint of Guinness in Cambridge's oldest pub. I missed Donald and Devon and worried about them, but could not discuss this honestly with Annette because she would have been threatened by my lonesomeness for my sons. Black, black, tones were thus painted on the bright medieval town centre. These would not be lightened for yet a week, when the arrival of Marianne Klopfstein, Annette's school chum from Geneva, for a four-day stay with us cheered Annette; and with a rented Avis auto, on the very day of Marianne's departure we picked up Annette's parents, who flew to London from Zurich. The Avis auto was rented on 2 August 1974 for a mere 17.70 pounds per week. We took it for the 17 days until August 19. The whole bill was but 79.94 pounds, after 1056 miles of driving; but even that small sum seemed expensive in comparison with my late 1960s rentals from Marshalls on Jesus Lane.

Xenology and Stardust in Meteorites

The NATO Conference began on Tuesday July 23, the morning after Annette's birthday party. In my kickoff talk, "Analyzing the Abundances of the Elements", I tried to explain how we sorted the elements and their isotopes into distinct nucleosynthesis processes, some in different stars and some in different regions of the same star. It was a purely introductory and pedagogical survey to facilitate that which was to follow during the three-week workshop. Next morning Edward Anders, reviewing element abundances within the meteorites, touched on the meteoritic evidence for the existence of decidedly different isotopic patterns within the element xenon. These patterns coexisted in the meteorites: perhaps in different minerals, or perhaps only released at differing temperature when heated. These differing isotopic xenon patterns became my second inspiration for the new kind of astronomy that I would introduce later in this conference. The story of anomalous xenon isotopes is somewhat lengthy and technical; but I include it for the better prepared reader.

Xenon is a gaseous element in the earth's atmosphere, a minor harmless component of the air that we breathe. About one ton of it is isolated annually on earth by liquefaction of air. Xenon is a chemical element that does not form solid compounds, and is one of those gaseous elements described as "noble gases" owing to its lack of chemical reactivity. The noble gases are used in gas-discharge tubes generating light having color characteristic of that element—so called "neon lights". But Edward Anders' topic concerned the nine stable isotopes of the xenon gas, isotopes that in air always reveal the same relative abundances. The atomic weights of those nine stable naturally occurring isotopes of xenon are A=124, 126, 128, 129, 130, 131, 132, 134 and 136. This large number of stable nuclear isotopes, more than any other element save tin, allowed for a large number if distinct isotopic abundance ratios. I took satisfaction that four of these, the ^{129}Xe, ^{130}Xe, ^{131}Xe and the ^{132}Xe isotopes, were at the second r-process abundance peak, as Phil Seeger and I had calculated with Fowler in 1964. ^{134}Xe and the ^{136}Xe isotopes were also synthesized by the *r* process, but with abundances less than in the peak. The ^{128}Xe, ^{130}Xe and ^{132}Xe isotopes were also *s*-process products in stars. But the two lightest isotopes, ^{124}Xe and ^{126}Xe, were produced only by the *p* process. What an intricately tangled web the solar mixture is! I had reviewed the purely nucleosynthesis aspects, these question of what nuclear processes produce which isotopes, in my opening introductory talk.

What Edward Anders was now reviewing was the existence within meteorites of some very distinct and special patterns of xenon isotopes that could be drawn out by careful experimentation with the xenon gas trapped within a meteorite. These patterns did not contain these isotopes in their usual proportions. A meteorite sample is a solid rock having many mineral phases within the rock, and this noble gas can not participate structurally in these mineral phases except in a limited sense; namely, the xenon atoms could be held within the rock phases by the relatively rigid positions of their constituent elements— xenon atoms trapped by their size and immobility within the minerals. But then, how did they get there?

Edward Anders presented the conventional interpretation of these exotic isotopic patterns. Each pattern was produced within the meteorite after the meteorite formed, allowing the crystal structures to be able to retain the xenon atoms, which would otherwise excape as a noble gas. One pattern was created by cosmic-ray collisions with barium nuclei in the rocks, collisions that produced certain xenon isotopes as nuclear fragments; one special ^{129}Xe excess[10] was trapped in the rock following the radioactive decay of ^{129}I, which could, before its decay, chemically bond within the rock as iodine in the mineral structures; one pattern resulted as fission fragments of radioactive ^{244}Pu nuclei, contributing to the abundances of four specific heaviest isotopes; and a final pattern that contained a much more abundant spectrum of heaviest xenon isotopes that Anders preferred to interpret as fission fragments of an extinct superheavy element, identity unknown, also once alive in the meteorite but now dead. What a stunning display of isotopic patterns from this one chemical element! Such are the detailed mysteries that can be scientifically extracted in laboratories from rocks that have fallen to earth. Xenon is in more senses than one a noble gas. I was then and remained fascinated by these patterns and by wondering what it was that accounted for the presence of those radioactive parent nuclei that gave them birth. What had been the cause for those short-lived radioactive nuclei, ^{129}I and ^{244}Pu and perhaps an unknown superheavy element, to have existed among the atoms of the cloud from which the solar system was born? Is that really the correct picture? Sometimes nature is rich with puzzles to the point of excess.

No sooner had Edward Anders set my mind churning with excitement than Robert Clayton from Chicago followed to describe the excess ^{16}O that I had learned of[1] but not concentrated on during Hoyle's visit to Rice. Lightning struck me near his talk's end when he surmised, as he had in his *Science* paper, that dust rich in ^{16}O may have formed near some presolar supernova that ejected a huge number of ^{16}O nuclei. I was all ready to object that his proposal could not work, indeed I did stand to remark that such large ^{16}O excess could not be assembled in circumstellar dust, when I promptly halted, needing to reseat myself. I had been stunned, as by a lightning bolt, by a new thought, though none in that room knew it. They thought my remarks had concluded with my one negative argument. I had seated myself with a sudden vision, an unexpected flash in my neurons. I suddenly saw where the pure ^{16}O-carrying dust could condense, could form very abundantly, almost must form, and even richer in ^{16}O than anyone had imagined. It had occurred not in the gas surrrounding some supernova remnant but within their very cores during the expansion that followed the explosion. The ^{16}O-carrying dust condensed deep within the bowels of every young supernova, when their still-hot gases were not yet out into interstellar space but only been started on their way. The refractory elements condensing from the hot vapor were bathed in pure ^{16}O. Such ^{16}O-carrying dust would not yet be visible to one viewing the supernova from the outside. The new ^{16}O-rich dust was buried deep within the rapidly cooling and expanding fiery furnace of the thermonuclear explosion! I knew that pure ^{16}O existed in supernova cores because of our calculations at Rice.

I said not a word, but sat the rest of the day and thought

this over. Ideas overflowed: The ^{16}O atoms would condense as aluminum oxides, the same gem as common sapphires, along with the very abundant aluminum that had just been created by explosive nuclear fusion of carbon. But wait!-not only ^{16}O atoms, also radioactive atoms, specifically ^{129}I and ^{244}Pu and perhaps an unknown superheavy element. These radioactive nuclei could condense when the supernova's center had cooled below 2000 degrees.

Each xenon miracle of which Edward Anders had spoken could happen here within the newly condensed supernova dust- -not *could* happen-- *must* happen. The supernova's dusty guts would contain radioactive ^{129}I that would decay later within the growing dust into which it had condensed, dust which might also contain fission fragments from new supernova ^{244}Pu nuclei that had condensed, still alive, within those solid grains, and might even contain fission fragments from Anders' advocated superheavy parent of "carbonaceous-chondrite fission xenon"[10].

Wait! There's more still! New connections flooded my brain, so I heard not a word of what was being said. The superheavy element did not need to live long enough to survive in interstellar clouds, waiting for our sun to begin its 100,000 year labor of birth contractions. Maybe eventual incorporation into our solar system meteorites was not required. A fissioning radioactive nucleus needs live only until the interior of the supernova had exhausted its heat with pushing. Then, within a year the dust could grow. The superheavy element needed live only a year. These realizations were stunning. Their correctness could alter all chronological interpretations that were being advanced using these extinct radioactivities. I suddenly saw the possibility of a whole new world- -a world of interstellar dust carrying the very isotopic patterns of xenon gas that the meteorites, coincidentally or not, demonstrably contained.

I said not a word. My first decision had been made when I instinctivel sat back down. This was an idea too good to share by casual public utterance. It was too good to toss into the rough and tumble of open discussion, which is the life blood of science. I remembered Willy's good advice upon my discovery of radiogenic osmium: "Publish it yourself; then talk." I remembered my questions over sharing the possibility of gamma-ray astronomy of ^{56}Ni with Stirling Colgate, questions which arose because I talked before publishing. So I sat there silently in the conference chamber, dumb, no longer concentrating on the spoken words. Others asked politely and futilely of Robert Clayton how the supernova dust could form outside of a supernova; and how it could cause 5% enrichment in solar-system rocks? No one knew. *Only I knew.* I remained quiet but agitated, distracted. I hardly heard the remaining talks. It is always like this at the moment of discovery. Scientists greater than I have written of the single mindedness of the moment of conceptual discovery.

Sharing the new idea

During the next week I spent as little time in the workshop as possible. I sought out my old office and sat at another's desk there, feeling comfortable again, and worked. Yes, calculations confirmed that my idea was very promising. Many outstanding isotopic possibilities should be concentrated in dust that routinely exists in interstellar space. I wrote a list of candidates on a sheet of paper as they occurred to me. Considering the small size of the established isotopic anomalies, I thought, *You ain't seen nothin' yet.* I did not yet care how their isotopic effects became inherited by the meteorites. It was enough for now that huge isotopic effects were routinely carried in solids scattered among the interstellar matter. That interstellar mix eventually would provide the meteorite-forming material to the birthing solar system. I was intoxicated with this new idea. It was new, fresh, maybe even capable of experimental test, and carrying abundant consequences. I could not see all of those consequences, but every day new ones appeared. I had enough discoveries to publish the ideas. The details describing how the interstellar dust seeded the meteoritic minerals to be made later could wait for independent scientific study. Yes, I was ready.

I began to write a paper for submission to a journal. In so doing, however, I made a huge tactical mistake. In the interests of time, and to allow me more time to mine out the other applications of supernova dust, I made the rather unfortunate decision to focus first on the possibilities for the element xenon by illustrating how the idea could address the puzzling "xenology"[10] within the meteorites. Xenology would constitute my inaugural application of this idea. Among the host of total possibilities I wanted a topic with sufficient focus and impact to be quickly digested. The element xenon seemed ideal and possessed that rich cohort of nine stable naturally occurring isotopes that sample the processes of nucleosynthesis within stars. Because it is almost chemically inert, xenon has a very small abundance within meteoritic rocks. There is very little there, with the fortunate consequence that an exotic xenon component trapped within a meteorite would produce a sizeable distortion of the relative proportion of those isotopes in the tiny amount of xenon gas within the rocks. .

The three anomalous isotopic patterns in meteorites that Ed Anders had been reviewing were all previously known to me. But my new realization was that radioactive iodine, radioactive plutonium and a superheavy radioactive element would all have been much more abundant in supernova interiors than they were in the early solar system. Those radioactive parents of xenon could have condensed within supernova dust, making supernova dust the interstellar carriers of more highly concentrated xenon patterns. The known isotopic patterns would be much larger in interstellar dust than within meteorites. My mistake was to overlook a combative human element in suggesting that the meteorite patterns, which were well known, might reflect a small component of interstellar dust in the meteoritic material. This would cause a huge uproar among meteoriticists. That uproar would divert attention from my original idea.

So feverishly did I work on this idea, one week thinking it through and a second week writing my paper, that by August 9, the final day of the NATO workshop, I had finished the handwritten draft and mailed it to my typist at Rice, asking that it be ready on my return. Relieved that it could then be promptly submitted after my return to Houston, I yearned to break my secrecy, to tell of my idea. So I decided to reveal it in a formal setting in which the origin of the idea would be easily traceable to me. On the last day I was to speak about the heavy elements, in a session called "*s* and *p* processes". In my remarks I said that although we in the solar system seemed to have only a mixture of nature's processes of nucleosynthesis, the interstellar dust can be expected to contain unmixed records of individual nucleosynthesis processes. I then described how supernova dust may carry the unusual xenon patterns, and, by contrast, how xenon's isotopic pattern in the *s* process, of which I was speaking in this talk, might even be carried in interstellar dust that condensed in *s*-process-enriched atmospheres of red giant stars[16], a different type of star altogether. There—I had disclosed it, the first mention of an idea that would revolutionize astronomy by creating a wholly new branch of it. Because I did not want the moment of origin to become lost, I carefully stressed within those public remarks that my paper introducing these new ideas would be submitted to *Science* within two weeks. To leave a further paper trail in my *Science* paper I added at its conclusion two final sentences:

> *These ideas were first publicly presented on 6 August 1974 at the NATO Advanced Study Institute, Origin and Abundances of Chemical Elements, in Cambridge, England. I appreciate helpful remarks on that occasion from E. Anders, R. N. Clayton and J. H. Reynolds without at all implying that they concur with this picture.*

I felt safe, mistakenly so as it turns out, at the thought of having stated this in print in *Science* magazine. On the final evening, August 9, we had a conference dinner in King's College Hall, the same Hall where I had served just three years earlier as master of ceremonies for the banquet honoring Willy Fowler on his 60th birthday. The same portrait of Henry VIII on its wall, and with which I produced a great laugh by intoning,

"We are honored to have a portrait of Willy Fowler painted for this occasion on the wall behind this podium."

How I loved the Cambridge setting. How I would miss it. How I wished Fred and Willy were also there, where, it seemed to me, they belonged more than any of these others. But they now hated the Cavendish Laboratory for its destructiveness toward Fred, and told me so, and did not now want to greet the Cavendish leaders in radio astronomy, especially Martin Ryle, now celebrated with fame of their own, but who had played a role in driving Fred from Cambridge and from his institute.

But tonight at King's College Fred Hoyle was absent, absent at the Institute, now renamed, built by and for him. He had left for good. Not even the statue that would later be placed in its garden to honor Hoyle was there. At this time I wanted very much to see him; and he and Barbara had welcomed us to visit them. In high spirits over the new scientific direction of my work, I asked Annette to promptly pack our things so that we could make a quick drive to the Lake District before finally embarking from Prestwick Airport near Glasgow for return to the USA.

Heavenly Ullswater

I savored auto trips in England as one might savor fine wine. This time I could not but wonder if it would be my last. With a 7:00 am departure from Cambridge we made such good time that stops with Annette were frequent and rewarding. I explained to her my love of England's timelessness. I tried to describe my tension over the paper that I was now planning to submit for publication from Houston. I tried to help her understand why Fred Hoyle had simply resigned the Plumian Professorship. Already in Yorkshire by noon, we paused to visit the parish church in West Tanfield, just off the A1. West Tanfield is, in its own timeless configuration, looped through by the river Ure, on the road past Jervaulx Abbey into the west end of Wensleydale at Leyburn. The stone church spoke with ancient voices as those in England seem to do, with brasses to be rubbed and monuments to those that died in her cause. Then our drive continued through incomparable Wensleydale, to Windermere, and thence up to Ullswater. We reached Fred's new home, Cockley Moor[11], high above Ullswater, at 5:15.

The Burbidges were already there; and a jolly time was shared in the sprawling vernacular Lake District farmhouse, part 200 years old, part 100 years old, and part just 50 years old, all renovated nicely by Hoyle. It was heavenly. Plans were laid for Fred's research visit to Rice in springtime, after which we all dined at Old Church Hotel on Ullswater. Tired but happy we retired to a sweet B&B that Fred had reserved for us in Patterdale, along the flats just south of lakeside. Quiet, stony, and sheep-flocked hills surrounded. Place Fell rose sharply to 2155 feet before us on the near eastern shore. It seemed heaven on Earth.

After bacon, egg and sausage, those especially delicious western-county pork sausages that are quintessentially English, a huge pot of breakfast tea standing ready under its cozy, a large fly buzzing around the window in the August 15 morning air, we dressed for our day with Fred. Our auto retraced slowly the A592 along Ullswater's western shore, past the stone Ullswater Hotel in Glenriding, wherein thirteen years later I would take my final night in Hoyle's neighborhood, past the sharp northeasterly bend along the lakeline to a glen mouth where the A5091 turns steeply upward for one mile and 960 feet of elevation, to Dockray village, then 400 feet more upward along a small lane, finally left on High Row, dead-ending a small way beyond Fred's house. The bright morning sun glistened off Ullswater most memorably in the cool morning air. Geoff and Margaret Burbidge had already returned from the hotel, and we talked until after lunch--mostly cosmology and the state of British astronomy, the inevitable topics when the Burbidges and Hoyle were in conversation. After Burbidges departed, Fred led us on a hike of modest effort in the hills south of Patterdale. From the 1424 feet of Arnison Crag we looked due north, over Patterdale, along the amazing glen that is the south terminus of Ullswater. There can be only a few more beautiful spots on earth. Barbara Hoyle had a teapot set out almost immediately after our return to Cockley Moor.

After tea Fred and I excused ourselves for his office [12]. There we analyzed further the situation on the surface of a white dwarf when the thermonuclear runaway occurs. The thermonuclear explosion is confined to the skin of the star, where hydrogen abounds, and does not penetrate to the hydrogen-exhausted interior. We had already published the opportunity for gamma-ray line astronomy of the radioactivity produced, but now we debated whether an *r*-process-like neutron burst could accompany the explosion. Although we could not be confident in the absence of hydrodynamic models calculated by computer, we decided to go forward with a second publication on that topic. It appeared[12] in *Astrophysical Journal*.

On Saturday, August 17, Hoyle emphatically displayed why this land had been chosen as his and Barbara's home. From Glenriding we trekked up past the Greenside Mine to Glenriding Common near 2000 feet, along Red Tarn Beck to a steepish climb to Striding Edge, across whose crags[13] we picked our way to the gentler slope up to the summit of Helvellyn, at 3113 feet the highest point in the Lake District. Lazy sunshine bathed our lunching at its top. We continued due north along a ridge of summits toward Matterdale Common, one of the more remote places in so civilized a wilderness. From the 2807 peak of Great Dod we descended to High Brow, from which Fred's house lay just below. I will not forget that loop. We talked along the way of our ideas and hopes; but the star attraction was the fells.

Barbara had dinner for us, lamb in casserole. One more glorious day, Sunday, and Fred and I walked alone, thankful for four days of impeccable weather, while Annette stayed with Barbara. We reclimbed to Blencathra's top at 2847 feet, and where we had climbed in 1970 as well. On page 150 of *The Dark Night Sky* I published a photograph, taken four years earlier in 1970, of Fred reading the Ordnance Survey map on this very spot. We talked and talked, of many things, all the day through, and I learned ever better who Fred Hoyle is. I count this one of the high honors of my life, one awarded by no committee, supported by no Dean, challenged by no peer review or referee, simply friendship and time with a great man. It was here on Blencathra that I had the time to describe my excitement over supernova dust suspended in the interstellar medium and being identified by the unusual isotopic signatures that I expected within such dust grains. Fred was very enthusiastic, even laughing once and musing,

"Why didn't I think of that."

He knew that he had been very close to this idea; but never, not once, did he remind me of his original publication in 1970 with

Chandra Wickamasinghe that had predicted supernova dust would condense within the adiabatic supernova-interior decompression that I now reformulated for him. Despite his enthusiasm over the likelihood of presolar aluminum oxides of pure ^{16}O, despite newly intensified interest in that meteoritic discovery of excess ^{16}O that we had discussed in Houston during November, he did not even gently remind me that the prediction of dust condensing within supernovae had been anticipated by Chandra and himself, with no mention of isotopes however. So, forgetting where four years earlier I had first heard that idea, I even described it again to him. This embarrassed me later when I recalled that I had learned it from him. I would submit my first two papers on this new astronomical technique without reference to their paper. I have found this oversight to be true of exciting scientific discovery--that in the flush of new insight scientists often forget the source of a relevant idea borrowed for a new argument. My excitement focused on my prediction that dust produced by this supernova mechanism would bear remarkable isotopic signatures that differ profoundly from those in solar material, and that the consequences of such dust might be found in meteorites. That was totally new, and Fred agreed that this could open a new research window into nucleosynthesis in supernovae.

Incredible Referees

Immediately after returning to Rice University I submitted my discovery paper, now typed and awaiting my arrival, to *Science*. I chose *Science*, a weekly scientific magazine, owing to the wide ranging general consequences of my new ideas. I anticipated broad interdisciplinary interest. I mailed it on 19 August, 1974. Robert Clayton's discovery of the excess ^{16}O had appeared in *Science;* and several papers on the putative superheavy parent of CCF xenon had also. *Science* seemed the right choice. To keep it focused enough and short enough I had concentrated only on the xenon anomalies, with just a word about ^{16}O. The work was intended as a first short paper in what I realized would open a substantial future. I was content to show just that the condensation of supernova dust within the deep interior of the supernova, while it was expanding and cooling, could make possible a reinterpretation of each of these xenon anomalies.

After about 2 1/2 months, a worrisomely long time for such a short and important paper, I received a letter from Editor John Ringle regretting that *Science* had decided not to publish my paper. I was absolutely incredulous! I quickly read the referees' comments with mounting anxiety. My heart pounded. For the first time in my heretofore placid career in nuclear astrophysics, I was to learn something that would follow me through remaining decades. Anonymous referees, referees for scientific publications and for grant applications, fail badly when confronted with totally new interdisciplinary ideas that confront their own prejudices. I was entering a decade during which almost everything that I wrote or proposed in this soon-to-blossom interdisciplinary science would be narrowly and wrongly dismissed, rejected or castigated by some erring referee, often venting his own hostility at having his prejudices questioned. I have stated this opinion strongly. I mean it strongly. I could almost feel my backbone toughening as I read these first reviews:

From Referee 1. *This is a speculative paper which should be exposed to public scrutiny. I am skeptical for one important reason-*-followed by his own incorrect reasoning for why dust could not nucleate in supernova ejecta. Nonetheless, his good judgment had recognized that this idea did need publishing. The editor should have published it on the basis of this positive evaluation by a leading scientist, A.G.W. Cameron, who later identified himself to me.

From Referee 2. *This author does not disclose the fact that fission tracks have been found in grains from meteorites. The probability that these fragile grains formed in the circumstellar envelopes of exploding stars and then floated around in the interstellar gas for billions of years without reaching track annealing temperatures is very small indeed.* The underlined (by me) words really stuck in my craw! What did this referee know of those "probabilities"? *Probability* is a scientific word, not a hunch! What he was admitting is that my idea seemed implausible to him. Nor did he realize that interstellar dust did not get hot. But I was getting hot. The fission tracks had not even been the main idea that I was giving to the world. So the referee seems to imply without actually saying so that I have overlooked fission tracks, thereby disqualifying my argument. His appeal to the word *probability,* without any quantitative assessment thereof, was unscientific.

From Referee 3. *The possibility that all these kinds of material could contain them (^{129}Xe-carrying and iodine-bearing grains) in the incompletely melted circumstellar grains no matter where you look in meteorite collections is so remote as to be an absurdity.* How does this referee know the "possibility is remote"? These are the words of referees who want to reject an idea because it counters their beliefs, so they appeal to undefined probabilities, remote, very small, skeptical, speculation without ever assigning numbers to back up their choice of words.

From Referee 4. *This paper has little to recommend it. It is an unscholarly and most casual document which purports to present a "rather revolutionary picture", but instead presents a speculation that is not deeply analyzed, that ignores much of the data and existing literature and does little if any predicting , although it is stated that "these ideas do have considerable predicting power".* But later, after really building up a head of steam with abundant small criticisms, he continues, *I could go on but it isn't worth it. A good term paper would have to be much better than this. The author was obviously shaving and got an idea which he wrote up while admiring himself. This is seen in the illuminating self reference,* "These ideas were first publicly presented on 6 August 1974 blah, blah, blah.."

I slammed the reviews onto my desktop and sat there, trembling with mounting indignation.

I feel the same reaction today, thirty-five years later, that I felt initially. My indignation has only been enriched by a sense of humor during the subsequent decades. In the first place these reports appear at first glance to be, except for the vitriolic and *ad hominem* remarks in the fourth report, what one would expect from a paper that truly had little merit. My paper is made to seem really bad, and the referees each include at least one expert-sounding statement. It takes careful and skeptical reading to see that there is not a single criticism that is quantitative and verifiable. But how is an editor to know? Any editor would reject a paper getting this collection of reviews. Part of my anger was that the referees understood that they were scuttling my paper. They were attempting to write plausible sounding criticisms that they knew would cause the editor to reject my paper without ever stating clearly why my ideas were either wrong, or not original, or uninteresting. These are the only three reasons for rejecting original work in science.

The first review was written by Al Cameron, an astrophysicist whom I admired for his pioneering work on nucleosynthesis in stars. He and I had been the "new boys" in 1957. My admiration for Al has grown over the years at the straightforward clarity of his opening sentence. But why need he, in a sensitive review, add remarks that express only his own lack of clarity about how dust can condense from supernova gases? A referee's report is no place for such things, for they only confuse the editor, causing him to interpret the referee's blind spot as a possible deficiency in the work of the author. Hoyle and Wickramasinghe had already published a paper in *Nature* showing why dust condenses within a supernova, and this referee should have known that their paper was more significant than his hunch to the contrary.

The fourth referee's identity was obvious to me at once, although I will not name him. His review reveals his love of spinning words as well as his sense of being a master of the universe, or at least of seeing himself as master of isotopic cosmochemistry. His anger and *ad hominem* fury has no valid place in science, and I am surprised that the editor did not simply discard his purported "review". This reviewer could not be a younger scientist, for they constitutionally welcome new ideas, whereas referee 4 showed fear and panic that some of his publications might have erroneous conclusions. This is not pretty, but science too has its ugly personal side alongside its sublimely beautiful method.

In his books extolling the beauty of evolutionary science Richard Dawkins lampoons the creationist who makes rhetorical remarks such as, "I can see no reason why the caterpillar would turn spotted green to match the plant on which he grows other than the hand of The Creator." Dawkins points out wryly that the truth is exactly what the would-be wise man has admitted; namely, *he can see no reason*. So, inspired by the creationist impulse, he invokes the hand of the Creator. In similar vein the astrophysicist referee negated his clear and positive opening remark with his own short-sighted worry, a worry that instead of being a basis of rejection was an expression of his own ignorance. Today astronomers know without doubt that abundant dust indeed does condense from the cooling gases of the supernova explosion. Hoyle and Wickramasinghe had foreseen this correctly and I utilized their reasoning as a basis for my paper. So Cameron's inessential and unknowledgeable remark took on the role of undermining his support that my paper should be published. I have witnessed this phenomenon countless times during my career, concluding that, in science as in law, gratuitous remarks and questions are better omitted in favor of the actual duties of the referee and juror.

Fomenting my Change

When the first volcanoes come, who knows where they will lead? Lava flows beneath the earth, this way and that, forging mountains of basalt.

These referee reports accelerated a change that was already underway within me. My personal challenge was whether I would quietly withdraw or get in the ring and box. For that reason those reports played a more significant role in my own life than on the course of science. This is my reason for dwelling on them. Hear me out.

Anonymous referees 2, 3 and 4 had very obviously been selected from well known meteoritic chemists, evidently leading scientists of that field, ones whose works have influenced the course of science.

Each betrays itself in one of the most common inadequacies of scientific specialists. Each preoccupies itself excessively with its author's own special expertise; namely, how well does an idea explain chemically the host of facts that the chemist has measured in meteorites? Repeatedly down the years have I seen this very short sightedness discard intellectual interest in a new idea by preferring to argue over how well that idea fits a host of unexplained details, usually chemical details that are primarily the referee's own bailiwick. Resistance may even blind him. Gratitude for expansion of the theoretical idea base might not even be mentioned. Turning his report toward his own expertise enables each referee to appear to be penetrating and simultaneously to turn attention toward his own measurements. At scientific conferences I have repeatedly watched the same human weakness plague questions following verbal presentations of a new picture or mechanism. Instead of asking the presenter to clarify his idea and topic, the questioner interjects bits of his own chemical knowledge, followed by asking the presenter, "How do you explain that?" Referees frequently abdicate their primary responsibility. With such an approach any astrophysical idea can easily be denounced by a referee. This can be done without ever coming to grips with whether the idea is new and might be useful. Such were the cases of my referees.

Lest I offend let me hasten to add something about the fields of meteoritics and of science editing. Meteoritics is no different than other scientific specialties except in that for decades, essentially pre-Apollo, its community had been more insular than most. It was dominated by chemists despite the science itself concerning astrophysical events about the origins of stars and solar systems. This occurs because the chemists make the measurements. They had grown accustomed to building astrophysical models from their measurements, like the blind man building an elephant from feeling only his final foot of trunk. The more expansive manner of thinking common in astrophysics had not made a significant presence among meteoriticists in 1975. That lack was very obvious as I attended conferences on meteoritics. For refereeing decisions by journals, good editors are aware of this short-sightedness of specialized referees. Their problem as editor, no matter how astute and well meaning they may be, is how to take this propensity of referees into account. Referees are excellent scientists, but they may fail to ask, "Is this an important new idea?" If the referees do not answer that question clearly, there is no way that an editor can be expected to recognize the value of a paper advancing a new picture.

Unfortunately, I too had committed strategic errors that inadvertently had stimulated the negative reports that I received. I could have avoided the entire brouhaha by writing my paper differently. I had not considered the frame of mind of the referees to whom the editor would send my paper. I had not paused to consider to whom the editor would send my paper for review. In astrophysics I was accustomed to reviewers recognizing that their responsibility in evaluating a work is to answer three simple questions: (1) Is it new? (2) Is it interesting? (3) Are its arguments logical? If the answers are *yes*, the paper should be published. My strategic error in judgment was choosing the venerable xenon-gas anomalies for my first paper on cosmochemical consequences of the existence of supernova dust and of its incorporation into meteorites. It had been naive of me to not realize that so rich an experimental topic would engender severe resistance to a new, unproven, and perhaps ultimately irrelevant approach to what that community of experts called *xenology*[10]. Furthermore, with so much data and so many published papers existing on the xenon isotopic anomalies, the editor was certain to send it to two or three of those who had measured and interpreted their own xenon data; and those authors were certain to carry vested interests in their own interpretations of that data. They would feel critical about a new physical mechanism that carried disturbing consequences for xenon isotopes. So my error had been to make the paper seem to concern xenon within meteorites instead of its new astrophysical idea; namely, that dust condensing within the supernova material will condense from matter carrying extremely unusual isotopic patterns, and that dust may perhaps either be found within meteorites or leave a chemical memory of its unusual isotopic composition in solids made from interstellar ingredients.

That error came back to bite me. My baby was being thrown out with the bath water. But I had done what I had done; and now I seethed with fury that it had not received a single proper scientific evaluation.

I mailed an indignant letter to *Science* editor John Ringle. On November 27, 1974, I resubmitted the paper with a few clarifications of the few appropriate criticisms. What I could not do without withdrawing the paper entirely was to refocus the attention of the referees away from the question *how possible is Clayton's interpretation for explaining the excess* ^{129}Xe anomaly? to the question *is supernova dust that carries within it large isotopic anomalies from extinct radioactivity of interest for cosmochemistry?* I could not rewrite my paper with a new topic without it becoming a new submission. But there exists a problem with withdrawing the paper; namely, once a paper is withdrawn or rejected, the scientific priority for any important result or idea contained within it loses its documentation. Indeed, other scientists actually did submit my ideas within a year as being their own ideas! Science is sometimes like that. Insightful ideas are sometimes published by reworking ideas first advanced by others in unpublished form. Such loss of priority became a significant consideration for me. I had absolutely no doubt that my ideas would prove to be very important. I wanted to publish a revised paper in *Science* in order to display its original "received date", which journals routinely publish along with the papers, documenting the date on which they received the paper for publication. A *Science* postcard acknowledged receipt of my revised version on 4 December 1974.

I fumed so with an anger that could not be satisfied solely by the tame process of resubmitting my paper for further consideration. Never had I been so indignant. Indignation was a new emotion to me, and under its stress an original counteroffensive surfaced, forced up like lava from the eruptive pressures. *I would expose the referees!* I would circulate their objections in the mail, widely, for all to see. This idea was not simply revenge on the referees, focusing primarily on the fourth; it was much more. By making a *cause celebre* of my controversy, the origin of the ideas could not be forgotten. This tactic had been used by Fred Hoyle and would be used by him yet again. Fred told me this amusing story concerning Cornell astrophysicist Thomas Gold's vitriolic public defenses of his 1960s arguments that the surface of the moon would prove to be very dusty. To Fred's question,

"Tommie, why do you provoke them so over your idea?"

Gold replied, "Because if I do not make them angry they will try to take the idea away from me."

On 3 December 1974 I mailed to about twenty leading scientists a letter beginning with the salutation "Dear Colleague" and accompanied by xerox copies of each of the four referee reports. After exposing the weakness of each report I wrote specifically of the fourth:

> *This referee seems to dislike me. Nonetheless, since he is one of our colleagues, I thought you might like to join me in playing America's favorite guessing game: What's My Line? Will the real referee please stand up! I have submitted the enclosed revised version to Science, and I would welcome your candid advice concerning whether I should continue my intent to publish.*

The first referee, A.G.W. Cameron of Harvard, responded to me promptly on January 3, openly over his signature. He was quite reasonable. I never heard from the others, confirming their desire to remain anonymous. None wished to claim his report. But you can be sure I heard many opinions identifying the fourth referee. Each offered the same identification. He would perhaps never forgive me for ripping off his cloak of anonymity. As Fred Hoyle put it to me on this issue, "The offender never forgives."

The second review by *Science* did not take nearly so long. On 21 January 1975 I received the letter from the editor, John Ringle, saying that they could not publish my paper.

"Unfortunately," he wrote, "the referee has advised us against publication and we must abide by his advice".

I could understand that, so abysmal was the new referee's grounds for dismissal. It falsely accused me of resurrecting an old and discredited idea by Eberhardt and Geiss. This too was scandalous, because my idea differed fundamentally from the one advanced by Eberhardt and Geiss. The adjudicating referee then applied the *coup de grace*, intoning:

> *The problem with this paper is that it is too superficial to treat the question fully in light of all available data. If it did so, it would be too detailed (and long) for Science.*

I was speechless at all three grounds: the first an ignorant accusation that would nonetheless convince an editor; the second that it did not treat all of the data; and the third that (although it might be possible to treat all the data) it would then be too long (and too detailed) for Science. My fury escalated. Had this lot never heard of publishing a new idea?

I wrote another "Dear Colleague" letter to accompany wide distribution of this adjudicating review. I mailed these around to most significant figures in isotopic cosmochemistry. It said:

> *My paper "Extinct radioactivities: trapped residuals of presolar grains" was again rejected by Science. The referee's remarks are so inane that you ought to read them. Such incompetence or maliciousness should be exposed in science just as much as in society. I am reluctant to believe that this distinguished scientist has willfully misrepresented my paper. But it is hard to see how he could fail to see that the stepwise release experiments are consistent with my picture. To claim that a full account would be too long for Science is beyond comprehension.*

> *I do not know which picture is right. My approach is that of a theorist presenting a new picture for interpreting our world. It is up to experiments and reasoned arguments to pin down the truth. The referees have consistently forgotten this, pretending instead that it is my obligation to prove the correctness of my new picture.*

For the remainder of my long career of refereeing, I personally refused to do so anonymously. I asked that each of my reports be mailed to the author over my signature.

During the ominously long silences that punctuated the reviewing I had been busy establishing additional evidences and other manifestations of my idea. I was determined to add coals to the fire. Firstly I had immediately reworded the *Science* paper and resubmitted it to *Astrophysical Journal*, where it was published[14] without a fight in August 1975. In its concluding paragraph I documented the two public presentations in Cambridge of its central idea. Secondly, I began outlining a second paper based on the general class of supernova-produced radioactive nuclei. Thirdly, as an Executive Committee Member of the American Physical Society's new *Division of Cosmic Physics* (now called *Division of Astrophysics*), I was expected to arrange invited speakers for cosmic physics sessions held at the American Physical Society meetings. The slate of speakers for the February 1975 Anaheim meeting, as prepared by me, included myself. This somewhat irregular step may have seemed rife with conflict of interest; but I saw it otherwise and did not hesitate. I submitted my slate of speakers to Phyllis Frier, Chair of the Division of Cosmic Physics. She saw nothing wrong with me on the program. On 24 October 1974 my letter of invitation for my own talk arrived from W. W. Havens, Jr., Executive Secretary of the APS. So on page 79 of the January 1975 issue of *Bulletin of the American Physical Society* appears the following slate of speakers :

A1. "Measurement of the gamma ray spectra of galactic and extragalactic objects". Robert C. Haymes, Rice University (30 min.)

A2. "Is the universe open?--is deuterium cosmological?" David N. Schramm, University of Chicago (30 min).

A3. "Recent observations of some objects with large redshifts", E. Margaret Burbidge, University of California, San Diego (30 Min).

A4. "Extinct radioactivities: trapped residuals of presolar grains14". Donald D. Clayton, Rice University (30 min).

Perhaps appropriately, this all occurred in the Disneyland Hotel, site of the February 1975 APS meeting.

The new research paper that I began writing immediately would be promptly submitted to *Nature*. I planned to emphasize isotopic effects of radioactive nuclei in interstellar dust having nothing to do with xenon. My strategy was to promptly illustrate many consequences of my idea without burdening them by connection to xenon's vast lore.

I was extremely well prepared for identifying extinct radioactivity in supernova dust. That scientific preparation had begun during my research papers with Bodansky and Fowler and with Woosley and Arnett on the explosive burning of oxygen and silicon in supernovae. Those explosive burnings synthesized the lion's share of very abundant radioactive nuclei. I was at the right place at the right time, rather far ahead of the scientific world in joining technical nucleosynthesis knowledge with growing awareness of experimental data from meteorites.

My initial case advanced the possibility of finding pure ^{22}Ne gas within supernova dust. Normal neon has three stable isotopes, A = 20, 21 and 22, with the ^{22}Ne being normally the second most abundant. The hint of purer ^{22}Ne within meteorites had already been published[15] by David Black, who discovered a gas component that he called neon-E and that was quite enriched in the ^{22}Ne isotope. I treated the ^{22}Ne excess as being atoms left within interstellar dust by the beta decay there of radioactive ^{22}Na. The ^{22}Na would be condensed into the supernova dust along with stable sodium during the first year of its expansion, and it would decay only later. I again had been prepared two publications that emphasized the explosive production of radioactive sodium, designated ^{22}Na, in stellar explosions: firstly the paper with Hoyle on gamma-ray lines from novae; secondly, the paper with my Rice student Mike Howard on ^{22}Na production in the shocked helium shells of supernovae. The ^{22}Na nucleus is radioactive, having halflife of 2.6 years, and it decays to stable ^{22}Ne. So if nova dust or supernova dust were to condense within their rapidly cooling expansion, prior to later mixing with the interstellar gas, the sodium-bearing dust would contain isotopically pure ^{22}Ne after the ^{22}Na decay inside the newly condensed dust grain. I suggested this cause of Ne-E in meteorites, and I chose "22Na, Ne-E, and extinct radioactive anomalies" for its title.

Fortunately this paper was sent for refereeing to David Black, the discoverer of Ne-E, and he accepted it. It was published[15] in the September 1975 issue of *Nature*. David Black later evolved into a big scientific presence by moving beyond his cosmochemical beginning, first into the astrophysics of star and planet formation and subsequently as Director of the Lunar and Planetary Institute in Houston, near Johnson Space Center; but his discovery of Ne-E remains of high historical importance. Although my explanation of the cause of Ne-E was not immediately accepted, Black saw that its idea needed to be among those of the cosmochemical lexicon, and he saw this despite this situation using the identical physical mechanism that I had used for the doomed xenon paper for *Science*. More sensitive experiments in subsequent decades slowly revealed ever higher degrees of purity of the ^{22}Ne in some meteoritic samples, which slowly caused a favorable shift to my idea.

But neon is nonetheless another noble gas. To avoid having stardust fossils regarded solely as a noble-gas effect, I searched carefully for important applications within refractory elements, those elements that participate in the basic crystal structures of solids. So I turned to an old nuclear friend, one of the six isotopes of calcium, namely ^{44}Ca, which was synthesized in the supernovae in the form of radioactive ^{44}Ti, having a halflife near 50 years. I argued that if the cooled supernova interior preferentially condensed the element titanium, the decay of the ^{44}Ti within that dust could be detected in the form of calcium having enriched abundance of its mass-44 isotope. This very suggestion did provide the smoking gun for identifying supernova grains about two decades later. Secondly, a radioactive isotope of calcium, ^{41}Ca, which is synthesized in the explosive supernova shells of oxygen along with the most abundant two stable isotopes of calcium, would decay later within interstellar dust to ^{41}K, the heavier of potassium's two stable isotopes. Based on that fact, I predicted ^{41}K-rich potassium within calcium-rich supernova dust. Both of these examples became important in the decades ahead as this new type of astronomy received experimental support. Wary now of possible myopic misunderstandings, I added at the end of that *Nature* paper a final sentence to emphasize that my focus was on the physical idea rather than on its manifestations: "Other possibilities suggest themselves once the key idea is in mind."

I kept pushing this astrophysical idea until it became absolutely clear that I had established a new and fruitful idea for cosmochemistry and for meteoritic science. Never in my life had I been so resolute in carrying the argument to the opponents of an idea. I felt at times like a general assembling his troops and weapons for an assault. The verbal presentations[4] that I began annually in 1975 at meetings of the Meteoritical Society and at the Lunar and Planetary Science Conference were aggressive, and delivered with an outspoken assertiveness and confidence that was designed to wither feeble opposition. I intentionally became much more combative than in my scientific life prior to age forty. This was not simply a staged show, but rather a pervasive change in my public persona derived from some intense maturation within me. There appeared counterparts within my personal relationships with others, which became more forthright.

A thought that came to me during this work is that science needed new scientific words related to specific formation mechanisms for differing components of *interstellar dust*. The wishy-washy term interstellar dust was much too vague for isotopic science. What's in a name? A great deal as it turns out, because names related to physical mechanisms not only clarify the confused terminology that I have almost continuously heard at scientific meetings, but they have the added bonus of associating for posterity the mechanism and its founder. One can in discussing stellar energy hardly say *the CN cycle*, for example, without thinking the name Hans Bethe. And I had witnessed at Caltech how Murray Gell-Mann had so brilliantly frozen his name onto the coming standard model of elementary particles by naming its basic constituents *quarks*, having *strangeness* and *color*. When the name becomes profoundly important, so does its inventor. So new words lurked among my volcanoes.

s Process Nuclei in STARDUST

The 1974 rejection by *Science* occurred during one of the intensely prolific two-year periods of my career, as did the meeting of the American Physical Society in Anaheim. Annette and I had returned from Christmas holidays in the Black Forest on January 7, 1975, one month prior to that APS Meeting. In the interim I discussed with my research student Richard Ward our paper calculating the abundance pattern of the *s*-process isotopes within the element xenon. That undiscovered specific pattern among the nine isotopes of xenon may remain physically isolated from the other isotopes of xenon by condensation of dust from stellar gas in red-giant stars. Because the atmospheres of evolved red-giant stars had been seen by spectroscopic astronomers to have been enriched in *s*-process nuclei, with barium being the easily observed abundance indicating the size of the *s*-process overabundance in the stellar atmosphere, my inference was that any dust that condensed in the stellar wind carrying gaseous atoms away from the surface would trap xenon atoms having the *s*-process pattern. That idea was to be the topic of the third of three quick new papers on xenon isotopes trapped in interstellar dust grains. Surprisingly, it would be the first nucleosynthesis pattern to be unequivocally identified in meteorites. Nonetheless, but amusing only in retrospect, this xenon paper was also rejected in 1975. I had mailed it to a cosmochemistry journal reasoning that it would be cosmochemists who would eventually discover the *s*-process xenon. I could not add another rejection to the list of wars that I was currently fighting. One war was enough. I would not resubmit it16 for publication until the predicted stardust was actually discovered three years later.

Dust that condenses in stellar winds needed a specific scientific name. Calling all of these separate components of *interstellar dust* by the single name interstellar dust was simply too vague. *Presolar dust* was also no good as a name because most of the interstellar presolar dust was not a condensate from a stellar wind. In a bold stroke I named all dust that condensed thermally within cooling atmospheres of stars *stardust*. What could be a better name than that? In early papers I routinely capitalized it, STARDUST, to make it clear that it is a scientific word with defined scientific meaning rather than simply a poetic word.

STARDUST literally falls to earth. Its delivery is poetic. Five billion years ago, before our solar system had come into existence, these solid pieces of stars had been suspended in the interstellar matter along with other kinds of interstellar dust and much more abundant gaseous atoms. The STARDUST had condensed out of the hot but cooling gases of dying stars. These ejected particles then had begun lengthy lives among the interstellar gas and dust. Eventually the STARDUST-bearing interstellar gas joined a massive molecular cloud, a cold, dark interstellar mass that then chilled the STARDUST for a subsequent 100 million years or so at a temperature of 10-50 degrees above absolute zero, about -400 degrees Fahrenheit. At a time of escalating crisis a dense portion of that cold, dense cloud fell onto itself under its own weight. It collapsed to form the sun and a warmed disk of debris orbiting it. Some of the STARDUST had also fallen into that disk. There rocks formed, and then aggregated into small planetary bodies. STARDUST was trapped, much of it unharmed, within these first rocks, which grew by accretion of more cold matter and of each other into small planetary bodies.

Some of these survive today as asteroids and comets in our present solar system. Within them the STARDUST resided for about 4.5 billion years. Then much more recently, about a million years ago on average, pairs of these asteroids in orbits about the sun collided. When they did so, countless small rocky fragments spewed into interplanetary space, launched by their velocities into new orbits about the sun. These orbit for about a million years until some of those rocky fragments fell to earth. We find these fallen asteroids around the world, and call them *meteorites.* They are most easily found in deserts and in Antarctica, where running water can do them no damage. A few are even witnessed in their fall to ground, great fireballs streaking toward ground. These are gathered, classified, and stored in museums of meteorites. There the STARDUST sat within the meteorites until the 1970s, when mankind first began to search for the causes of anomalous isotopic material within them. My papers predicted what the isotopes in STARDUST should be like, and chemists began actively searching for them. In the 1980s they would be found. They would be scientific manna, falling by grace to ground and nourishing mankind's knowledge of natural history. The stardust gems themselves are literally falling fragments of stars.

A related poetic coincidence lubricated the discovery. A large infusion of scientific investment by NASA into cosmochemistry laboratories to enable them to better study samples of the moon, first returned in 1969, also empowered them to discover STARDUST. This entire providential good fortune happened within my lifetime, allowing me, for all mankind, an unparalleled opportunity.

A Blizzard of Ideas

My fortieth year (1975) was a prolific one in my life. Ongoing battles may sometimes drain us of energy; but somehow I was bursting with science energy despite, or perhaps because of, my battles.

At this time my imagination was captured once again by measurements by Raymond Davis, Jr., of a deficit of neutrinos coming from the sun. The theory of thermonuclear power in stars, a theory that I had explain well in my textbook, required huge numbers of neutrinos to be passing through every small area of earth. Almost 100 billion neutrinos pass through one's fingernail each second! They are unfelt because they interact so incredibly weakly with other nuclei that they simply pass through. Only once in a great while does one of these neutrinos actually cause a reaction. Davis's scientifically romantic experiment was located deep in the Homestake Mine in Lead, South Dakota, where cosmic rays could not penetrate. On the surface of the earth the cosmic rays cause many more reactions than do the solar neutrinos. Davis had calculated that the radioactive argon atoms that he could count in his huge tank of chlorine would be made at the great depth of the mine only by neutrinos from the sun reacting with the chlorine nuclei. The mine was too far underground for other particles to reach.

Davis's experiment had been in my thoughts for many years. Its pull on nuclear astrophysicists was very strong. I quickly found that my new frame of mind seemed to induce radical leaps within me. I sensed myself thinking outside the box. Two bold new ideas occurred to me that seemed to provide possible explanations of the observed deficit. Both challenged the accepted scientific wisdom in a manner similar to that of my stardust ideas.

The first idea lay in statistical physics. Supported by our computer models of the sun, I submitted to *Astrophysical Journal* my paper with Ray Talbot and my students Mike Newman and Eli Dwek explaining the deficit of solar neutrinos by a non-Maxwellian distribution of relative velocities for the interacting nuclei within the sun[17]. Eli had worked with me for his Master's thesis on a mathematical representation of a progressive deficiency in the rate of the most energetic collisions at the solar center. Those collisions occur with a continuous spectrum of collision energies in thermal gases. I had seen what I thought to be a physical possibility within many body-physics when the interparticle forces have such a long range that they involve more than two particles in a collision, such as the electrical repulsion that interacting ions must overcome in thermonuclear reactions. The paper was accepted for publication, but its radical motivation never became compelling.

On January 27, 1975, we submitted a second paper[18] to *Astrophysical Journal* explaining the solar neutrino deficit in terms of the existence of a planetary-mass black hole at the center of the sun! The black hole would have existed before the sun formed around it, and its gravity would have stimulated collection of the sun's mass. Owing to the angular momentum of matter trying to fall toward that black hole, it was not possible that it fall directly into the black hole. It instead formed a disk orbiting the black hole, which kept growing by continuing infall until it became our sun. I had been formulating that idea with Ray Talbot, my postdoc supported by the Air Force Office of Scientific Research, and my student Mike Newman. We constructed computer models of the sun to evaluate the central temperature and energy balance within the sun. A black hole too massive would have led to the sun's disruption before today, and a considerably smaller black hole would have negligible effect. Its mass would have to have initially been comparable to that of a planet. This new idea was, I believed, the first paper to be published of a model of a star having a central black hole.

The black hole at the center makes for a hot disk that actually liberates heat to the star as matter spirals on shrinking orbits into the black hole, slowly making the black hole larger and more massive. I had to invent a physical model for the accretion power liberated as heat at the solar center. I decided that the matter must accrete slowly enough that the radiation pressure caused by photons from the hot disk did not prevent further accretion. We found that if a significant fraction of the sun's total power results from the gravitational settling of matter into the black-hole horizon, the solar neutrino puzzle could be solved. This new work would also be accepted for publication and remained for almost two decades another possibility for the mystery of the sun and stars.

Each of these solar-neutrino papers were, in my opinion, highly original works, and ones that had been competing with my raging battle over supernova dust for my attention. Fueling my outrage was the amazing fact that I could publish a paper about a black hole in the sun among astrophysicists easier than I could publish a paper about supernova-condensed dust in the interstellar medium among chemists! Something was very wrong in cosmochemistry; and I resolved that, once my stardust papers were finally published, I would personally and forcefully participate with the cosmochemistry community to introduce astrophysical perspectives into it.

Also on the negative side, I learned that scientific work can stimulate hard feelings between old friends. When my friend John Bahcall, who had shared with me the first two years of the new Caltech program in nucleosynthesis, later wrote his technical monograph, *Neutrino Astrophysics*, he cited Stephen Hawking as source for the idea that a black hole could solve the sun's problems with neutrinos. This was pure hero worship. I wrote to John to object that he should not cite a celebrity of science for a casual remark that did not even address the effect of a black hole in the sun and that was so obscurely reported as a casual remark as to be virtually without influence. It is not a casual comments that lead to science progress, I added, but rather a competent evaluation of the impact of a new idea, and why. John did not like my criticism, so this strained our friendship slightly. But such things occur in science as we defend our contributions.

Both of my suggested solutions of the neutrino problem lie

abandoned today because successful neutrino experiments would later place the solution of that puzzle at the feet of the neutrino itself. There exist three kinds of neutrinos, but unknown in 1975 was their quantum oscillation from one type of neutrino to another. Of these three only the electron neutrino could trigger Davis's experiment. By the time the original electron neutrinos had left the sun they had changed their identity. This chameleon-like change became the accepted resolution of the solar neutrino problem.

But what I realize today is that neither solution submitted by me in 1975 would have been advanced by me five years earlier in my personal evolution. The personality eruption occurring within me during 1974-75 led me far outside the box of conventional thinking. I had heretofore operated within that scientific box, for there existed ample room there too. In that spirit, these two solar neutrino papers of 1975 were much more significant within my own scientific growth as a man than within the world of knowledge. In fact, the fresh idea of the black hole at the solar center so captured my imagination that I made it the basis of a scientific novel that I would publish ten years later[8].

Still more new work simultaneously contended for my attention. In January 1975 I had been able to return to the *Astrophysical Journal* editor the revised version[15] of my paper "^{22}Na gamma rays from young supernovae". This work continued my ten-year-old goal of laying out the prospects for a gamma-ray astronomy that could test explosive nucleosynthesis in supernova explosions. Radioactive ^{22}Na became the isotope involved in two new 1975 applications to astrophysics. The same week's outgoing mail carried the last of the photographs for publication in my mid-career memoir *The Dark Night Sky*. And on 24 January, 1975, just as the *Science* rejection letter arrived, Annette and I drove to Dallas in order that I might serve as best man in the wedding of my brother, Keith. I took along my Cambridge tuxedo, tailored in 1967 for the St. John's College Ball, for the 8:00 pm wedding in Bay Shore Presbyterian Church, followed by reception at The Hilton Inn. Throughout this family affair for that brother with whom I had played through thick and thin, my mind spun alternately to memories of our childhood, my divorce from Mary Lou, my wedding to Annette in Germany, and to my rage over the referee's rejecting reports. So I laughed and cried at the same time.

A Trip to Disneyland

1975 was still a time when driving across the country seemed romantic, especially with a young German wife who found it so. It was also just right for a tennis player with a new flair for adventurous science and a glamorous image. My yellow Fiat Sport Spider convertible, purchased during my first sabbatical in Cambridge in 1971 when Annette had lived with me as a language student, provided a visual symbol for this romance. We loaded the cooler with minimal road supplies and left Houston. Dallas, my parents' house, and my brother's wedding, was to be just the first stop. Disneyland was our goal, with many diversions along the way. I did not question whether taking this auto trip at this time was sensible or not, as I would have were I ten years older. My spring-semester teaching load was especially light, so it was academically possible to get away. It is Rice's policy to allow light academic loads for star researchers that also become star teachers of their specialty.

At 4:00 pm on the afternoon of the next day after the wedding, we pulled away from my parents' home on Melshire Drive, just north of Forest Lane in north Dallas. We breezed happily along US 287 with our windswept hair and sunglasses to steak dinner in Wichita Falls, where evidences of that great tornado that had recently swept down tornado alley through Wichita Falls lay scattered about. Annette fancied my hair long, in the fashion of the times, albeit for the most part for men twenty years my junior. And she moved its part from its lifetime right side to new-age left. Like a Viking astride his ship's poop deck, I sped along with shoulder length blond hair[19] blowing in the wind, getting too much sun, hirsute with chin goatee and lengthy mustache, my contact lenses hiding my bookish interior, with a beautiful but often scantily clad young woman at my side. Hotel managers sometimes smiled knowingly during our registrations. I enjoyed this at the time but it is not a comfortable recollection today.

From an overnight at Childress, the gateway to the panhandle, we flew along the highway to lunch in a very attractive roadside park near Santa Rosa NM. I drove too fast. We ate rye bread, the dense Volkornbrot style loved by Germans and by me, with cheese and fruit while we gazed northwest to the Santa Fe Mountains beside Las Vegas NM. Then more speeding past Clines Corners, where I had bought snow chains for my first visit to Phil Seeger in Los Alamos, to our 5:00 pm arrival at Hotel Winrock in Albuquerque.

James Renken, a fellow graduate student from Caltech's Kellogg Lab, had invited me to revisit Sandia Laboratories to continue their hopes of designing a gamma-ray-line telescope having much more sensitive energy resolution than the sodium iodide scintillators used by Haymes. My first visit had been in May 1974, when I presented the physics colloquium entitled "The Astronomy of Radioactivity". Basically it came down to management's decision at Sandia concerning to whom the research dollars would flow. Jim hoped I could help the flow into a gamma-ray telescope. I had not seen Jim since graduate school, and he had been at Sandia Laboratories engaged in diverse defense related technologies. I gave another talk there and worked all day, while Annette enjoyed the swimming pool. I stressed the advantages of the scientific yield they might obtain by using germanium gamma-ray detectors. Dinner followed at La Placita on Albuquerque's old town square. On the next morning I worked only half the day in order to promptly hit the road again. But I assisted with the writing of a prospectus for a study of gamma rays from novae, a goal based on my recent paper with Hoyle, and for a study of the emergence delays for gamma-rays from supernovae, as described in my 1974 paper on ^{57}Co lines. Twelve years later their instrument would be among the first to detect the gamma lines from supernova 1987A.

By 12:15 on that last day we were rolling on to Flagstaff. We took the memorable drive through Oak Creek Canyon south of Flagstaff, standing out frequently to walk to better outlooks. It is a canyon of remarkable beauty. Near Prescott, Arizona we surprised an eagle with roadkill as we flashed into view over the top of one of a sequence of undulating hills affording considerable driving sport. The eagle took flight along the roadway, and we gained on him as

362

his great wings, slowly flapping, seemed to span the entire two-lane highway, and then he rose up and was away. It left me breathless. I love wildlife. January snow lay on the mountains, and we stopped to hike upward through the snow until exhausted by the heavy going, panting, with sweaty bodies. With such fun, and with what Thomas Wolfe would describe as a master-of-the-universe feeling, I put aside my anxiety over the rude treatment that supernova dust had received from referees.

As I spoke with Annette to again clarify that battle with the referees for *Science*, a battle that I made clear I intended to win, I realized that I needed to clarify another sensitive situation. The secrecy about my two sons had now gone on for five years, since the first occasion on which I had met her parents. *This is a bunch of damn nonsense*, I thought to myself. The same backbone that I was surfacing for the science battle now asserted itself in this important matter of family.

"We need to tell your family about Donald and Devon", I declared. "In fact I think we should arrange for them to visit Wutöschingen."

After a noticeable silence Annette agreed that this could now be done. Her family could not object, not now after three years of marriage. I reminded her that since Mary Lou's paralysis my boys needed me more than ever. Together we agreed on a plan to invite Devon, whose morose side continued to worry me, to join us the following summer in Cardiff. We had agreed to spend much of that summer there to be close to Fred Hoyle and Chandra Wickramasinghe, two pioneers of the origin of interstellar dust. After some discussion, Annette agreed to that visit too. *What a relief!* I drove on with a happiness of spirit and a greatly lightened burden on my heart.

Finally in Disneyland Hotel, we consumed the first two days in fun-loving tourist activities. This was for Annette, who wanted to see those famous tourist sites. So I abandoned my habit of attending science sessions continuously. On Thursday we visited Movieland Wax Museum & Museum of Living Art, and I found myself surprised to have such a good time. Annette loved it. On the next afternoon we toured entertainment Mecca, Disneyland itself.

That was undeniable fun. Willy Fowler had suggested we meet for dinner in Shipyard Inn, which we did, and Dave Schramm, one of the speakers within my session next day, dined with us. Willy invited Annette and me to come after the meeting to their house on San Pasqual in Pasadena in order to spend a couple of days with Ardie and him. I loved Willy for that. Fun with a great and dear colleague was my kind of fun, science fun.

The session of invited talks on astrophysics occurred on Saturday, February 1. There was a large audience, certainly not owing to my talk but rather owing to a rapid growth among physicists of excitement over cosmological issues. Each talk was enthusiastically received. Dave Schramm was always an engaging advocate for big-bang nucleosynthesis. He reviewed how the abundances of the isotopes of H and of He, both isotopes of both elements, provided strong evidence in favor of a Big Bang. Dave, one of the great operators in astrophysics, wasted no opportunity to set himself up in public talks as the champion of the Big Bang. I never knew a scientist so determined to promote himself. He established a surprising degree of fame by those efforts.

During my own talk, entitled "Extinct Radioactivities: Trapped Residuals of Presolar Grains" chosen to match exactly the title of my now rejected *Science* paper, I noticed, or perhaps I should say "heard", that my presentation had not gone unnoticed by the Caltech laboratory that measured isotopes in meteorites. Two of its workers, denizens of a laboratory that proudly called themselves The Lunatic Asylum as a witticism on their measurements made on lunar samples, sat middle center. I could see clearly who they were. They were close protégés of the lab's famous director, G. J. Wasserburg. What they did next astonished me. They shouted remarks during my talk, a rudeness unheard of at American Physical Society meetings. From their shouted interruptions--no hands raised or anything like that--simply shouted confrontational remarks-- I could see that they had been sent to do some dirty work. This was a planned disruption. I spoke doggedly on as the audience turned to identify them, just as a politician might when confronted by a heckler, but I regret to say that I could not clearly discern at the podium while delivering my own thoughts exactly what specific objections it was that they shouted into the room; but their target was clearly me and my objectionable ideas. This outburst was probably retaliation for my effrontery in circulating the reviews of my doomed *Science* paper. And if the entire thing seems rather silly, as it probably was, their retaliation nonetheless warned me that I was in for a long battle. Later I felt some sympathy for these lab workers, one a research associate and the other a graduate student, when I realized that they had been injected over their heads into the world of astrophysics. What's more, it was unlikely that my Science paper would have been sent to them personally to referee. Probably one of the reviews was written by another researcher in their laboratory. Nasty stuff.

After the meeting Annette and I drove to Pasadena to accept that overnight invitation from Willy and Ardie Fowler. After eggs and toast with marmalade with English tea in the Fowler dining room next morning in Pasadena, I asked Willy about his reaction.

"Very unusual rhubarb", he agreed, "but you need remain aware", he cautioned, "of the importance that that community attaches to their published positions".

It is so. Only in meteotitic science had I heard so much meeting time expended by individuals defending their past published works rather than reporting their new results. Edward Anders was the champion of this technique. I routinely heard such self justifications at the Lunar and Planetary Science Conferences and the annual meetings of The Meteoritical Society. So I went through both of my scientific papers on supernova dust with Willy, explained all of their ideas, clarified the exaggerated isotopic abundances that I predicted to exist in interstellar dust, isotopic features that undeniably revealed counterparts in documented known facts from meteorites, facts that surely, I argued, existed routinely within the interstellar dust. Willy was totally absorbed, even excited, privately supportive, and promised to try to calm them down over in the Lunatic Asylum. He loved my ideas as much as I did because they followed so naturally from coupling the science of nucleosynthesis in stars with Hoyle's argument that supernovae will surely make a lot of dust while their interiors expand from their initially dense hot state. The next day we resumed this theme, and recorded[19] with photographs our first private time together since 1972 in Cambridge. Annette took the photographs of us with my camera. Seated at the table in their dining room, we smile happily in these photographs, like two boys at play, which is just what we always were together.

Having heard enough of controversy, Fowler was eager to hear of our recent visit with Fred in the Lake District. Fowler had previously visited Hoyle there and, by his own admission, loved it without compromise. I described Fred's optimistic frame of mind and his lack of backward looking. Full descriptions of the climbs that we had taken were mandatory.

"When will we go back to Scotland," he asked with a good natured wink.

At this opening Annette was able to pitch in with an exciting report of our Scottish climbs prior to the Cambridge workshop. How Willy loved to show his envy! He was a ham, feigning agony at every top, but his envy was sincere. How pained he was to have not been there with us, how eager to hear details of our climbs of his previously bagged Munros, how he wanted to know exactly if we had taken the same routes up and how the weather on each had been. Willy loved that stuff, and made everyone love it with him. He pulled out his copy of *Munro's Tables* to check on the dates when we had climbed those same mountains together. He also wished to hear about Cambridge, his beloved but now renounced Cambridge, and how his house on Oxford Street had looked.

"Forlorn," I replied.

"Were Ryle and Hewish and Pippard there?" he asked, adding, "I never want to see the bastards again!"

Eventually we returned to the controversy over my paper in a more jolly humor, and he agreed emphatically with my resubmission of the *Science* paper to *Astrophysical Journal*. I was so grateful, so relieved, for his encouragement. From that day until his death in 1993, Fowler privately supported my ideas, even hoping that they would prove correct. But he did not always show that support to others. It was Fowler's lifelong manner to avoid any public endorsements of positions that might be incorrect. He cautiously protected his own stature. So his private support could not translate into strong actions in my behalf in the controversial years that were dawning. The question in all of this was not whether I was or was not correct in my proposed details for xenology. It was whether these new ideas that I had introduced, and their new possibilities for meteorites, and their crystal-clear new predictions were valuable additions to scientific discourse. Fowler possessed no compass to decide whether the ideas were correct--nor could anyone else do so at this point in time. Faced with the tidal wave of sentiment against my suggested reinterpretations of xenology, Willy could privately admire its arguments, and even encourage them, but he could not help me advocate them. I was then, and remain now, somewhat disappointed in his lack of public support to follow. But after Fowler was awarded the 1983 Nobel Prize in physics, he openly stated his fascination for them in his 1983 Nobel lecture, where he wrote:

> *[Wasserburg and Papanastassiou] espouse in situ*
> *decay for the observations to date, but Clayton argues*
> *that the anomalies occur in interstellar grains preserved*
> *in the meteorites and originally produced by condensation*
> *in the expanding and cooling envelopes of supernovae*
> *and novae. Wasserburg and Papanastassious write,*
> *"There is, as yet, no compelling evidence for the presence*
> *of preserved presolar grains in the solar system. All of*
> *the samples so far investigated appear to have melted or*
> *condensed from a gas, and to have chemically reacted to*
> *form new phases." With mixed emotions, I accept this.*

The underlining in the final sentence of this quote from his Nobel address is my own. It openly calls attention to Willy's conflict. It was the closest he could come to visibly saying that it was only with grave misgivings that he had allowed himself to be persuaded of the incorrectness of my suggestions. When, four years after his Nobel Prize, the stardust grains bearing the huge excesses in their abundances of the daughter nuclei of radioactive nuclei were finally discovered, Fowler would be excited for me, and would remind me that he had been sufficiently convinced by my arguments to hold out hope of their correctness in his Nobel address. He pointed excitedly to the same quote from that address that I have reprinted above. I too would be happy about this; but I would have been even happier had he had been able to support the usefulness of my ideas against those who would damn them prematurely, and who did so for more than a decade.

The total experience created a new credo for me; namely, if a new idea is scientifically original and scientifically interesting and might correctly explain some phenomenon, it is almost always wrong to try to prevent its publication on grounds that it does not apply to some other phenomenon. The reason is very, very simple. The fundamental principle of science is skepticism. Being by nature skeptical, scientists test new ideas by experiment and reject them when they are demonstrated to be wrong. Scientific rebuttal in the open literature, which is the correct scientific procedure, will quickly kill off wrong explanations. But publication of the new idea that is not in itself demonstrably wrong allows it to come to light. It allows others to consider its consequences. There is a great difference between a new physical idea and a new explanation of an old phenomenon in terms of that idea. The idea and the specific explanation have independent lives. The history of science is one huge rubble pile of false ideas or false explanations. The great size of that pile matters not at all. In the scientific method, it is the continuously purified edifice of science that survives.

The idea storm year during 1974-75 had been momentous in my life. Many consequences would follow me forward in time. I can not fathom how a person like myself remains happy, productive and good natured during extended times of severe stress and worry. But it is one of my characteristics to remain so even in the darkest of times. At the same time a fiery new public persona had been forged. It was to be further honed by the science controversy that would follow inexorably from these first papers on cosmic chemical memory. My personal stress was so great that I remain forever grateful for the new fire within my belly, for the stoking of it helped me through this difficult period. The year 1975 closed with a new me.

CHAPTER 15. CARDIFF AND HEIDELBERG: STRATEGIES FOR STARDUST

In March 1975 at Johnson Space Center for the sixth annual *Lunar Science Conference* I spoke with an energetic young German scientist named Till Kirsten. I had raised in my rejected paper to *Science* the general issue that the excess abundance of the special ^{129}Xe isotope found in meteorites might have arrived in supernova dust instead of from live radioactive ^{129}I in the solar system. I suggested that although radioactive ^{129}I was the source for this excess ^{129}Xe atoms found in iodine-bearing minerals, it had decayed while residing in the interstellar material in dust formed in supernova explosions. If that were so, ^{129}I need not have existed alive in the early solar system. All meteoritic scientists firmly supported live ^{129}I, so my proposal had placed me in very hot water. Kirsten was nonetheless interested in my argument, and we discussed the xenon isotope measurements that his team was making in his laboratory at the *Max Planck Institute for Nuclear Physics* in Heidelberg. After the discussion I remarked that my German wife and I were interested in spending a year in Germany. Kirsten leaped at that remark, saying that he would welcome us to do that in Heidelberg. Furthermore, he proposed nominating me for an *Alexander von Humboldt Senior Scientist Award* to support such a stay. I was floored! That would be serious support.

Kirsten's idea was more attractive than my own notions. It would enable me to work in a center for meteorite research to develop my theory of isotopically heterogeneous interstellar material. And it might at the same time provide relief for Annette's homesickness. I was confident that if we returned to Germany for a year the United States would suddenly look more attractive to her. That experience had happened to many Germans that I had known in the United States, to return home after a year struggling critically with American society only to find that their views of both the United States and of Germany had changed. I was determined to give this marriage every chance to succeed. Annette was quite excited when I told her of the verbal invitation. So I returned the nomination forms with a copy of my vitae to Kirsten a month later. He made my nomination.

I needed a wider strategy to bring the implications of stardust to the attention of the scientific world. Rebutting referee reports may or may not have been a good idea, but those actions do not carry much scope. I was plotting a decisive change of career in order to aggressively develop that new astronomy. This is no small investment. To do so would amount to placing my career in nucleosynthesis and stellar evolution on hold. Such a drastic step back from my scientific identity required courage and confidence, but my decisive new personality did not flinch. Heidelberg would be a good idea for this evolution; furthermore, I initiated a parallel astrophysical thrust to my stardust idea with summer research visits to Cardiff, U.K. during summers 1975 and 1976.

Several overlapping reasons inspired my Cardiff plan. Chandra Wickramasinghe, a friend from Cambridge and a past student with Hoyle, had become Professor of Astrophysics at *University College*, Cardiff. Chandra had established himself, along with Hoyle, as a pioneer of the origin of interstellar dust. Fred Hoyle took an honorary research affiliation with *University College* in order to better pursue their pioneering research into the nature of the observed carbonaceous opacity of interstellar dust. They envisioned some type of reproductive interstellar chemistry as a mechanism for evolving a large fraction of interstellar carbon atoms into a form resembling bacteria. This flirtation with the origin of life was to bring down much scorn upon them. I resonated with their plight owing to the rough treatment that my papers on isotopic stardust were encountering. Wickramasinghe and Hoyle were essentially tracing the origin of life to chemical reproductive memory that had its beginnings within the interstellar matter. This thrust seemed a useful cousin to my own isotopic chemical memory within stardust. Secondly, my Cardiff plan allowed me to maintain close contact with Hoyle. Never yet free from my emotional yearnings for some idyllic locale that seemed to beckon me back, it enabled me to continue my love of England in a Welsh venue. Thirdly, this was the first occasion when Annette and I took one of my sons along, in this case my younger son, Devon. I instinctively wanted to shore up Devon's emotional pessimism. This seemed successful when it appeared that Devon had been extracted by the Cardiff visit from his depressed emotional shell.

Chandra and I struck an agreement for this plan to begin immediately after the special conference *Frontiers of Astronomy* in Venice to celebrate Hoyle's 60th birthday. I was one of its science organizing committee. I housed there with Annette and her mother, who had driven us from the Rhein to Venice, in an old hotel at Piazza San Marco. The conference site was the beautiful monastery on the Isola St. Giorgio (St. George Island), which is immediately in front of the Palazzo Ducale and the Piazza San Marco in Venice. What an inspiring and appropriate venue for this renaissance man!

Organizing Committee for conference honoring Fred Hoyle on his 60th birthday, in Venice, 1975. front, Geoff Burbidge, Margaret Burbidge, Fred Hoyle, Barbara Hoyle, Willy Fowler; rear, Clayton, Martin Rees, Elizabeth Hoyle

The title of this conference borrowed from Hoyle's very influential 1955 book *FRONTIERS OF ASTRONOMY*. Following on the heels of a golden era for nucleosynthesis, the conference program in 1975 still had a largely nucleosynthesis flavor, with talks on that topic by (in sequence delivered) Truran, Arnett, Fowler, Barnes, Pagel, Woosley, M. Burbidge, M. Shapiro, Wickramasinghe, Schramm, and myself. My talk "Dust as Tracers of Chemical Evolution" presented the cosmic-chemical evolution theory of isotopic anomalies in presolar dust. The chemical memory derives from its stardust content. The setting was very inspiring. Religious paintings adorn the walls of the monastery on the Isola San Giorgio[3].

Following the honorary conference for Hoyle we arrived in Cardiff on July 24, 1975. Chandra and his wife Priya met us at the Post Hotel in Pentwyn following our arrival by train from London's

Paddington Station. A lovely house at 3 Heath Park Crescent had been booked for us by Priya Wickramasinghe. We resided there during both of these summers while its owners, Mr. and Mrs. Dark, were away constructing their holiday house in the south of France. For a very modest fee they also loaned us their *Rover* for our transportation. After I presented a couple of department colloquia, one on isotopic expectations for stardust and the other on astronomical measurements of grain condensation and growth following nova explosions, Chandra and I quickly selected the latter topic for our summer research. I wanted results, fast. So did Chandra. A photo was taken[2] of us on a warm June day planning this calculation in the garden at 3 Heath Park Crescent.

We quickly decided to buttress my recently submitted paper with Hoyle[1], "Grains of Anomalous Isotopic Composition from Novae", with numerical calculations of the growth of carbon dust from hot expanding gas following a nova explosion. Chandra and I constructed a numerical model[2] for that grain growth and the associated growth of the infrared emission from novae. Our paper was submitted to *Astrophysics and Space Science* on 16 September. This excellent pathbreaking paper confirmed once again how quickly a good new idea can become publishable. I would estimate that today most research papers in astrophysics take several years from conception to publication. The single most frequent exception occurs when papers such as ours spring from a new idea for explaining a new astronomical observation. This particular paper would shore up astronomical aspects of stardust rather than the meteoritical aspects. It set my new tack.

It was the custom at *University College* that each summer Chandra Wickramasinghe and his department would host a summer astrophysics conference on a topic of current research. These were held in a splendid estate in central Wales that had been given to *University College* by the Davies family as a useful and appealing meetings outpost for the university. The country house was named *Gregynog*. My visit of it thrilled me and inspired a new strategy. My aggressive plan would use *Gregynog* as a venue for an international conference on isotopic anomalies within interstellar dust and within the meteorites. I proposed this to Chandra for the following summer and he approved it immediately. It was right in tune with Chandra's interests. I told Hoyle and Fowler of this plan and obtained their enthusiasm for it and their promise to attend. Annette and I even took a quick drive up to the Lake District to discuss the deeper strategy with Hoyle. He reinforced my courage for my career redirection by admitting that he had found it useful to change his research emphasis every ten years or so. Hoyle and Fowler agreed to serve on the Science Organizing Committee, which usefully added the luster of their names to the attractiveness of such a meeting. In his typed letter from his Caltech office dated August 8, 1975 to me at *University College* pledging his support, Willy warned with following words that Jerry Wasserburg and other Caltech researchers would probably not attend a conference being planned by me:

> *I will definitely affiliate as an organizer for workshop at Great Egg Nog. I will work on Wasserburg et al. but must warn you that they are pretty bitter! The more said in any brochure about the experiments and the less about grains and other explanations the better chance we will have to get them to attend. I say, lure them into the trap and then lay it on.*
>
> *All the best,*
> *Willy*

I regarded the confluence of factors as perfect: a pioneering conference featuring a new astronomy; an historic and graceful large country house where attendees would be lodged; Hoyle and Fowler joining Chandra and myself as its science planning committee; and my opportunity to personally plan the program and extend the invitations. They allowed me free rein. This was my baby!

During fall 1975 and spring 1976 I was constantly busy at Rice with selecting the invitees and handling all correspondence about the science program. My initial invitation list included many prominent astrophysicists: Jean Audouze, David Black, Al Cameron,

Dave Dearborn, Freeman Dyson, Mike Edmunds, George Field, J. Geiss, T. Gold, K. Hainebach, Mike Howard, Mike Newman, Henry Norgaard, Buford Price, Hubert Reeves, Ed Salpeter, Phil Solomon, Dave Schramm, Jim Truran, George Wetherill, Stan Woosley, and we four organizers. Almost all participated. It was a concentration of astrophysical power that had never been seen at a meeting of meteoriticists. I invited an equally large number of distinguished meteoriticists, and constructed the program largely having them describe their measurements of anomalous isotopic compositions in meteorites. A strong European contingent was invited to spread the new ideas to Europe. I hoped that the glamour of the meeting and of its illustrious attendees would stimulate the desire among leading meteoriticists to attend a remote countryside conference, even if they doubted that its chief planner understood much about meteorites.

When summer 1976 arrived Annette and I took Devon with us. We took him along to Scotland, hiking in the highlands, to a return visit to Cambridge and our homes from 7-9 years earlier. We shared a nice tour of western counties en route to Cardiff. Devon seemed to have a wonderful time, so that I felt an inner glow when we took him to his departure flight from Heathrow on July 15. We had seen little of the silent depressed look that always worried me about Devon. Visits followed from my sister and her husband and by Annette's brother and parents. Everyone wanted to visit Wales. So I did little work prior to the *Gregynog* conference.

By early August the *Gregynog* participants began to arrive. We invited Stan Woosley and his first wife to spend the preconference week with us. Stan began and remained a special friend. Willy Fowler arrived on August 6. We picked him up from Cardiff airport, ready as always for science action. Willy and Stan and I spent two weekend days going over research problems together. A photo of Stan and Willy in our home on Heath Park Crescent can be seen at <1976 Fowler and Woosley in Cardiff>.

We invited George Wetherill and Freeman Dyson for afternoon tea on Sunday August 8. I enjoyed having famous guests. George Wetherill was a prominent geochemist who won even greater fame later with computer studies of the assembly of small rocks into larger planetary bodies in the early solar system, the prelude to formation of the large terrestrial planets. I had previously discussed isotopic anomalies with Freeman Dyson, a brilliant physicist who played a major role in the mathematics of quantum electrodynamics, during his visit as the 1976 Houston Lecturer at Rice University. He twice had dinner with me during that visit. I told him of my plans for the *Gregynog* conference and he surprised me by saying that he would like to attend. Freeman Dyson was also feeling an attraction to his own life history, as was confirmed when he mailed me a copy of a letter written from *Gregynog* by Walford Davis to Dyson's father in 1929. Many personal lures contributed to the splendid participation in this workshop. On Sunday afternoon many participants called to confirm that they had arrived and were ready for the Monday trip by autos to *Greynog*. Getting there was half the fun. Who would not relish the task of finding a fabled remote location in central Wales for an exciting assembly of intellectuals? Monday was Willy's 65th birthday, an occasion to be noted, so armed with birthday presents of whiskey and cigars, he and Freeman Dyson rode along in our car. As we arrived at *Greynog*, all participants were settling into the rooms that I had assigned them. The excitement built during dinner for our beginning on Tuesday morning.

My talk on the first morning was arranged to follow John Reynolds' keynote address on "xenology", which described again the several isotopic patterns of xenon that had been identified in meteorites. This was, as all present were well aware, the unfortunate topic of my rejected *Science* submission. So some tension mounted in the room for my talk, "Theoretical Issues: a Preview Focus". But I had planned carefully. Eschewing xenon in meteorites entirely, I presented a sweeping and fresh view of how dust condensation during mass loss from several distinct types of stars was expected to sprinkle the interstellar matter with dust carrying distinct isotopic patterns from their host stars. Xenon within supernova dust was but one pattern among many that could be expected. And if stardust could not be found in meteorites, or if it had been thermally absorbed in forming new minerals within the meteorites, I nonetheless predicted a chemical memory inherited by those minerals in the form of diluted isotopic patterns of that stardust.

I was confident as questions were asked by the participants that my first goal had been achieved. I could be seen as champion of a new isotopic dimension within interstellar dust. That was my platform.

Three exciting days featuring presentations about unusual isotopic ratios, or hints of same, or speculations about possible causes, followed. A Thursday evening session after dinner, scheduled after a free afternoon to enjoy the surroundings, featured isotopic surprises from the lunar soils. As stimulating as these talks were, in truth we knew very little about isotopic anomalies at that date. This lack of firm knowledge was what made this conference so prescient. Researchers had discovered only a few special examples in a backdrop of overwhelmingly normal matter. The hope of a science bonanza was often pooh-poohed by scientists who did not attend, notably the residents of Caltech's *Lunatic Asylum* and by Edward Anders. But their powerful dismissals of my ideas thereafter lacked force among those who had participated. Despite the paucity of indisputable isotopic anomalies, the *Gregynog* workshop was nonetheless an historic event, the first of its kind. Most participants retain fond memory of it as a special moment when an emerging issue of great importance was first addressed by a distinguished interdisciplinary audience. I made several good friendships that would be important as I entered phase two of my strategy. One of the group photos before the main entrance to *Gregynog Hall* can be seen at <1976 Group at Gregynog workshop>.

Paying the research bills

Professors in research universities are expected to seek grants to support the costs of their research program. Universities are not in the business of supporting scientific research, although many critics of universities seem to think so. What universities do provide is an infrastructure, an adequate salary, the opportunity of an intellectual life devoted to science knowledge, and graduate students. But we must bear the costs of the research, including the graduate students if they are not teaching, travel funds, and publication costs. Back at Rice University in the fall, I therefore had to find new sources of funding for my career reorientation.

I submitted a research proposal to NASA to the program that supported meteoritic research. The proposal was rejected by referees that were hostile to my initial publication on xenon isotopes, which had by now been accepted and published in *Astrophysical Journal*. I had grown accustomed to having my research rejected by meteoriticists; but that did not deter me. A happy ending for me occurred during a visit in Washington with program director William Quaide at NASA Headquarters. He told me quite simply and directly that he had been doing his own homework on my case and he was going to fund it despite its rejection by the referee. I believe that he had spoken to Al Cameron about my negative referee report and had been advised by Cameron that my ideas were well worth NASA research funding. So Quaide simply put me in the program! NASA was to fund my research in that area for the following three decades. A new research proposal was required and sent out for review every three years.

I discovered a private foundation in Houston, *The Robert A. Welch Foundation*, which supported chemistry research within the state of Texas. I described in my proposal to them the novelty for chemistry of studying solids that condensed during mass outflow from stars and within the deep interiors of supernovae. They were impressed enough to grant their financial support, which they renewed for the remainder of my career at Rice University. Renewals every three years in both programs, by the way, required significanty positive evaluations by my peers of my follow-on proposals.

So serious was I about my career reorientation that I took out membership in *The Meteoritical Society*. It is an international society devoted to the scientific study of meteorites. I had never even looked at a meteorite, much less studied one; but I saw their study as a key for my goals, which included oral presentations annually for the next two decades. Its large European membership fit my Heidelberg plans. Heidelberg and Mainz in particular boasted many German members. So I submitted an abstract of my talk for the 1976 program in Bethlehem, Pennsylvania. On

October 16 Annette accompanied me on a flight to Allentown, PA through Pittsburg, where we changed to a small Allegheny Airlines flight. The host hotel was the historic Hotel Bethlehem, which was then subsidized by Bethlehem Steel Corporation. After checking in, we had a drink in their nice bar with Paul Pellas and Frederik Begeman, two distinguished European meteoriticsts that we had met at *Gregynog*. Our lively chat as we talked of *Gregynog* and of the papers that we planned to present at this meeting was relaxed and natural, giving me the sense that I was going to like this society very much. They showed motivation to search for the stardust that I had predicted. I was gratified that when I described my impending Heidelberg plan, both were kind and generous in their enthusiasm. *This strategy is going to work,* I thought to myself.

My first oral presentation at my first meeting of *The Meteoritical Society* presented a model of the changing isotopic composition of the xenon gas in the planetary disk owing to the progressive release from gradually heated interstellar dust of a large fraction of the ^{129}Xe daughter isotope. This trapped ^{129}Xe was abundant, I claimed, owing to the supernova dust in which much radioactive ^{129}I had condensed initially —my rejected *Science* submission! I daringly used the stardust basis of my rejected paper to advance a model that could impose temporal order on the sequence of meteorite formations. I avoided asserting that ^{129}I had not been actually alive in the meteorites. I thus avoided the specific assertion that had so upset meteoriticist referees of that paper and that now, three decades later, seems to have been incorrect. But I hoped for a good reception for the idea of using stardust by showing that it can lead immediately to observable consequences, whether my model prove correct or not. So my first observable possibility seemed to be that the differing ^{129}Xe/^{130}Xe isotopic ratios within trapped xenon observed in different meteorites might in part be caused by the releasing of a significant fraction of the ^{129}Xe isotope from interstellar dust. This left the straight line correlation the result of the decay of live ^{129}I in the metorites but interpreted its absolute level, measured by the intercept of the straight line, as fossil. I admit I was relieved, however, when my talk did generate considerable interest—which had been my goal—including a surprising defense of the presented idea by G. J. Wasserburg of Caltech's *Lunatic Asylum*. I always enjoyed Jerry Wasserburg and admired him for his science. Everyone did. His science was always brilliant, winning him early fame, and his wit was outstanding. But my suspicion that he inwardly seethed despite his kind remark was confirmed at a subsequent meeting when he admitted to my face, "I hate your guts!" I feel sure that his animosity had started with my submission to *Science* and with my circulation of the referee reports that rejected it.

Edward Anders, who was to be an even more implacable foe, saved his biting remark for the discussion period following his own verbal presentation. In answer to a question about a puzzle in his xenon data he replied, dripping sarcasm,

"Perhaps Clayton's *rugged interstellar travelers* could do the job."

This strategy may not work, I thought to myself amid the laughter. An irony of Anders' dismissal of my work was that his brilliant chemical techniques would soon lead to the discovery of stardust. His paper at this meeting in Bethlehem was a dramatic step in that direction, and I was spellbound by his findings. His lab had begun dissolving meteorites in strong acids. Stones dissolve in strong acids; and a meteorite is a very complicated stone. Anders reported that when they did this a residue of solids that would not dissolve was left behind. Although the residue was but one percent of the mass, it contained eighty percent of the noble gas content of the entire meteorite! The noble gases were concentrated within that residue. I think he was well aware that he had found the trail that would eventually identify carriers of the noble gas atoms. He probably hoped that those carriers would *not* prove to be be stardust.

All in all I felt that my attendance at this society meeting had successfully initiated my personal goals within a new community of colleagues. Although the details of my proposed scenario in the early solar system would not prove to be correct, the stardust upon which those details relied would prove to be correct. I had also been very encouraged by the open-minded curiosity of the Europeans. This bade well for my Heidelberg effort. I was so encouraged by the enthusiasm and questions of the large audience that I thereafter made it part of my strategy to present annual talks at both *The Meteoritical Society* meeting and the March *Lunar and Planetary Science* meeting at Johnson Space Center, and to make them sufficiently original and interesting that I would attract a large and curious audience. This proved to be the case as the legacy of stardust unfolded. Stardust and its possible roles in cosmochemistry was proving to be a prolific concept.

In May 1976, prior to *Gregynog* meeting and *The Meteoritical Society* meeting, I had received the official letter from the *Alexander von Humboldt Foundation* appointing me as *Senior Scientist* in Heidelberg. This was a quite high honor in addition to providing ample financial support for a leave of absence from Rice University. Annette appeared to be thrilled. I was cheered by that. So we excitedly made plans to leave Houston in time for Christmas 1976 in the Black Forest with Annette's family and then to reside in Heidelberg until September 1977. Eli Dwek and his wife Florence, always helpful, drove us to the airport for our flight to Zurich on December 19, 1976. Then we settled in for eight months of German life. First stop of course was Wutöschingen in the Schwarzwald, where we charged up our drained batteries with two weeks of festive family time. Enjoying snow-flecked trees, foggy-breathed hikes, Christmas decorations, and delicious meals, Annette and I loved one another, and in that period of happiness our daughter was conceived. On New Year's Eve we drove to the high Schwarzwald and had dinner with a bottle of *Spätburgunder*, a worthy German pinot noir, and danced at Hotel Adler, where our wedding feast had been four years earlier. My spirits were high with the sense that negative emotions were being dispersed like snowflakes in the winter wind.

To reside seven months we would need a car, so I spent some time in shopping for used cars and taking advice about them. I settled on a 1970 Peugeot 204 from Autohaus Fink in the neighboring village of Degernau. It was clean, looked nice, and most importantly, still had an approved TUV inspection making it legal to drive throughout 1977. German auto inspections are quite strict by American standards. I paid 2200 DM for it, about $1000. So just like that we had a car, and it would run well without trouble until we sold it to a Heidelberg student in August.

GALLEX

Till Kirsten had alerted me that he hoped I could participate in discussions in early January 1977 contemplating new underground experiments designed to measure the rate of arrival of neutrinos from the sun. My arrival in Heidelberg was timely in that regard. I was expert in its hydrogen fusion reactions near its center, and I had published two possible explanations[4] of the observed deficit of solar neutrinos by Ray Davis's experiment, so I had complementary knowledge that Kirsten welcomed for their assessment of alternative experiments.

With high hopes we settled in Heidelberg on January 7, driving straight to Kirsten's house on Berghalde. We met his Japanese wife, Emiko, amid a cordial time of greeting and becoming acquainted. But he soon recommended guiding us to the Hausmeister of the apartment building in which they had reserved for us a nice one-bedroom apartment. This was in a seven-story apartment house at zur Forstquelle 6, located on the back side of the mountain that looms above Heidelberg city center, and near which the *Max Planck Institut für Kernphysik* (Institute for Nuclear Physics) was situated. When that apartment building was constructed, the *Max Planck Institute* had purchased many of its apartments to lease to scientific guests. This bold plan, and the deep pockets of the *Max Planck Society*, had solved the knotty problem of institute guests having to seek living accommodations in a medieval town dominated by university students. I had seen that same problem before, in Cambridge. So we four went to our apartment. It was fine. We lived for eight months in apartment 9. For us it was perfect. We were to reside in this same building for several years, and I enjoyed the forest walk from it to the institute lying higher up the mountain.

The assessment of new solar-neutrino experiments amounted to a contemplated career plan for Kirsten. If he were to undertake such an experiment, which he wanted to do if it proved feasible, its demands would gradually terminate his present career using xenon gases from meteorites to learn of their history. Xenology was the general field that Anders had spoken of in Cambridge during my moment of inspiration there, and which John Reynolds had spoken at *Gregynog* in the keynote talk. My enthusiasm for research in this area made the thought of Kirsten pulling out somewhat disappointing; but the excitement of a possible new solar-neutrino experiment more than assuaged any disappointment. Kirsten and I were in the same boat, contemplating entirely new career thrusts, and this kinship would bond us strongly. So after some days of shopping for our apartment and sightseeing about the enchanting *Altstadt*, the old city, I moved professionally into my Heidelberg office. I was pleasantly surprised to find that Kirsten wanted me to share with him the spacious and elegant *Joseph Zähringer Zimmer*, a director's room that had been named for the scientist who had previously directed the cosmochemical research and who had recently been killed in a tragic auto accident. The room had large, openable picture windows along the outside wall, two large desks located in the middle along that wall, two sofas and a conference table in one corner, bookshelves, file cabinets, and an industrious Frau Heidi Urmitzer working in the adjoining staff office. I liked this office very much, loved it actually, and would use this office with Till for seven years. Most of all, it kept me in daily contact with Till, so that our only problem was to not spend too much time talking over all the science ideas that would arise over the time to follow. You can imagine that such opportunities arose often, whether about solar neutrinos or about stardust.

Kirsten had invited Raymond Davis, Jr. and Oliver Schaeffer to come to Heidelberg for this initial discussion. Ray Davis had been running his Nobel Prize winning experiment in the Homestake Mine and Oliver Schaeffer was an expert in environmental production of background radioactivity. Kirsten had met them when he took his first sabbatical leave at Brookhaven National Laboratory. Also participating was Wolfgang Hämpel from Max Planck Institute, Kirsten and me. On January 11 we conducted this first team meeting in the *Zähringer Zimmer*. A photograph of this first meeting can be seen at <1977 First meeting of GALLEX in Heidelberg>. During the daily discussions we considered each of several new experiments that might succeed in clarifying the deficit of solar neutrinos. This was complicated by three differing sources of solar neutrinos, each having its own energy threshold for the needed detectors. It was necessary to choose stable element targets that could be made radioactive by absorbing neutrinos from these distinct neutrino energy ranges. The main problems for each were avoiding the production of radioactive nuclei by cosmic rays or by natural terrestrial radioactivity and the challenge of counting the numbers of radioactive nuclei caused by the neutrino absorptions. These would require a deep underground site. Where underground could such an experiment be housed? Would it be possible to assemble a sufficient number of the designated target nuclei that would absorb the neutrinos? Would it be possible to measure and identify the radioactivity caused by the absorption of neutrinos? How could the small number of radioactive nuclei be isolated from the great mass of the target? How would the radioactive counting be done? Relatively few questions about the sun arose, so I was relatively silent, absorbed by the many issues concerning low-level radioactivity. Ray Davis was a font of information about the problems of recovering and counting the radioactive atoms. Only from time to time did we turn to suggested explanations of the deficit that Davis was measuring. At that time I suspected that the problem lay with our understanding of the sun. It asked a lot to believe that we could understand completely what was going on in the sun's thermonuclear reactor located within the central one percent of the sun's volume.

Kirsten had already done a lot of work when an enlarged team met again[5] in September 1979 at the Max Planck Institute. I could easily participate during my annual residences there. A preference for a neutrino absorber target of gallium atoms was by then clear. The experiment would be called *GALLEX* and be housed in an underground Italian tunnel for automobiles through the Gran Sasso, where Kirsten had discovered that the Italian government hoped to develop a laboratory for underground physics experiments. Kirsten patiently surmounted many difficulties and technical problems in turning his dream into a reality.

*Till Kirsten and the author at Kirsten's home
on Berghalde in Heidelberg.*

Going with Gamma Ray Observatory

Kirsten and I discussed the question of my relationship to
the GALLEX team. I reluctantly decided that I would have to
withdraw after the 1979 GALLEX meeting. I decided that my
big team effort would lie instead with *Gamma Ray Observatory*
(GRO). Two scientists trained in Bob Haymes' Rice group, James
D. Kurfess and W. Neil Johnson, Jr., had submitted a proposal to
NASA to produce an instrument for that proposed space mission.
They would be *Principal Investigator* and *Project Scientist*,
respectively. The instrument would follow the Haymes design,
albeit with generous improvements. To produce the strongest
proposal for that big competition I decided to commit my time
to gamma-ray astronomy rather than solar neutrinos. Certainly
this decision was influenced by my predictive role in gamma-ray-
line astronomy. I saw in it a rare chance to possibly carry through
from prediction to observational discovery of radioactivity in

supernovae. As a result I became *Co-Investigator* on the proposal for a shielded scintillation spectrometer to record the gamma rays. That experiment, called *OSSE* (Oriented Scintillation Spectrometer Experiment) in a typical NASA acronym, was one of five initially selected for the space observatory. To later save costs one experiment had to be removed, leaving OSSE as one of four. The expectations for gamma-ray astronomy had been such a prominent goal of my career that I decided to give my full effort to it. In a strategy session at Goddard Space Flight Center in April 1979 it was decided that Reuven Ramaty and I would co-sign letters to prominent scientists asking them to write to NASA with their endorsements of the importance to NASA science of the GRO mission. Although I withdrew from GALLEX, I followed the experiment with intense interest and was very proud of Till Kirsten when its successful measurements not only confirmed the deficit of solar neutrinos but also suggested that the problem lay not with our understanding of the sun but with our understanding of fundamental properties of neutrinos.

*Reuven Ramaty and the author in Ramaty's office
at NASA Goddard Space Flight Center planning a
strategy to influence congressional support for
Gamma Ray Observatory, April 3, 1979.*

Becoming a Meteoriticist

Then in late January 1977 I focused my energy on my main goal for
establishing my Heidelberg affiliation, namely, the reorientation
of my science career toward cosmochemistry. I tackled this
change aggressivly. Not only would I continue predicting the
isotopic fingerprints of the solid pieces of stardust, but I also saw
the stardust as carriers for an isotopic memory that could remain
when that stardust was chemically reorganized along with other
atoms in the growth of new solids in the solar system. I would call
this combination *cosmic chemical memory*. My strategy called for
publishing a series of papers displaying the abundant predictions
that were explicit or implicit in the central ideas of my papers; and
it called for oral presentations at cosmochemical conferences,
where meteoriticsts would listen in person to my ideas whether

they read my papers or not; and it called for discussions with Heidelberg experts that would help advance my understanding of the complicated world of the mineralogy of meteorites. I dove zestfully into these.

This flurry of activity in the first half of 1977 in Heidelberg had been preceded by my submission of a paper[8] to *Icarus*, a solar-system studies journal. Entitled "Solar System Isotopic Anomalies: Supernova Neighbor or Presolar Carriers?" this paper compared the decay of the growing number of known extinct radioactive nuclei to judge whether their decays had occurred within the meteorite itself or within interstellar dust prior its inclusion in the meteorites. These were identified by data from meteorites showing that, although the radioactivity is now extinct in the meteorite, an overabundance isotopically of its daughter nuclide provides evidence that the radioactive nucleus was once alive. It might have been alive either in the meteorite as it is found today or in precursor interstellar solids that once contained the live nucleus and that have carried the daughter abundance into the formation of the meteorite sample. I was arguing in favor of condensed grains from stars being responsible for much of the fossil evidence, evidence that was interpreted by others to show that those radioactive nuclei were present and still alive in the meteorite when it formed in the early solar system. I envisioned that my extinct-presolar-carrier model would become the best single explanation for extinct radioactivity. In this I overreached the mark, but the exercise showed convincingly that competing interpretations might be possible. Publication of this paper established me as a key player in the science yet to unfold.

Almost daily after the GALLEX meeting I had conversations that opened my eyes to the chemistry of meteorites. Ahmed El Goresy helped me with meteoritic mineralogy, and brought to me his new measurements of the mineral structure of meteorite inclusions and explained their relevance to me. Jim Jordan explained the details the xenon mass spectrometer that he and Kirsten used to measure the xenon isotopes in meteorite samples. Elmar Jessberger clarified many points of isotopic cosmochemistry that arose in his measurements. Hugo Fechtig and Eberhardt Grün introduced me to their work and knowledge of dust in space. And Till Kirsten offered a critical sounding board for my outpouring of speculations. Understand that this crash course continued to educate me for five uninterrupted years and two follow-on summers. My construction of cosmic chemical memory was encouraged beyond measure by this support. For the next few years I was struck more and more by the good-willed curiosity of the Europeans and its contrast to supercilious and obstructive reactions of some senior American meteoriticists. Slowly I learned that the young American meteoriticsts were largely with me, however. They listened to my presentations at meetings even when assured by their superiors that they were listening to crap. They told me personally that they were grateful for my radical ideas. And I had seen evidence that this was the case, because for three decades, beginning with 1975, my proposals to NASA for support of this research were recommended for funding by referees and by the grants panel.

I continued without pause with papers and presentations. The submission of Abstracts for oral papers to be presented at the 1977 meeting of the *Meteoritical Society* was immediate. This would be my second such meeting. I submitted not only one but two abstracts, and was allowed to present both at the 40th annual meeting in Cambridge, July 24-29, 1977. In the published abstracts in *Meteoritics* , I advertised dual affiliations with Rice University and Max Planck Institute. In the first of these, "Astrophysical Implications of Isotopic Anomalies", I introduced astrophysical ideas that were foreign to the chemical and geological backgrounds of the overwhelming majority of participants; namely, that the early solar system was cold, not hot, and that the presolar dust would consist of accreted mantles from interstellar gas surrounding refractory STARDUST minerals from stars, and that, as I stated, "chemical alteration of presolar condensate aggregates is the key to the early solar system." I described four different types of isotopic anomalies to be expected in meteorites. This first presentation was devoted entirely to astrophysics motivated concepts and was designed to reach non-astrophysics scientists.

In the second Cambridge presentation, "Origin of Ca-Al Rich Inclusion in *Allende*", I turned my attention to the origin of the calcium-and-aluminum-rich inclusions (*CAI*). These are several-millimeter-sized whitish rocks that were prominent in the recently fallen *Allende* meteorite and that had been shown by several in the audience to contain isotopically abnormal material. Their mineralogy is dominated by oxides of aluminum and calcium. I stressed that calcium and aluminum were created in supernovae and that their first chemical bondings following their creation would occur while supernova matter expanded and cooled. Therefore, those two elements would condense into Ca-and-Al-rich oxide minerals that are stable at high temperature, and they would do so while still *within the interior of the supernova.* Moreover, I added, astronomical observations of Ca and of Al in the interstellar gas showed that those elements are depleted from the gas phase. That is to say, individual atoms of Ca and Al are rare in the ISM. This is because those elements reside overwhelmingly in dust.

"Such supernova condensates will be very hard gemlike minerals, especially Al_2O_3 sapphires", I said, "because they condensed from a vapor that was too hot for less refractory minerals to condense".

I named the SUperNOva CONdensate formed initially in this way the *SUNOCON*, a name constructed from the capitalized letters This was offered as a new scientific word, one that is needed for intelligent discussion of the components of interstellar dust. But the use of that word was resisted by objections to my general approach, despite the fact that calling such wonders of nature "interstellar dust" or "presolar dust" amounted to tepid pabulum.

Resistance would not be muted until SUNOCONs were actually found within meteorites, which did not occur until the 1990s; but even then the moniker "supernova dust" was preferred by most authors despite that name being equally applicable to several differing types of dust that can be made in association with supernova explosions. I stressed in my talk that the oxides of calcium and aluminum would be highly enriched in the [16]O isotope of oxygen, similar but much more extreme than the [16]O richness of CAI that Bob Clayton was describing at this same meeting. I described how the interstellar SUNOCONs would become coated by mantles of other elements as they coagulated into icy dustballs during aggregation processes within the early cold planetary disk surrounding the infant sun. I concluded with a proposed origin for the CAI found in meteorites; namely, subsequent heating in the disk of these mantled SUNOCONs, causing evaporation of most of their mantles and fusing their remainders into the observed CAI.

"The CAI inherited their ^{16}O-richness from the SUNOCON component", I claimed, resting my case.

The large and curious audience at the Cambridge meeting of *The Meteoritical Society* sat through these presentations with evident interest, some questions and some criticisms. It was evident to me that many were, frankly, stunned at being asked to consider notions that were totally foreign to their experience, curious but noncommittal in the face of so much that we did not understand. They certainly did not flock at once to my banner! Quite to the contrary, the most outspoken opinions stated diverse reasons to doubt what I had said, mainly based on the dogma that the solar system began in such a hot state that all elements, calcium and aluminum included, must have been gaseous rather than within solids, and that the CAI had condensed first from that gas as it cooled. My Abstract stated,

"The usual interpretation of the CaAl-rich inclusions as high-T condensates from the solar gas cannot be correct."

I maintained this repeatedly in heated discussions at many meetings over the next five years, almost as often as my opponents stated that my theory cannot be correct because the solar gas would have begun totally vaporized by great heat. The latter ultimately gathered a pale of doubt when in subsequent years the CAI were found to contain puzzling isotopic patterns within many elements, not only oxygen but especially calcium, titanium and aluminum. A hot solar gas contains no isotopic anomalies because it would be homogeneous, and that homogeneous mix is our standard for *normal isotopes.*

Nonetheless, my thesis that the interstellar aluminum oxide SUNOCONs provide the ^{16}O richness of the aluminum-rich minerals had provided an understandable example of cosmic chemical memory, allowing many in the audience to grasp a concrete feeling for the theory. Later, however, I came to doubt the correctness of my explanation of the ^{16}O richness. I always maintained that it is the theory that is important, not the speculative explanations based on that theory. Opponents on the other hand continued to attack the specific applications and to ignore the general theory. What later would come over our horizon to offer the best explanation of the ^{16}O-richness had not even been thought of! We were pitting one flawed explanation against another. Science is sometimes like that until the correct explanation arises.

A cornucopia of isotopic-memory

I have no explanation for the breathless pace of my work at this time, especially having a competing priority, namely, my life with Annette in Germany. And if I had many irons in the fire, each was glowing red hot and extracted quickly.

Before traveling to my much anticipated return to post-Hoyle Cambridge and to my presentations at the 1977 *Meteoritical Society* meeting there, I had worked day and night to write papers for submission to solar-system journals rather than to my accustomed astrophysics journals. On March 25, 1977, one week after my 42nd birthday, I submitted a full paper[6] detailing the ideas of the second presentation to be delivered in Cambridge, "Cosmoradiogenic Ghosts and the Origin of Ca-Al-Rich Inclusions". The *ghost* concept was the most elementary application of the stardust idea. Namely, if the initial mix of isotopically anomalous stardust and solar gas is defined in bulk to be isotopically normal, the solar gas in the absence of stardust would have a negative anomaly where the stardust has a positive excess. I called that negative anomaly a *ghost*. I intentionally chose catchy names during this period despite knowing that they would be ridiculed by some. But I realized that the physical concepts were valid and would have applications. Even the ridicule was useful for fixing my role at this time in the minds of meteoriticists. I submitted this paper[6] to *Earth and Planetary Science Letters*, a European journal having an editor in nearby Mainz. It contained several examples of ghosts, especially the low ^{129}Xe/^{132}Xe that appeared to exist in early solar gas owing to much of the ^{129}Xe being in the stardust, and a table of magnesium isotopic anomalies expected in the initial minerals Al_2O_3, $MgAl_2O_4$, and

MgO. Each example derived from a simplified numerical model of stardust condensation in supernovae. To make my general case for the existence and relevance of both stardust and SUNOCONs, I began composing a comprehensive position paper[6], which was published in 1978. It would be my manifesto, and would remain one of my favorite papers throughout my career.

I also constructed a model having the potential to invalidate the claim of the Caltech *Lunatic Asylum* to have proven that a surprisingly large abundance of radioactive ^{26}Al existed in the early solar system. Naturally, this alternative aggravated Jerry Wasserburg, who felt that he was the discoverer of ^{26}Al-derived ^{26}Mg in meteorites. But exacerbating that unfortunate rift between Jerry and me had not been my goal; rather it was to demonstrate with a specific model how the correlation of excess ^{26}Mg with the element aluminum could be generated chemically from presolar dust without need for live ^{26}Al. My model used a concept that I gave another fanciful name[10], *charmed* ^{26}*Mg*, and which further exercised ire at Caltech even though the concept had interesting possibilities. *Charmed* meant in my picture that as the initial Al_2O_3 structures partially melted and rearranged, they carried their internal ^{26}Mg atoms along with the aluminum in unmelted cages. I vigorously defended this possibility for years in other meetings, but finally dropped it in silence when chemists presented more advanced arguments for believing that the ^{26}Al had been still alive in the meteorites.

The differences in philosophy of science that underlay this rhubarb while it raged would be described by me as follows. Chemists saw that the simplest explanation of their data is that the ^{26}Al had been still alive and that the material melted totally. They wanted the simplest *chemical* explanation. Being an astrophysicist, I wanted the truest picture of what physically occurred. What seems simple chemically may impose grave difficulty for a physical description of natural events. I did agree that if charmed ^{26}Mg would not be possible, a neighboring supernova must have injected live ^{26}Al into the molecular cloud from which the solar system formed[10]. This would have made the birth of our solar system atypical of starbirth, however, because far too few supernovae occur for them to pollute each forming new star. This position violates our Copernican instinct, namely that our solar system should not be special. Nor did I go quietly, expressing for years my physics reservations that supersonic supernova ejecta, which is made very hot after about 100 years by what is called *the reverse shock wave*, could physically be admixed into a cold interstellar molecular cloud. Mixing hot, low-density gas into dense cold gas is not easy.

The reader surely notices that the technical details of my work at this time are difficult to follow, but their thrust captures my pace during this special time of life-altering goals. I wish to share the excitement of new technical possibilities as they emerged at that time, even if difficult to understand. I wish to share that they came like a bracing rain, smacking the face and mind.

By June 30, 1977, still prior to the Cambridge *Meteoritical Society* meeting, I had finished and submitted a second paper[7] called "Interstellar Potassium and Argon" to the same highly respected journal. It involved a description of the presolar natural history of the two elements that were central to the potassium-argon dating technique. Could presolar history of stardust account for age estimates for distinct meteorites that had been based on a model of the *in situ* decay of radioactive ^{40}K to ^{40}Ar? The ^{40}K nucleus has a halflife of 1.28 billion years, nearly four times shorter than the age of earth but not so short that it is all gone today. ^{40}K is *not* an extinct radioactivity. The decay by radioactive potassium in the earth, and the subsequent escape of daughter noble gas argon from the earth's interior into the earth's atmosphere has made ^{40}Ar the most abundant of all argon isotopes in our atmosphere despite being the least abundant of argon isotopes in the universe! This provides one of the great clues to the history of the earth, a subject of growing scientific importance. In this new paper I also defined STARDUST to be a scientific word, designating by it those refractory condensates that originated within the atmospheres of red-giant stars. In those stars the *s* process had made perhaps half of all radioactive ^{40}K in the interstellar gas by a single stellar neutron capture by stable ^{39}K. A collection of varied STARDUST should include overabundances of *s*-process isotopes. I emphasized that much of the ^{40}Ar would still be contained within interstellar stardust that had condensed its radioactive parent nucleus, ^{40}K, and I gave several possible manifestations of this in early-solar-system materials.

But the story of interstellar potassium and argon by no means ends there. I added a discussion of excess ^{36}Ar, another stable abundant isotope of argon, noting that ^{36}Ar will be carried in stardust owing to decay within it of radioactive ^{36}Cl made in nucleosynthesis. I also returned to more details of my prediction from 1975 that the element potassium would be marked by excess stable ^{41}K within calcium-rich interstellar dust owing to the nucleosynthesis of its parent, radioactive ^{41}Ca. These details do not bear repeating here, because they are difficult; but the point is that I did this paper swiftly in the hopes that the systematics described by me would accelerate experimental identification of presolar stardust in meteorites, a goal that I stated explicitly in its final paragraph. It was a breathtaking time.

Subsequent decades would show that in these papers I had gone overboard with enthusiasm for the possibilities inherent in stardust; but the eyes of the world were nonetheless opened to a presolar history that they had never dreamed existed. All of these papers and more to follow were submitted by me as their single author, in contrast to the multiauthor papers with my Rice colleagues in the early 1970s. The ideas flowed to me too quickly to entertain even the possibility of coauthors!

Family issues; getting acquainted with Germany

Today I am surprised that I was able to initiate such an energetic *blitzkrieg* into my new world of meteoritic science during the first half of a year that was also full of major personal events. I was simply so intensely motivated that ideas arose and were developed even while taking a busy family-oriented look at Germany. A pregnancy test at the Heidelberg University Women's Clinic confirmed on January 24, 1977 that Annette was pregnant. This was joyful news, but anxiety causing within Annette, and therefore within me as well. Morning sickness commenced in her almost at once and was frequent for a few months. But we enjoyed reconstructing the timetable for conception and placing it during Christmas holidays in Wutöschingen. I experienced some hope of redemption when thinking of creating a new family, and it revealed itself in helpful optimism.

In February we had a lot of travel, beginning on February 4th with a visit home to Wutöschingen to share the pregnancy news with her family. That was an exciting priority, and we were toasted by the Hildebrand family. Two days later we drove off to Munich for my first visit to *Max Planck Institute for Extraterrestrial Physics*, where I met many who would be colleagues in gamma-ray astronomy. We had dinner with Klaus Pinkau, Volker Schönfelder, J. Trumper and George Gloeckler, all regarded names in gamma-ray astronomy, following my invited seminar entitled "Prospects for Gamma-Ray Astronomy". A close relationship to them was forged and, by extension, to the instrument that they were proposing for *Gamma-Ray Observatory*. Then back to Heidelberg where on February 16 we saw our first ultrasound images of our daughter. Then on February 20 we were off again to Rottach-Egern on the Tegernsee south of Munich, where we were honored guests of the Humboldt Foundation for their annual welcome to new Senior Scientist Awardees and where I delivered my inaugural address, "Dust to Dust: Controversy over the Origin of the Solar System". Then back to Heidelberg to work on finishing my papers, then back to Wutöschingen, and so on, and so on. These dates and venues are recorded in my daily diary entries, as was my custom. How was such an energetic pace possible? Somehow the poetry of living in Germany, now with a German family, augmented the excitement of introducing cosmic chemical memory into cosmochemistry. So miraculously, its applications sprinkled from me like salt from a salt shaker.

Significant visits to other German science centers continued unabated. In late April we were invited to Darmstadt, where a historic confluence of talent in nuclear astrophysics in Europe was temporarily employed at the Technische Hochschule. I met several in that single visit who would become world leaders in nucleosynthesis: Mounib El Eid, Ewald Mueller, Marcel Arnould,

Wolfgang Hillebrandt, Friedl Thielemann and Koji Takahashi. I snapped an historic photograph of that remarkable group[9]. Few other photographs capture so many that were to mold the future of nuclear astrophysics in Europe at a moment of its rebirth.

We followed that trip with one to the leading meteoritics group in Germany at the Max Planck Institute for Chemistry in Mainz. Very genial discussions occurred with Fred Begemann, who was conducting an experiment to search for stardust using my prediction of excess ^{44}Ca in titanium-rich rocks, and with Heinrich Wänke, a leader in the compositions of the planets and who as group director invited me to a wine garden following my seminar, and with Herbert Palme, Ludolph Schultz, and many more. Their demeanors eschewed the contentiousness that I experienced among American meteoriticists, preferring instead to understand the nature of my predictions. Their good will assured me that my European plan was the correct one. My invitation to Bern in June by the leading Swiss research group reconfirmed this good feeling. There I met Johannes Geiss, Peter Eberhardt, a future president of *The Meteoritical Society* who was zeroing in on the ^{22}Ne-rich neon that I predicted to have resulted from ^{22}Na decay in stardust, Hans Balsiger, who would later become a researcher at Rice University, and Kurt Marti. Following my seminar we also had a friendly *Nachsitzung*, that lovely European hospitality of gathering for drinks and food at a local restaurant or wine garden. I did not meet a single European scientist that I did not like or who castigated my ideas the way prominent Americans had done.

Next day I visited a great shrine, the apartment that Albert Einstein had lived in as a young man while he worked in the Bern Patent Office and, incidentally, created the theory of relativity. This was one of the magical moments of creativity of the 20th century. I stood unspeaking for ten minutes, seeing only a nearly bare room and an open window to old-city rooftops. I was humbled, almost in tears, and, for once in my life, at a loss for words.

The Meteoritical Society meeting in Cambridge occurred so late in the summer of 1977 that I worried about Annette flying home to Houston in time for our daughter's birth. *What if the airline would not let her board*, I worried. The pregnancy continued to make her sick and tired. Nonetheless we were able to enjoy the return to Cambridge while I presented my contributed papers. At the banquet of the society, held in St. John's College, hosting everywhere for me ghosts of Fred Hoyle, Bob Clayton turned to me, wine in hand, and said "I don't think you have a ghost of a chance of being right"; but his criticism was good natured, and in the happy boisterous setting I was unsure which of my many predictions and models he was speaking of! I presumed he meant the model for the origin of the CAI that I had just presented in my talk.

On August 11 we were allowed to board our flight to Houston without facing questions about imminent birth, perhaps because Annette's tall and slender physique did not show at a glance that she was eight months pregnant. Initially homeless in Houston, we were invited to live with my Rice astrophysics colleague Ray Talbot and his wife Prudence until the baby was born. That generous invitation allowed us to focus on prenatal care, to have several welcome reunions with Donald and Devon, and to buy a new house on Bartlett, about a half mile north of the Rice campus. After a final ultrasound image on September 2, Annette's contractions began at 3:30 AM on September 3, and one hour later we went to the hospital. There I was allowed to dress surgically and to participate in the delivery. I cherished that so much! I could not but remember how isolated I had felt at the births of my sons, and how I could not suppress an emotional disappointment at their Caesarian deliveries. Annette and I had discussed this, and she was resolute that she intended to have a natural childbirth, not for me but for herself. At 6:51 AM our daughter Alia was delivered, a full-term baby. Completely conscious throughout, Annette received Alia onto her breast for the first ten minutes of her life. Only then was Alia removed for cleaning and rest in an incubator. I watched all this with an inner peace and with my own instinctive sense of the rightness of it all.

The author with daughter Alia, age 1 yr, in September 1978.

Donald and Devon came individually several times to speak with us about their concerns for their crippled mother. This was very difficult for me because I was grief stricken over Mary Lou's irreversibly bad luck. Donald worried that he could not cope with her outbursts of anger. She was emotionally on the edge in addition to being paralyzed from her waist down. It was heartbreaking. On September 15 Mary Lou was placed in St. Joseph's Hospital because of her unstable mental condition. So, reluctantly in some ways but enthusiastic in others, we began legal papers to claim custody of my sons. And eventually we moved Donald and Devon into our new home on Bartlett. It fulfilled my wish that when we returned to Heidelberg on May 7 for summer 1978, the boys could fly over to spend time with us following a few months of having Mary Lou's parents live with them in their home while their school year concluded. We took both boys to Wutöschingen to meet Annette's parents for the first time. That all happened without manifest problems. But the entire sequence of events had produced another problem. Years later, Annette told me that our marriage began to fail "when you took those boys into our home." Surely I had no other choice. What other course would be consistent with human love and responsibility. But I can sympathize with Annette's inevitable emotions. Annette was a young twenty-five year old mother of a new baby daughter, in love with her older husband who shared her youthful feelings, whereas welcoming two teenaged sons into that home introduced undeniable aspects of middle age. These included my arising to drive them to Lamar High School, for example, and of their teenage presence at our dinner table. I loved it because I was reunited more completely with my sons. And I sympathized with Annette; but what else could we do?

The telephone rings in *s*-process stardust

During fall 1977 I received the exciting telephone call telling me of the discovery in an acid residue of a meteorite of one of the stardust-carried isotopic patterns that I had predicted three years earlier. The *s*-process xenon isotopic pattern predicted to reside in red-giant-star stardust was concentrated in Edward Anders' laboratory in Chicago by dissolving the meteorite in acid and by examining the xenon content within tiny bits that would not dissolve. Ed telephoned me to ask,

"Whatever happened to your paper with Ward on *s*-process xenon? I want to cite it."

I told him the story of its withdrawal in 1975 following expressions of doubts by the referee for the cosmochemistry journal to which I had submitted it. He then startled me with an unprecedented surprise:

"Well, we have found *s*-process xenon contained in carbonaceous particles within the meteorite residues that will not dissolve by acids. It is nearly pure *s*-process xenon. "

Edward Anders and colleagues had not actually isolated and identified the specific structure of stardust, but they had isolated a collection of rare meteoritic grains that contained a xenon isotopic pattern that asymptotic-giant-branch stardust was expected to carry. And it was not simply *enriched* in *s*-process xenon; it was essentially pure! I realized that this would be a key victory in my war with American leaders of meteoritics. The isolation of the pure stardust was subsequently inevitable. A sweet irony surrounded this, nor was it to be the last irony that was to involve Ed Anders. Ed had been an outspoken critic of my thesis of stardust within meteorites. He called my ideas "crap" on more than one occasion, which always struck me as strange coming from one so cultured. He had with biting sarcasm referred to the stardust idea as "Clayton's rugged interstellar travelers" from the podium at a *Meteoritical Society* meeting in Bethlehem PA. And his criticism of me and my ideas did not end with this shocking phone call. They extended into the 1990s and included his opposition to my nominations for at least two high honors. Yet here he was on the telephone revealing quite forthrightly that he had discovered one of the stardust predictions that I had made. And he wanted to cite our prediction in his forthcoming paper! The important moral here is one that I love about science. When the data speak we scientists are scrupulous in citing the work of others that either led to it or explained it. It is a code of honor that Ed Anders holds as dear as I do. So my rejected paper with student Richard Ward was quickly submitted instead to *Astrophysical Journal*, where it was published[11] in 1978.

The following four years in Heidelberg continued in much the same spirit as that of these eight months there in 1977. Annette and I lived on zur Forstquelle, and for my office I shared the Joseph Zähringer Zimmer with Till Kirsten, one of my best friends for life. I attended wonderful meetings of the *Meteoritical Society* held among the finest romantic destinations in Europe—Cambridge, Heidelberg, Toulouse, Bordeaux, and Bern—as I gained maturity and understanding of meteorite science. Each year new evidences of stardust were reported at these meetings. I worked with focus on stardust and cosmic chemical memory as a new branch of astronomy, never once doubting that it would one day be seen to be just that. I maintained that it would retain growing importance in astronomy long after the bickering over its relevance for meteoritic science had been forgotten. So when I was honored by being selected as the 1981 *George Darwin Lecturer* of the *Royal Astronomical Society*, I not only accepted but chose for my title

"Cosmic Chemical Memory: A New Astronomy"[12]. I began writing this during my full year there in 1979-80 as a Fulbright Fellow.

Suddenly I was shocked back to precious memories of my Iowa childhood. I was living in Heidelberg in May of 1979 when my grandmother Kembery died at age 92. My mother telephoned to share the news. I became tearful as we talked. This dear woman and her lovely Iowa farm had lived in my memory since my earliest years. She was my German ancestor. She had ridden beside me in the car at age five during our move from Iowa to Texas, when I had suffered loss as one of life's bitter pills. As I went about my research into cosmochemistry at the Max Planck Institute, I was struck by how frequently thoughts of her and of that farm rushed back over me. This profoundly sad event in my life cemented my early insight from my fifth year: *Things will never be the same again.* On a positive note, I found that I became, at long last, ready to let it go. Grandma had taken her leave. The grip of that Iowa infancy slackened, and I became more maturely free. It had taken a long time.

Strike Two

This year would be my last residence in Heidelberg with Annette. Our marriage deteriorated badly during the first half of 1980. She seemed to feel almost frantically that she must have more freedom. There was so much experiential growth that she needed. She wanted to go to discos and dance with young guys. Perhaps her perceived need arose because I was simply not young enough. I am certainly no prize on the dance floor. I was no longer seduced by the motivations of youth, although I always tried to invite activities she loved--going to the beach, sunning, playing tennis, listening to rock music, taking auto trips with the top down, picnic concerts in Herman Park. But even with my complement of youthful zest, our baby daughter restricted activities that I found inappropriate. I was burdened by cares associated with middle age. From early childhood I lived strongly in my own mind, interpreting my experiences with thought; and sex was no exception. That now lacked the spontaneity of twenty-year olds. I am not sure that a true reason can be found; but my instinct is that it lies much deeper than such issues. So we divorced after returning to Houston in September 1980. She chose her liberated new life to be in Houston rather than in Germany, which said quite a bit to me. She too could not go back again, not to Germany, to experience what she had missed during her first time through.

I suffered painful *déjà vu*. Divorce is serious psychological business. I grieved once again over the loss of love and its dream. I was suddenly a defeated man. I worried over the daily life of my young daughter, now, in her third year, to be a victim of divorce. I worried about Annette's future because she had little training for life. I again sought out a church-sponsored help group for divorce. After taking several bad pitches earlier, this was definitely *strike two*.

Within the next half year, however, I glimpsed a changed perspective on what had happened. I had gambled on a vision of sexual love and happiness, and I had learned that many factors must be overcome by improbable unions such as ours. But I *had tried*. I had learned that sexual love with a person who seemed increasingly immature did not seem appropriate. It is not easy to love someone and support her insecurities at the same time-- unless she is your child! We regrettably did not earn a happy marriage. By not valuing what I personally brought to her life and our marriage, Annette failed that test. I should not feel blame for that. And I could not take the blame for her regarding me increasingly as her hated father. I came to admit that from the beginning I had suspected that it was not working, but, instead of calling it off early, I had nonetheless pursued my course out of not wanting to lose this pretty young woman who had once showered me with affection. I saw that I had been immature in trying to preserve that. So, by late 1981 I could confess what I had known all along but suppressed; namely, marrying Annette had been *my* mistake. Within months I noticed a growing sense of relief that it was over.

CHAPTER 16. THE BIG DATE

The date was October 3, 1980. I was in the throes of divorce. I had gone for lunch to *Sammy's*, the small cafeteria in the Rice University Student Center. Stepping with my tray from the cashier, I could not help but notice a young woman sitting alone at a table nearby. Her head was shaved bald, and it glistened in the light of the ceiling fixtures. She was smartly and imaginatively dressed. At first glance, her confident appearance did not suggest either tragedy or accident, which intrigued me. Why then the shaved head? I paused uncertainly, as my mind raced. Should I speak to her? After Annette's explosive termination of our marriage, I certainly wanted nothing more to do with women; and yet I did too. Looking for a place to sit, I carried my tray to her table and asked,

"Do you mind if I join you?"

"Please do," she replied.

"I just had to ask what caused you to have a shaven head," I said while noticing a beautiful, deer-like face, having large dark eyes.

"My boyfriend dumped me," she said bluntly, "and by the time my hair has grown back I hope to be over him."

Her determination to bounce back appealed in my own situation. I had seated myself half prepared to empathize even with a health tragedy, but instead I found fresh determination embodied in that old Broadway hit song "I'm gonna wash that man right out of my hair". I was not actually looking for a girlfriend. And I needed a kooky one like I needed a hole in my head. But I was in pain, and I was discouraged, and I found emotional kinship with this stranger. My immediate thought was that her boyfriend had made a poor decision. As time came to return to work I realized that I had to suggest something or we would not meet again. So I invited her to attend a play with me at a Houston theater. She accepted! That gave me a lift. When I arrived to pick her up at her apartment, she emerged dressed in a beaded-satin 1950's party gown. She flaunted not only her heartbreak, but also nostalgia for times past. I viewed her with some combination of humor and amazement. She had become quite memorable already.

This first date initiated a fast evolution into what would be the seminal friendship of my life. We had significant things in common. I learned that Nancy McBride had been born in Pennsylvania and high-schooled in London, England. Before returning to the US to enroll in college, she had lived and worked in the small village of Tongue on the north coast of Sutherland, Scotland

"I have been to that village", I exclaimed.

"I worked tables in the Tongue Hotel", she clarified.

She had returned to the United States to study art and architecture and had recently returned from an architecture internship in New York City to achieve her Bachelor of Architecture degree (a second, professional degree in architecture). I shared with her that I had lived in Cambridge, England during the very years of her London experience. In the same space of time we had hiked on some of the same, remote Scottish mountains. The similarities of emotions and experiences with England were uncanny. Today I am unnerved to look back and realize that the course of our subsequent 27 years was made possible by one fleeting chance encounter.

Our coincidental histories in Scotland made a special bond between us. We spent frequent, friendly time together. Our passions for England and Scotland made us feel that we shared a longer, richer history. After a few months those feelings inspired an adventurous plan that made sense only to our increasingly romantic friendship. While contemplating whether to participate in another NATO Workshop being held at the *Institute of Astronomy* in Cambridge, the idea arose to go together to the north of Scotland to rent some simple cottage for a long holiday following that Cambridge Workshop (on Supernovae). While in tongue Nancy had lived in a very simple cottage and loved it. My schedule allowed exactly a month during summer 1981 between Cambridge and *The Meteoritical Society* meeting in Bern. Captivated by our idea, we located a house near the tiny village of Achiltibuie on the northwest coast. It was offered for summer periods by its owner, a Mr. MacLennan. We sent money to reserve it from July 15 to August 15. Making that booking reignited the special emotions we shared. I had, by the way, arranged in the divorce for Alia and Annette to continue living in our house on Bartlett Street. Somewhat comforted by that arrangement, I found it to be a time to bury my daily worries about them.

I left Houston for the Cambridge meeting on June 26, where for two weeks I had a room in historic Clare College. I would deliver my talk "Supernovae and the Origin of the Solar system", rekindling at once my fire for the isotopic puzzles within the early solar system. Nancy flew into London on July 11 and took the train to Cambridge. From there we drove a rental car to Tongue, where Nancy had worked as a waitress in the Tongue Hotel. She had sought out that experience in preference to a forced return to the USA with her family, to their consternation, at the end of her father's London stint for Gulf Oil. She had been quite determined, unwilling to simply pack up and return to Houston. During our revisit to Tongue, a postage stamp of a village, Nancy relished staying in the Tongue Hotel and being on the other side of the service. Next day we walked up the hill to the simple stone cottage about one mile from the hotel in which Nancy had lived during that time. For quite a pause we savored beautiful desolate views of the stony hills and of the azure *Kyle of Tongue* penetrating inward from Tongue Bay in the north, past Tongue itself for three miles, until it narrowed to the Kinloch River that drains into it. I watched Nancy in silence as she recaptured her feelings as she silently looked about. *Ben Loyal* loomed visibly several miles further to the south. I had hiked to its summit with Hoyle and Fowler and radio astronomer

John Bolton in 1968 during my second trip to the Highlands. An uncanny bond between Nancy and me grew stronger.

On Saturday July 18 we arrived at Mr. MacLennan's cottage, *Blairbuie House*. Its name sounds grander than the house actually was. But it was perfect, a light yellow stuccoed two-story house standing alone in a hillside of sheep. We enjoyed four weeks there, in isolation from the world. I wrote on my novel based on the solar-neutrino puzzle, and Nancy edited and criticized the writing with an unerring literary eye. And on alternate days, if the weather was fine, we drove to the base of one of the many nearby mountains and hiked to its top, packing our lunch. We first climbed gentle *Cul Beag*, and cooked a leg of lamb afterwards to celebrate Nancy's maiden climb in these western highlands. One by one, day by day, we hiked lovely mountains-- *Canisp, Cul Mor*, where in the fog we encountered botanists searching for rare plants, up magnificent *An Teallach* on the day of Prince Charles's wedding to Dianna, which we picked up on the car radio after descending, over the highly eroded baroque looking pinnacles of *Stac Pollaidh*, up long *Beinn Dearg* where five deer majestically leaped a stone wall in front of us, fogged in on *Beinn Allegin* amid the geologic wonders of Torridon's pre-Cambrian quartzite, and trekking up massive *Conival*. We were irrepressible, reaching the optimism of each pinnacle. It was four weeks of sheer poetry. It was four weeks of talking through our disappointments in life. In this special time, our friendship turned to love.

In Fall 1981, as I was trying to find the way to financially rebuild after my second divorce, I invited Nancy to accompany me to look at a small house in northern Montrose District that I thought I might afford. This 1930s bungalow at 901 Bomar had original hardwood floors, set on pier and beam, in a setting of contrasts. The quiet little street corner seemed bucolic, without traffic, old fashioned, peaceful, yet the view through its live oak trees showed clearly the skyscrapers of downtown Houston only a few blocks away. Nancy looked it over with her combination of imagination and architectural eye, leavened by her strong practical inclinations. She liked it.

Essentially cashless after the divorce settlement, I asked my father if he could loan me $20,000. I hated asking, but I have always known that I can depend on my parents. My salary was adequate for ongoing costs of home ownership, and I promised repayment within a year. I could save $20,000 in the year ahead. With large down payment (about 30%) I could buy the home at a good price. My father never wavered, perhaps both of us remembering that Grandpa Clayton had twice loaned him $1000 for each of his first two airplanes. This sweet modest house became home for me, and after our marriage in January 1983, two years after our meeting, it became Nancy's first home. My daughter Alia loomed large in my thoughts. She remembers this Bomar house as one of her first homes, because as a four-year old she spent most weekends with us there beginning in winter 1981 when I bought it. Nancy helped me decorate her bedroom colorfully and imaginatively, and that sweet little four-year old daughter settled into our lives. She proudly repainted a kitchen cabinet door to help us remodel its interior. I watched all of this in stunned amazement. Nancy never resented my time with Alia. Nancy never tried to sequester me to herself. Nancy embraced Alia, and Alia her. Nancy became a tender, inspiring stepmother before she became a wife. And in one touching moment, obviously feeling sorry for me being wifeless, Alia suggested, right out of the blue,

"Daddy. You could have Nancy."

How Nancy and I loved that sweet sentiment—maybe even childhood wisdom. Alia was, from the first day, very aware of the great differences between Nancy and Annette, her mother. The amazing stability of this new friendship at my time of upheaval enabled genuine fathering to resume. Alia had a collection of stuffed animals in her room, each of which she gave a name. One of these, usually Mun-chi-chi, the monkey, was invited at day's end to share the evening highlight, story time.

"Story time!" I would announce with an air of something very special about to happen. Alia would scramble with her invited animal to her bed where the nightly sign-off occurred. I read, repeatedly, the Beatrix Potter series to her, one book nightly, starting with *The Tale of Peter Rabbit*. I had done exactly the same thing with similar enthusiasm when Donald and Devon had been preschoolers, even using the same early editions that I had purchased for them in Cambridge in the 1960s. Each child liked their idiomatic Lake-District language as much as I did, preferring these "difficult books" to the routine books written for children.

von Humboldt reprise to Heidelberg

In summer 1982 I accepted the invitation, backed by substantial financial support, from the Alexander von Humboldt Stiftung to return to Heidelberg as a follow-up visit to their original award in 1977. Their goal was that each Senior Scientist Awardee establish and maintain where appropriate a significant relationship with his German host institution. Leaving my new home on Bomar was not as difficult as leaving Nancy for a long period. I had misgivings about doing that. This was six months prior to our eventual wedding. Unsure of my own feelings, I expressed my reluctance to leave her, and she proposed to a solution. Namely, she would take two weeks off from work in an architecture firm, June 13 until June 26, to come to Heidelberg to be with me during that time. I was relieved, because, although I had not yet surmounted my previous losses to see clearly my intent with Nancy, I did not want to lose her during three months of absence. My overriding personal doubt at this time was whether I had soured permanently on marriage. I just could not resolve my doubts from two failed marriages, or whether I might win Nancy permanently. Her promised visit allowed me to tread water while I tried to figure it out. A second exciting visit was expecting my two sons, Donald and Devon, now 22 and 20 years old. They would visit me for six weeks in Heidelberg. I rejoiced at that prospect. I had never come as close to them as I had wished after my divorces, first from their mother in a traumatic time for them and later from Annette in a traumatic time for me. My personal life had hovered in a state of near ruin, so I needed time with these three important people, to rebuild if possible.

To accommodate these visitors I again obtained a nice two-bedroom apartment on Zur Forstquelle in Boxberg, in the same building that I had known now during five years of life there. It was thus in a happy frame of mind that I resumed work at the Max Planck Institute. But first I took a quick train to Munich to pick up the BMW that the Humboldt Stiftung had made possible for its awardees and which I leased for the summer at a very friendly rate negotiated by them as a BMW advertising campaign to influential Americans. What a sweet deal!

Sharing again an office with Till Kirsten, conversations with him and others largely concerned the puzzling abundances of extinct radioactive nuclei in the early solar system. Their numbers had grown. Meteoriticists paid lip service to a waiting period of perhaps 50 million years (Myr) after the last nucleosynthesis event until interstellar material could contract to higher density on the road to forming the sun. This so-called *waiting period* would allow the radioactive ^{129}I nuclei to decay from its large interstellar concentration to the low level it was observed to have when the meteorites assembled. I speak here of the isotopic ratio ^{129}I/^{127}I. But the halflives of all such extinct radioactive nuclei that were then known differed greatly. To be specific, they were ^{26}Al (0.7 Myr), ^{53}Mn (3.7 Myr), ^{107}Pd (6.5 Myr), ^{129}I (15.7 Myr), ^{146}Sm (103 Myr) and ^{244}Pu (81 Myr). In a free-decay interval preceding solar birth, if such a waiting period actually occurred, the abundances of those now-extinct nuclei would have been halved in a time equal to each halflife. With sufficient waiting period to get the ^{129}I abundance right, the shorter-lived nuclei would be essentially gone, already decayed to their daughters before the meteorites could assemble in the solar system. *Something is wrong.*

I began at once to question that notion of a dormant waiting period after the last nucleosynthesis and before the solar birth. It slipped off the tongue, easy to say, and made for transparent mathematics; but it did not work. Meteoriticists seemed happy with the simple concept; and astronomers, in their ignorance of meteorites, offered no help at all and very little interest. As was often the case, I decided it was up to me to resolve the problem. I did this alone. I continued to publish papers as sole author of cosmochemical arguments precisely because astronomers are ignorant of meteoritics and meteoriticists are ignorant of astronomy. Being knowledgeable in both, working on the boundary between those sciences, I realized that I was way ahead in the interface of these fields of research. And so my next original resolution followed swiftly.

While I was reviewing the physics of the interstellar medium, I saw a previously unknown physical condition that would greatly change thinking on this problem. The interstellar gas consists of three phases; a hot phase heated by shock waves, a warm phase of neutral hydrogen, and a cold dense phase of molecular hydrogen. New stars form only within the last gas phase, the *cold molecular clouds* as they were called, whereas new nuclei from supernova explosions would enter the dilute, hot phase. Both of those extreme phases had to mix with the warm atomic hydrogen phase. This occurs by thermally inducing a change of gaseous phase within a fix sample of interstellar gas. The mixing of these three phases in the ISM occurs by something akin to diffusion wherein portions of hot gas are cooled and become warm atomic gas, and cold-cloud gas is heated to become warm and neutral by the new stars born in the clouds. The three phases exchange matter as this occurs. This renders them identical for stable isotopic ratios, but different in their gaseous concentration of radioactivity.

Excitedly, I formulated a steady-state mixing model for these three phases. My model was designed to apply to the average ISM. The paper[1] was submitted to *Astrophysical Journal* on July 15, shortly after my sons arrived in Heidelberg. Each of the three phases contain average radioactive concentrations that differed from the other two phases. The concentration of each radioactive nucleus was greatest in the hot-gas phase, intermediate in the atomic-hydrogen gas phase, and least in the cold molecular-cloud phase. To my satisfaction, the concentration of each of the various known radioactive nuclei in the average molecular cold cloud depended roughly to its proportion to halflife, rather than *exponentially* with halflife. This important insight helped rationalize the similar concentrations of radioactive nuclei that are evidenced by early solar-system material from the meteorites. On August 11 I presented this new model in a colloquium in Munich at the *Max Planck Institute for Extraterrestrial Physics*. My son Devon went along with me for that trip. We had a very enjoyable time together. I had begun during this summer to love Devon more and more for who he had become, even as I saw more evidence of his depressed nature and knew that he was gay. I had no problem with the latter and enjoyed meeting his Los Angeles friends, while he was attending UCLA. His depressed nature, now subdues by the excitement of Europe, was my worry.

The Wedding

I almost risked losing Nancy owing to my doubt about marrying again. But she understood what I had been through, and she supported me as a friend. What gave me the courage and assurance was not growing love, for all contemplating a wedding think they are in love, but what I would characterize as *devotedness*. To those to whom Nancy pledged she remained completely devoted. I saw this repeatedly with her friends. And so she was with me. Realizing that my first two marriages had lacked precisely that ingredient, I found courage to face the third strike. Whatever my personal failings in those previous marriages, I knew that they had been to women who were not able to let devotion to our marriage be a guiding principle. Asking Nancy to marry me turned my life around. We set our date as January 15, 1983, in the Rice University Chapel. There with my three children and my parents and siblings, we had a simple but ceremonial service. Nancy eschewed the white dress in favor of a smart wool pantsuit, another statement of her independent nature. I wore my Scottish wool tailored suit. My first son Donald, who had turned 23 years old just one day before, was my best man; and my 5-year-old daughter, Alia, was our flower girl. But by the time that ceremonial day arrived, our emotional marriage was already in full sail. Our devotion had locked into place the day I found the resolve to ask her to marry me and to believe in her reply.

I can not say enough about Nancy. She is imagination combined with practicality.--not as unlikely a combination as it might sound--from the distance, standing out, something opulent; up close, practical, determined, patient, steady-- all knit together with the vision of an artist. Her nature was the key that transformed an unlikely pairing into a lasting and deep commitment. Nancy is so interesting and charming as a person that I prefer being with her to any other person. I had never previously been able to say that. It was our growing love and her exceptional nature that also freed me substantially from the sexual preoccupations that bind men. Her noteworthy beauty was not the main point. Nancy was friend before lover, helper before wife, stepmother before mother, guardian of contentment rather than an agitator. Our marriage was totally unlike my first two. Nancy recognized and preserved my thinking time and my creative work, and she came more and more to protect me from the daily demands of our complicated world. She took over our financial affairs because she was so good at it. She handled family situations with sensitivity and generosity. More important was our growing love. We have loved one another unabated for twenty seven years. We also remain best friends to one another. As simple as this message is, it had taken me decades to glimpse it. She taught me life's lesson as no other had. I wasted decades pursuing sex first and getting to know the person later. I had been unable to handle my life with the same sure footedness with which I handled my career. Nancy enabled me to see my problems differently, and so viewed, they tend to vanish.

I loved Nancy's one liner about our marriage, "It's like a *Big Date!*"

A personal tragedy attended our wedding, however, like an unwanted guest. My father suffered a severe heart attack during our wedding ceremony. Thinking more of his devotion to us rather than his own fear, he said nothing except that he and my mother needed to return to Dallas promptly. The attack was confirmed, and heart surgery was initiated at Baylor Hospital in Dallas to insert bypasses and to cover with a feltlike patch a weakened spot on his heart. The surgery seemed a success. Nancy and I visited them with high hopes of his recovery. But the felt patch covered a bacterial intrusion, and a lengthy downward spiral was evident by the middle of summer. Even replacing the patch could not halt it. One touching memory lingers. While hope for recovery still ran high I repaid, in person, the $20,000 loan, and my father sprang up to get the IOU and wrote "Paid in Full" on it. In late summer, in dire straits and unable to converse owing to the oxygen equipment, we had our final conversation. Alone I asked if he was afraid, and he squeezed my hand once for "No." When I asked if he was worried about anything in particular, he squeezed it twice. "Are you worried about Mom?" I asked. He squeezed twice. I promised that I, and all four of his children, would care for mother attentively, and he grasped my hand and did not let go. His heart gave up on the last day of September, 1983.

It is normal for adult men to lose their fathers. Though a sad day, most of us move on with reassurances about their full lives and with understanding that this sequence for life is normal. But I was devastated. The full force hit me at the funeral, at which I, the oldest child, could not quit my sobbing as grief knotted me with pain. I was trying to navigate the loss of something of huge importance to me. More than a father, he was a god of my early years that I had temporarily forgotten, and his words from Cram Field came back to me: "OK then. You fly it." I lost about ten pounds during fall of 1983.

Springtime in Paris

Before my father's downward spiral, Nancy and I had seized on the opportunity for a special honeymoon trip during spring and summer. I contacted my old friend from Caltech, Jean Audouze, Director of the *Institut d'Astrophysique* in Paris, who had asked on several occasions if I could visit. Jean responded with enthusiasm, even after I told him that the three-week visit might be as much romance as work. French understand. Jean offered to us the third-floor apartment atop the institute. This was wonderful. Nancy touched me deeply by repeating that as nice as a honeymoon sounds, every day together was like continuing "a big date". We still carry that sweet thought in our hearts.

Before that was my son Donald's commencement ceremony at Rice University. Devon flew from Los Angeles for the occasion, despite ambivalent feelings for his brother. Don had made Devon's childhood rough, and Devon felt that they could not be close because they were so different. But it pleased Don very much that

Devon came. On Saturday May 7 we all attended a reception and lunch for the graduates of Will Rice College, my son's college. Don was graduating with a BS in geology. I felt very fortunate that he had been able to gain admission to Rice University, a highly competitive admission, and that Rice accepted him free of tuition as the son of a Rice professor. I paid only his college residency costs. This Rice degree was perhaps the best education per dollar ever obtained, and I had but few dollars available at that time. I think his four years cost me only $8,000. The commencement ceremony took place at 7:00 pm, out of doors, shaded from the hot setting sun by Lovett Hall, in whose outer balconies the brass band played "Pomp and Circumstance" and other traditionals. His maternal grandfather, Roland Keesee, drove to Houston from New Braunfels with his new wife to attend the service. It was my first time to greet Roland since my divorce from Mary Lou (Keesee), whose funeral I had been unable to attend, and I was nervous until ascertaining that he was full of friendly good will and pride. But we said little above the awkwardness. Each student crossed the podium to receive his degree from President Norman Hackerman, and I rose in salute to Don.

Another student in that graduating class was to become very important to me later. I had met Bradley S. Meyer earlier in the year when he dropped into my office with the request that I suggest a senior thesis topic. Brad was one of our incredibly talented physics majors that we have enjoyed at Rice University over the years, so his request was a no-brainer. I suggested that he work on the r process of heavy element nucleosynthesis by constructing a steady-state analysis of all branchings of the nuclear-abundance flow so that their effect might be measured. This would be a lot of work, so we met weekly for a progress report and new ideas. Brad took to it like a duck to water. We needed neutron-capture cross sections, beta decay rates, and neutron-separation energies along the r-process path. We obtained photoneutron rates by inverting the neutron-capture cross sections. Brad and I cobbled these together from various sources, including some guesses. I have retained my copy of his thesis, which was very interesting to me, allowing me to understand details than had not been possible in my 1965 paper on this subject (with Phil Seeger and Willy Fowler). Not surprisingly I was eager to recommend Brad strongly to the University of Chicago for his graduate work (inasmuch as Brad wanted to get away from Rice, a motive I supported). There he continued studies of the r process with Dave Schramm taking over the advisor role. Dave was extremely happy to have a student who already knew about the r process.

Let me add another word about the scientific growth of Brad Meyer. Brad made many long trips from Chicago to Livermore to learn more nuclear theory and nuclear systematics. There he became good friend to an earlier Ph.D. student of my own at Rice, William M. (Mike) Howard. They also shared interest in the r process. By that path Brad began, became, and remains a world leader on r-process studies. And here is my point. Brad was later to become my close colleague during the last decade of my career at Clemson University. I had no inkling of this significance for me as he obtained his own B.S. degree that lovely May evening alongside my son.

On May 9, after the close of classes and before my father's severe decline was clear, we flew to Paris. On Boulevard Arago, near the historic Paris Observatory and the Luxembourg Gardens, we set up house in the apartment of the institute and had a marvelous time. We shopped the open market on nearby Rue Daguerre, where we purchased food for meals that we could cook in the apartment kitchen. French restaurants are wonderful, but not nightly. Some of our own cooked meals were memorable—a whole cauliflower, raw calamari that we sautéed in butter, countless baguettes and cheese, French wines, and so forth. We discovered couscous in an Algerian restaurant on Rue Daguerre, and it became a favorite of our later home cooking. On a daytrip to the Cathedral of Chartres we lighted candles in ceremonial supplication for my father's struggle. But most of all, we enjoyed our intimacy with the streets of Paris and with each other, unburdened by life at home. It was a romantic time.

Jean and I met every few days for a discussion of the chemical evolution of the galaxy resulting from ongoing nucleosynthesis, and I presented one institute colloquium[2] "Discovery of s-process Nd".

On that day Nancy took a nice photograph of Jean and me that can be viewed on my web site for the history of nuclear astrophysics <1983 Audouze and Clayton at IAP>. My institute colloquium illustrated the astrophysics of cosmic chemical memory by using my recent discovery[2] of s-process neodymium isotopes in a carbonaceous residue of the meteorites. This was the same residue in which Ed Anders had found s-process xenon. These discoveries were news to everyone present. As usual in my experiences, astrophysicists had at that time not yet heard that cosmic chemical memory had been discovered in meteorites, and that the s-process isotopes could be seen not only as a theoretical component of the average solar composition, but also as standing alone within some dusty interstellar carriers of cosmic history. My friend and colleague Guenter Lugmair, a leading nuclear chemist, had discovered[2] isotopically anomalous neodymium in a carbonaceous residue of the Allende meteorite (which had fallen in Mexico in 1969, just as cosmochemical labs were tooled up for the Apollo 11 sample returns, and which launched many new discoveries in isotopic anomalies). Using his choice for normalization of his data, Lugmair suggested at the 14th *Lunar and Planetary Science Conference* that the anomalies had resulted simply from the radioactive alpha decays[2] of ^{146}Sm and of ^{147}Sm to ^{142}Nd and to ^{143}Nd, respectively. As usual, the situation was quite unclear. In the first place I renormalized Lugmair's data in a different way. By normalizing to somewhat different mass-dependent isotopic fractionation, I argued that the pattern of isotopic irregularities was instead the s-process Nd pattern carried in presolar carbonaceous grains. I attributed this to stardust collected into the meteorite at the time that it formed. In my published paper I wrote explicitly: *The astrophysical scenario is the condensation of dust from a red-giant atmosphere, perhaps a carbon star, where s-process abundances are expected in the presence of abundant carbon.* This was radical at the time, being four years before the mainstream SiC grains were discovered, experimentally isolated, and shown to carry s-process Nd as I had predicted. To obtain more exposure I decided to submit an abstract on the same topic to the 1984 *Meteoritical Society* meeting to be held in Albuquerque.

One should not be surprised that astrophysicists were unaware my cosmic chemical memory ideas. Active scientists can often be unaware of important new developments in their general discipline because we so single-mindedly pursue our own immediate specialized goals. And that occurs for good reason. That is what it takes to contribute on the frontiers, where the jack-of-all-trades falls quickly behind the technical expert on a research frontier. My personal satisfaction was that my technical insight lay in joining astrophysics disciplines in unforeseen ways into an emergent boundary discipline.

Before departing Paris for Heidelberg, we rented an Opel Kadett for touring the Loire valley and visiting its historic towns, Blois, Chenonceaux and Amboise, spending three charming nights in that fabled valley. In Amboise Nancy described her fascination with Leonardo da Vinci, who had lived there late in his life and who had become her hero when she studied drawing at *The Art Students' League* in New York. I purchased several bottles of wine in Chablis to take along to Heidelberg, where on Sunday June 5 we settled into our apartment on zur Forstquelle. That very apartment house had been my second home during these seven years.

Nancy at coffee on the Heidelberg Marktplatz.

That evening we were invited to Till and Emiko Kirsten's home for a welcome dinner. I took along one of the bottles of Chablis, chilled just right. And as usual, Till challenged me to a game of chess, about which he is a lifelong fanatic and which he usually won, but into which Nancy interjected amusing remarks; *e.g.*

"It seems to me that your queen in useless, Don!"

We visited often together until our departure on August 9; and as always I shared Till's office with him. We debated the GALLEX solar-neutrino experiment and talked about isotopic anomalies. As was our custom Till and I sought music[3] throughout the summer, finding *Cosi fan Tutte* at the Mannheim Opera House, more Mozart, the *Knabensänger*, in the Rokokotheater at Schloss Schwetzingen, where we often attended the excellent music festival, a Bach Cantata in the Peterskirche, a harpsichord recital in the Alte Aula, the Brahms Requiem in the Peterskirche, and, not least, several of Peter Schumann's Bach organ recitals at the Heiligegeistkirche on the market place. Heidelberg is a music paradise in summer insofar as performances by dedicated musicians in intimate settings are concerned. I love that about Germany.

I had been really proud to introduce Nancy to Till and Emiko. She charmed them immediately, as I knew she would.

"You made the right choice this time!" Till blurted with his usual delicacy.

I described my planned work on publications concerning cosmic chemical memory. The Max Planck Institute, through Till's sponsorship of me since 1976, had done much to support development of that theory. I described ways in which our team for *Gamma-Ray Observatory* was confronting our promising future in astronomy of radioactivity. In the first days in residence I participated in the annual celebration of the science year at Max Planck Institute, which featured invited talks and food and wine— *lots* of food and wine. Such occasions are delightful in Germany, which has an extremely gracious knack of celebrating science.

The *Oriented Scintillation Spectrometer Experiment* (OSSE)

The proposal from *Naval Research Laboratory* to *NASA* advocated building a scintillation spectrometer to measure the rate of arrival of gamma rays from a collimated view of the heavens and to measure their energies with about eight percent precision in the energy region where nuclear gamma-ray lines would exist. It was an evolved version of the balloon-borne gamma telescope that Bob Haymes had flown for his pioneering Rice University research, and both its Principal Investigator and Chief Scientist had been trained at Rice in Haymes's program. It was a traditional design and NaI detector that was known to work. The characteristics of this instrument featured capability for discovering my targets for supernova-produced radioactivity[4], which was one of its main advertised goals. Our proposal was funded and became one of the four instruments on the *Gamma Ray Observatory*, the second of NASA's great space observatories. My role was theorist and spokesperson for the role of supernova radioactivity. But first the funding had to be obtained from Congress. The price tag was too large to be absorbed by the NASA budget without a line-item augmentation of NASA's budget for that purpose. My letter-writing effort in 1979 with Reuven Ramaty was successful, eliciting many forceful replies from prominent scientists in favor of that funding. It worked.

Funding success was followed by bimonthly team meetings at the *Naval Research Laboratory* throughout the 1980s. There the hard work of building and testing the telescope and of planning for the observing strategy and for the required software for the mission was done[6]. This program provided a postdoctoral job for one of my best Ph.D. students at Rice, Mark Leising, who moved to *Naval Research Laboratory* following his Rice Ph. D. thesis on gamma-rays from nova explosions in order to continue work with the *OSSE* team.

Before settling in Heidelberg to a few weeks intense work, I took Nancy for a townhopping tour along the famed *Romantische Strasse* in northern Bavaria, visiting the fabled medieval towns Dinkelsbühl, Nordlingen and Rothenburg ob der Tauber. We spent one night in each and had a glorious experience doing so before returning to Heidelberg through Würzburg. In Würzburg I reflected on my now discarded plan of becoming more German in my own life. Nancy and I looked at *Germania Haus*, standing majestically across the "lion's bridge" over the Donau (Danube), where Annette's father had included me in two reunions with his university fencing club (with sharp foils, the scars from which were worn as badges of honor). Nancy could see from this short tour why I loved Germany; but over one dinner with Franken wine we agreed that Germany was not in our future. I no longer entertained any thought of making half a life there.

On June 29 I made my last visit to Karlsruhe, to the experimental *s*-process research program that I had helped initiate six years earlier. Our large joint paper summarizing *s*-process science in the light of their new experimental neutron-capture cross sections had just recently been published[5]. This hugely successful research program had made me very proud, although my role had been only that of consultant. Their measurements moved *s*-process nuclear data to a new level of precision, replaying my similar experience at Oak Ridge two decades earlier when they had begun the first experimental program of measuring neutron-capture cross sections for the *s* process. It has been said that one loves the topic of his PhD thesis for his entire life, and, for me at least, that has been the case. It was first love, the sweet awakening of science passion. I seemed to be examining my scientific life, turning it over in my mind, viewing it from distinct angles, while my love for Nancy deepened its insight as well.

Surprised by gamma rays from interstellar ^{26}Al

For the next month in Heidelberg I feverishly studied the published paper[7] reporting the detection of aluminum radioactivity in the interstellar medium. A NASA space mission measured the gamma-ray lines emitted radioactive ^{26}Al decay. I was in considerable consternation over this exciting discovery, and it would not let me go. On the positive side I was thrilled that the new astronomy advocated by me now had its first successful measurement. On the negative side, I had not predicted successful finding of radioactive ^{26}Al in the interstellar medium. This caused me actual anguish, because I had long made it my mission to lay out the observable gamma-ray targets for that community of scientists. But I had rejected the first radioactive emitter to be discovered! I had considered it and rejected it. Even in small things, I hated to make a mistake. But such a big thing was too, too much.

What was this event that changed nuclear astrophysics? The third *High Energy Astronomical Observatory*, a satellite dubbed *HEAO 3*, had made the measurment of the rate of arrival of 1.81 MeV gamma rays from central directions of the Milky Way. The ^{26}Al halflife is 0.74 million years (Myr), long for the life of a man but short for the Milky Way galaxy. This first detection of radioactivity in the interstellar matter opened a door of gamma-ray-line astronomy as an observational science more unequivocally than had Bob Haymes's pioneering research at Rice University. It was a moment which I had long awaited, awaited with nervous intensity.

I had considered this possibility before rejecting it. I had done so seven years earlier, when nuclear chemists were able to show from aluminum-rich rocks found within the meteorites that radioactive ^{26}Al had been present in the early solar system. Its abundance within the early solar system, as measured within meteorites, had been found to be 45 atoms of ^{26}Al per million atoms of stable ^{27}Al. That abundance was very large, so large that I quickly demonstrated that it could not possibly reflect the level of ^{26}Al radioactivity in the general interstellar matter. Instead, some special event must have created that ^{26}Al abundance during the process of formation of our solar system. Because the meteorite concentration of ^{26}Al could not be universal, it could not imply that gamma rays from it could be detected. Discovery of interstellar ^{26}Al radioactivity seemed to not have anything to do with the large concentration of ^{26}Al in the early solar system. My initial conviction remained correct. Where then had I gone wrong? Because this was so significant in my life, I here recount the resolution in some detail.

It would take many years to learn that my errors had been of three kinds, each devilishly conspiring to raise the expected observability from that of my first estimates. Firstly, the supernova models that I had used underestimated the amount of ^{26}Al that could be ejected from a single supernova. Secondly, astrophysical misunderstanding of galactic chemical evolution as metal-free gas falls steadily onto its disk underestimated the size of the $^{26}Al/^{27}Al$ ratio that should on average exist in today's ISM. Thirdly, ignorance of the ways that galactic supernovae cluster together in large numbers, thereby making the emission from that cluster uncharacteristic of the galaxy as a whole underestimated the amount of ^{26}Al that could be expected within localized regions of active star formation. So I had underrated and discarded the first galactic radioactivity to be found. My joy at the end of my own seventeen-year wait was dampened by chagrin that I had discounted what was to become the most important single nucleus for studies of the Milky Way nucleosynthesis. This siphoned wind from my sails. Because science involves the cooperative community of scientists, the self-centeredness of my initial disappointment is somewhat embarrassing. I had hoped to predict the entire field prior to its discovery, and the initial detection had put me in my place. Science is humbling.

I now made it my goal to clarify what had been wrong with my initial rejection of this possibility. About six 1.81 MeV photons strike, on average, a one-meter square surface per second from the general direction of the galactic center. This is a small rate of arrival since the actual detector was much smaller than 1 m^2. Even the small observed rate implied a lot of interstellar ^{26}Al. To achieve the observed flux requires that about three solar masses of radioactive ^{26}Al nuclei be spread throughout the interstellar gas. My previous calculations had estimated that no more than about 1/40th of that amount could be maintained by the ejection rate of ^{26}Al from supernovae. In light of this conflict I made calculations in my Max Planck Institute office for a paper[7] showing those contradictions. They exposed serious flaws in optimistic arguments suggesting that the observed three solar masses of ^{26}Al can be maintained from the 20,000 supernovae that are thought to have occurred during the last million year lifetime of ^{26}Al. My initial estimate of galactic ^{26}Al abundances based on their supernova creation rates for both aluminum isotopes showed that that idea did not work at that time. Subsequent calculations of supernovae by others showed that hydrogen burning in presupernovae created more ^{26}Al than did my initial estimates based on the aluminum producing cores of massive stars.

The invoking of collected historic supernovae was not novel, as that situation had been addressed in my 1971 paper[7] showing that ^{60}Fe radioactivity would have the same problems—too slow in its decay to be seen from a single supernova but visible from the 20,000 collectively expected during the halflife of the emitting nucleus. A single supernova, on the other hand, would have to have been very close, only about 14 pc distant, in order that its ejecta as a single event could be seen. Faced with these logical obstacles, I argued in this paper[7] that the ^{26}Al was the instead the result of nova explosions, which are 1000 times more frequent than the supernovae. In effect, I did what one should not do, namely, to argue that something observed can not happen! My counterarguments were, however, correct. The community could find no error in my conclusions. Another decade would pass before the resolution of this conflict could be understood. My paper[7] was submitted to *Astrophysical Journal* from Heidelberg.

An aviator spins down, but a second father soars

Nancy and I came home to my father's final chapter. On August 31, after our return from Heidelberg to Houston, reluctantly skipping the *Meteoritical Society* meeting in Mainz, my mother called to report that my father needed another surgery owing to infection under his heart patch and to an aneurysm. So on September 4 we were off again by car to Dallas's Baylor Hospital. He made it through that surgery and looked quite good on the day after, so my hopes returned to their usual high state. The doctors explained his status to all, and expressed hopefulness that they had excised the seat of infection that had been requiring a constant heavy dose of antibiotics. We returned to Houston optimistically. On September 19 my brother Keith called again to report on Dad's good condition. On September 24 I called Mom and Dad at the hospital to congratulate them on their 50th wedding anniversary. He was gowing weaker. Then unexpectedly my brother Keith called me on September 30 to say that the doctors had concluded that Dad would not make it. My father died on September 30, 1983. My mother called me saying simply, "He's gone, Don". The infections caused by the bypass surgery had steadily weakened him so that heart failure took him.

There then followed an intense inner emotional readjustment within me. Miraculously I soon found that I was accepting a new maturity, as if old conflicts had been buried with my father. Like a "born-again" man, I took on responsibilities of the eldest family male, maybe even a replacement of Dad in my psychology. It is known that a large psychological readjustment on the death of one's father is not uncommon. As example, it was especially commented upon in that circumstance by Nobel Prize winning physicist, Wolfgang Pauli, in a letter to his friend and correspondent, psychiatrist pioneer Carl Jung (as described by Pais's biography of Pauli, which quotes it):

> *On November 4, 1955, my father died at an advanced age. That leads to considerable change in the subconscious, and I suspect that for me this means a change of the shadow, since for me the shadow had been for a long time been projected onto my father.*

Twenty-five years earlier, in 1958 (coincidentally the year of Pauli's death) I had written from Caltech a long, impassioned, anguished letter to my father in Dallas. In it I confessed my loss of religious faith, and followed with an emotionally charged confession that he (my father) had always provided my image of God, and that now I was lost. This was at a time of near depression about my graduate work owing to sexual problems in my first marriage. I regret that this letter to my father seems to have been lost. In later years after his death I learned that my mother had no recollection at all of such a letter; therefore I concluded that for some reason my father had not shared it with her. And at the time he either did not reply to it or replied gently and obliquely to its main issues.

In October, just three weeks after my father's death, a coincidentally momentous event occurred with my surrogate father, my "Doktorvater" as Germans say, William A. Fowler. On the

19th of October the leaders of nucleosynthesis assembled at Yerkes Observatory for the opening of the conference *Nucleosynthesis: Challenges and New Developments*. It was for just such opportunities for us together that Nancy had quit her architecture job after we married. Our life together offered so much more. We were housed in a resort hotel in Geneva, Wisconsin, and I wanted all of my colleagues, especially Willy, to get to know Nancy. Willy and I had agreed to meet at 7:00 for breakfast on that first day. Willy always preferred eating with a beautiful woman. The fall setting at Yerkes was quite striking as a distinguished body of nuclear astrophysicists looked forward to the beginning of an exciting conference. I was, however, still in daily grief over the death of my father, images of whom interrupted my thoughts at unpredictable times.

William A. Fowler at breakfast with the author on 19 October 1983, just two hours after learning of his selection for the 1983 Nobel Prize in physics.

At 6:30 A.M. our room telephone rang. It was Willy who said

with noticeably trembling voice that he had been awakened at 5:30 by a telephone call from Sweden. He was to share the 1984 Nobel Prize in physics! And he wanted to meet us beforehand so that we could enter the dining room together. Such excitement! I was so proud for him, and happy for what it also would mean for me. I was proud that he wanted to be accompanied by us in the dining room. My spirits soared at this a moment we had long contemplated, hoping that it would come to pass. And now this man that I loved as a second father was to receive this most visible of high science honors. As always I was happy basking in his good natured aura. When at 7:15 we walked together into the dining room, a growing wave of applause began building as conference attendees, who had been alerted by the organizers, noted that Willy had entered. They stood and applauded continuously. It was the headiest breakfast of my life. Dozens of people, including resort-hotel guests that were not attending the conference, walked to our table to offer their congratulations and to share their own excitement for Willy. Nancy took two or three photographs of that breakfast, one of which is displayed on my web site for the history of nuclear astrophysics <1983 Fowler Nobel Prize>. This photo was published in my later, invited obituary[8] for Fowler.

On short notice Dave Arnett and Jim Truran, co-organizers, were able to locate a sweatshirt shop which immediately printed a sweatshirt for Willy bearing the identifier "Nuclear Astrophysics 1", as on a football jersey. Wearing that sweatshirt, Willy spoke to an adoring and sometimes rowdy group of admirers on that first morning. See the group and the sweatshirt at <1983 Group at Yerkes Conference for Fowler>.

Later that day I presented my talk "^{26}Al and Galactic Chemical Evolution". In it I presented the reasons for my neglect of ^{26}Al as a gamma-ray astronomy target. One might justifiably have called it "Clayton's alibi"; but no one did. No one could correct my reasoning. I showed that ^{26}Al *should not have been detectable according to current theory*. In science it is always an exciting moment when it can be shown that a discovery does not fit current theory, because it implies that something new, perhaps an unexpected correction to the theory, is coming about. No one could find an error of my arguments, however; and so we all accepted the ^{26}Al birth of gamma-ray-line astronomy as a puzzling affair—an unexpected gift from scientific providence. This has always been the main reason in science for performing new experiments or observations— that the results may surprise current wisdom and lead to new understanding.

I pressed onward into that theoretical breach. To the conference proceedings I would submit a quite different and new theoretical framework for the problem of interstellar ^{26}Al. In that 1985 book[9] I found a way to construct analytic mathematical representations of the chemical evolution of the galaxy in which the rate of star formation was chosen to be proportional to the mass of interstellar gas, which I therefore called *linear models*, and in which all quantities evolve as time passes, including the infall rate of metal-poor gas onto the disk of the galaxy. Sensing its great originality and utility, I titled my proceedings paper "Galactic Chemical Evolution and Nucleocosmochronology: A Standard Model". I was able to represent with simple mathematical functions the time-dependent solutions to the differential equations of chemical evolution. I was surprised that, for this venerable problem, I had discovered overlooked analytic solutions that were much more versatile and physically plausible than previous oversimplified steady-state representations. My solutions showed how the interstellar gas mass and the star formation rate first grew as matter settled onto the galactic disk, until both later passed through a maximum and then declined with time as the infall rate declined. These solutions included the interstellar abundances not only of stable species (*the metallicity* as astronomers called it) but also of the radioactive chronometers of astronomy. These chronometers displayed a much more correct and physically realistic interpretation of the relationship between the age of the galaxy and the radioactive clocks—the uranium isotope ratio in solar matter, and also of the ratios U/Th and ^{187}Re/^{187}Os. Fowler and Hoyle and many others had tried to treat those radioactive chronologies with their arbitrarily parameterized models of the evolution of those radioactivities, models that were in fact physically unrealistic.

My analytic standard model[9] for galactic chemical evolution would not be thought of as a realistic depiction of the many local details of galactic chemical evolution, which is an exceedingly rich topic. What it reveals explicitly are the properties of radioactive decay in an interstellar medium that is being reduced by star formation and growing by return of matter from stars (fresh radioactivity) and from infall onto the galactic disk. The solutions reveal generic properties of that interplay. My achievement in this was similar to the start of my career with the s process, in the sense that I interjected time dependence into problems that had been thought of in a stationary steady state. The time-dependent mathematical treatments exposed unforeseen properties of an old problem.

This topic took me back with joy twenty years to my early fame with radioactive chronometers, a topic that has motivated much of my career. My analytic standard model[9] showed one novel and noteworthy feature about extinct radioactivity, those radioactive nuclei that were alive in the ISM but whose halflives are too short to be found on earth today; namely, the interstellar ratio $^{26}Al/^{27}Al$ was increased by a factor (k+1) over previous calculations, where k is about 3 or 4. This provided a factor of 5 in the hunt for the reason for the surprising detectability of ^{26}Al in interstellar matter by *HEAO 3*! The physical reason for that factor (k+1) is the dominant sequestration of the old stable ^{27}Al within the interiors of old stars. That effect provided amplification of the relative amount of new, radioactive ^{26}Al in the gas phase. That effect is more subtle than it at first appears, because it does not occur in simpler models having no infall of gas onto the galaxy. Because its beauty was poorly understood by astronomers, and is even poorly understood today, I would publish ten years later[9] a focused analysis that showed how that factor (k+1) and the gradual increase by a factor six in Stan Woosley's supernova models for the production ratio for $^{26}Al/^{27}Al$ conspired to make the interstellar mass of ^{26}Al about 30 times greater relative to that of ^{27}Al than had been expected in 1983.

So I thereby solved the very problem that had misled me, and which had caused me to discard ^{26}Al as an observable radioactivity for current or planned gamma-ray astronomers. All of this would receive yet another new shock after the launch of *Compton Gamma Ray Observatory* in 1991.

Reading this story one may be struck by my foolishness in pretending to be able to predict all details of gamma-ray-line astronomy and in my self vindication that partly motivated my search for the new standard model of galactic chemical evolution. There is no denying it. I was trying to show that I had been justified in my neglect of ^{26}Al as a gamma-ray observable even while I had struggled since 1975 with the equally surprising puzzle of its large initial abundance in the solar system. My new analytic models of galactic chemical evolution were essential in surmounting, by the factor (k+1) resulting from the infall of metal-poor gas onto the galaxy, one of my obstacles to supernova nucleosynthesis of the observed amount of interstellar ^{26}Al. This was as much satisfaction as I could salvage in this time of my chagrin, to have turned my oversight into theoretical innovation. One sometimes hears it said that because computer codes are required in order to evaluate details of chemical evolution, the analytic models are not really significant. That is, however, misleading. Computer codes yield only tables of numbers, sterile in their ability to explain key physical phenomena, whereas good analytic models bring their causes into sharp focus. Mine was such a new theoretical tool.

Coming home

I was in such stress over my father's long downward spiral during all of 1983 that it is evident in retrospect that some regular activities throughout the year were invaluable to me physically and emotionally. The first of these was running, most often the three mile dirt track around the perimeter of the Rice University campus. Usually I ran it once (taking about 24 minutes in 1983), but sometimes twice on those days when I sensed that I especially needed it. That Nancy was a fitness freak who would often run with me made it better and easier. One acquires a special all's-right-with-the-world feeling in aerobic running, a fact known to all aerobic runners.

Second was manual labor on our house at 901 Bomar in the

Montrose area of Houston. This 1920s bungalow was quite charming in its down-in-the-tooth ways. With Nancy's architectural eye to guide us, we stripped extraneous additions from the walls and from the kitchen and bathroom, and we repainted the entirety. We tore up the rotted floor (termites) of the add-on back office, treated for termites, and replaced it with nice oak floorboards. Nancy produced visually modern and appealing interior color plans. Such work often consumed weekends, sometimes with my son Donald helping and usually with my daughter Alia joyously doing suitable tasks that Nancy defined for her with sensitivity and sweetness. She painted an entire kitchen cabinet. This modest three-bedroom house will forever be endowed in my mind with a sense of reconstituting and reorganizing life after personal and financial catastrophe.

Thirdly was the adjustment of Alia herself. As a result of careful negotiations with her mother, it became possible for Alia to spend most weekends with us. I could fetch her from Kindergarten at *Poe Elementary School* in Southhampton (just north of the Rice campus), where my sons Donald and Devon had also attended elementary school 10-20 years earlier. Owing to concerns about her daily life with her mother when she was home from her stewardess job with *Continental Airlines* but with sitters when she was not, it was a blessed relief to have her at home with us.

Daughter Alia with Nancy and me at the Bomar house.

Nancy never once objected that her presence restricted more adult things that we might otherwise have done, and my gratitude for her attitude would last a lifetime. After the degree to which Annette had resented my two sons, it was such a relief that Nancy embraced and made the best of the analogous situtation. But more than that, Nancy became a special friend to Alia, inventing projects that they could do together. Their special bond has persisted into Alia's adulthood. In high spirits on those weekends, I often cooked a totally mediocre dinner that Alia regarded as a special delicacy—chili, beets and rice; and every night, announcing "Story Time!" with an undercurrent of excitement and anticipation before reading to her one of the Beatrix Potter story books.

The emotional meaning of *story time* was not always evident to me; but it is now. In the first place it was my way to give back the love that I had received abundantly from my parents. It was not only an act of fatherly love and intimacy; it was also one way in which I held on to the dream of a loving and contented family. Twice now that goal had escaped me. The goal had also become entwined with my emotional embrace of *the shire*, of my own concept of England, where Donald and Devon had lived and played in gardens every bit as English as Mr. MacGregor's, whether backing onto the Trinity Hall Sports Ground behind *White Cottage* or behind *The Grange* in Great Wilbraham, alongside *Little Wilbraham Waters*. Beatrix Potter's complete Englishness, with which I had so bonded during my years at Hoyle's Institute, seemed to me to describe an orderly world having everything in its place, and that place very near to God's design-- though few Englishmen would concur today with that rosy view. No matter that this was my romanticizing. No matter that I emotionally sought the family farms of my childhood. My yearning and my love for family were real feelings. They were and are genuine. Their force had enabled me to grasp the threads of my own life in 1983 and turn it miraculously around, finding my family goal at last in the arms of the best friend I ever had.

CHAPTER 17. A BIG STAR FALLS HARD

The bigger they come, the harder they fall

—Robert Fitzsimmons, prizefighter

On January 28, 1986, shuttle *Challenger* exploded on its way toward orbit. The rubber O rings that were supposed to seal the joints in the solid-propellant rocket boosters failed, allowing exhaust to burn a hole in the fuel tanks. Seven astronauts were killed almost instantaneously, including Christa McAuliffe, NASA's first "teacher in space". Prior to the launch children around the USA watched Christa McAuliffe explain the science teaching experiments that she was prepared to demonstrate. A mere 73 seconds after liftoff, 8.9 miles above the Atlantic Ocean, her life ended before the eyes of many an idealistic child. Ten years later National Public Radio was to conduct interviews with those children, now in their twenties, and we heard their reactions-- first confrontations with death, first doubts of NASA or of the US government, first awareness that life can end in an instant. It was indeed a great tragedy. But one thing is clear. There was from the first, on every shuttle launch, a chance of disaster. The one certain thing that Americans had to confront was our cherished hope in infallibility. Collateral damage affected me; namely, that disaster also delayed the launch of NASA's *Compton Gamma Ray Observatory*. I am sometimes asked what we missed by its not getting up on schedule in 1987. We missed the astronomical event of the century—Supernova 1987A (*SN1987A*).

What we all feared while preparing GRO for launch was not a shuttle disaster but that no supernova would occur sufficiently nearby during its mission to provide an observable flux of gamma rays from the radioactivity. Their rate of arrival at earth declines as the inverse square of the distance to the supernova. Standard astronomical estimates are that a supernova occurs in our Milky Way Galaxy about every thirty years, although unpredictably spaced in time. Therefore the most likely outcome was that during a five-year mission we would have no suitable supernova within our Galaxy. We could live with those odds because young supernova remnants will still emit detectable gamma-ray lines for a decade at least. Then, as if by miracle, fate arranged that such a star would explode at just the right time.

Explosion may not be the best word to suggest that its bright visual display actually happened as consequence of its central mass falling onto itself because of the strong gravitational pull of its own mass. Its core collapsed onto a hard center having the huge density of the atomic nucleus. The star fell hard, leaving a gravitationally bound neutron star or a black hole at the center and expelling the overlying mantle. Nineteen neutrinos observed over about ten seconds after the collapse of SN1987A attest to the ferocity of that core collapse. It is amusing to be reminded that SN1987A actually exploded about 170,000 years ago, during pre-human prehistory, in the Large Magellanic Cloud, a small satellite galaxy attached to our Milky Way; but its gamma rays, traveling at the speed of light for 170,000 years to cover the 170,000 light years of distance to the Large Magellanic Cloud, reached Earth in the very year scheduled by a maturing mankind for the launch of *Compton Gamma Ray Observatory*. That cosmic coincidence is the irony. I often joked, "God was kind to arrange this. But a leaky O ring fouled it up!" Gamma Ray Observatory would not be launched until April 5,

1991, more than four years after the light from supernova 1987A reached earth.

The most intense predicted radioactivity was that of ^{56}Co, which is produced when the radioactive but short lived ^{56}Ni decays to it. The ^{56}Ni nuclei are the nuclei created by the explosion. The supernova is a dense ball, more massive and thicker than the sun, so those gamma rays cannot escape from the deep interior in the early months. The radioactivity is initially imbedded too deeply within the explosively expanding mass. One can no more detect it immediately than one can detect gamma rays from natural uranium decaying in the center of the Earth. But the exploding supernova is, unlike the Earth, flying apart at speeds near 5000 km/sec, or 3000 miles per second. That is fast! According to calculations, that expanding mass would become thin enough for the cobalt gamma rays to emerge after about one year, which would have been in early 1988, near the time of our scheduled launch. So the timing had indeed been almost perfect! We expected[1] the ^{56}Co gamma rays, especially those having energies of 847 keV and 1238 keV, to be visible for about three years. But as the *Challenger* hearings were unfolding, it became apparent that we would not get up in time to seek them. I felt deep disappointment. The gamma ray detection would have provided the first unambiguous proof that the iron of our world had indeed been created as radioactive fallout from ancient supernovae. Fortunately, that proof would emerge from observations of supernova 1987A by other research groups.

But there was a second, more subtle, loss associated with our being grounded in 1987. We could have measured details of the structure and mass of the exploding envelope above its collapsed core. By the year 1987 most astronomers no longer needed proof that the elements were synthesized in supernova explosions. They accepted that. The circumstantial evidence for that belief had mounted since the 1960s. Detecting the gamma-ray lines would, in that regard, simply prove it. Nor was it so pressing to demonstrate that iron was ejected in the form of radioactive nickel rather than as iron itself. Stable, rather than radioactive, iron isotope production and ejection had been championed during two decades, 1946-1966, by Hoyle and Fowler. The correctness of radioactive ^{56}Ni as nature's parent for ^{56}Fe had grown gradually more secure. In the first place, supernova theory demanded it unless our ideas were wrong. Secondly, the 77-day halflife of ^{56}Co matched very well the halflife for the decay of the optical brightness of one specific class of supernova, the so-called Type Ia supernovae. It also matched the halflife of the decay of the brightness of supernova 1987A. It had become apparent that supernovae would become dark after their first week were it not for radioactivity reheating the cooling interior and thereby keeping it shining. That 77-day match for SN1987A was in the last analysis a surprising and fortuitous result of the particular kind of supernova that SN1987A was. Being twenty times more massive than the sun, its precursor star had aged in just such a way as to be compact and blue at the time of its collapse. If it had been distended and red, as most supernovae of this type are, the visible effect of the radioactivity on the light output would not have been so clear. But this was not all known at the time when both the star and the *Challenger* exploded and when *Gamma Ray Observatory* was therefore held back.

The reorientation in my 1974 paper[1] had emphasized measuring *the structure* of the exploding stars with the gamma rays. In that work the rate of rise of the gamma ray intensity as the object expanded was argued to be more informative than its intensity after maximum, because the rise time measures where the radioactivity is located within the exploding object. A new goal would measure the star's radioactive structure with gamma rays, using not only the many strong lines from ^{56}Co decay but also lower energy ones that were emitted following ^{57}Co decay, which is the parent of the mass-57 isotope of iron, ^{57}Fe. The less abundant ^{57}Co would become the brighter of the two gamma-ray emitters after two years. Differing gamma-ray lines have different transparency within the supernova owing also to their different energies. In the energy region of about one MeV, the more energetic the gamma ray, the more penetrating it is. Those from ^{57}Co were less able to penetrate the supernova mass, so their breakout was retarded somewhat in time relative to those from ^{56}Co, which emits more energetic gamma rays. Likewise the strength of the 847 keV line from ^{56}Co would be seen to have been retarded with respect to that of the 2599 keV line, even though both are emitted at the same rates and from the same locations. The supernova structure could be mapped in this way. A rather poor analogy might be the use of X rays in medicine to map the interior of the human body. I began to advocate that interior mapping as another strong reason in support of gamma-ray spectroscopy of supernovae. These opportunities were lost in 1987 because *Compton GRO* was not up.

I suspected that mixing of the radioactivity into overlying portions of the expanding star should cause the gamma ray intensities to be detectable sooner than would be observed in unmixed models. I found it likely that the violent explosion would send fingers of radioactivity outward, penetrating through the overlying matter. This was to prove to be the case in supernova 1987A. But our inability to launch frustrated each hope to measure these effects myself, using the good time dependent precision that GRO could have provided.

For all of these reasons and more, the *Challenger* fireball was not only a national tragedy but also wrecked some of my personal scientific goals. Had my own observations with the *OSSE* team of the gamma light curves been able to confirm and utilize my own predictions and motivations from 13-28 years earlier, it would have been the luckiest outcome of my lucky scientific life. The issues are dramatic and fundamental for the natural philosophy of our material world. But slight frustration over this is of no consequence for science, and is truly of no concern to the scientist. For the most part, the desired understanding was achieved anyhow, by other observations. What is exciting to me is that we scientists did reach, during the supernova 1987A event, a profound new level of understanding of supernova nucleosynthesis and structure. That supernova spawned a flurry of theoretical papers looking intensely into the explosion mechanism of Type II supernovae. It continues unabated to this day.

It was just prior to SN1987A that Nancy and I decided to take my last sabbatical leave from Rice University. Waiting out the *Challenger* inquiry was no fun, except for the exciting brilliance of my hero from the 1960s, Richard Feynman. Here was one of the greatest minds of the twentieth century dunking O rings in freezing water! II had mistakenly thought that having a genius on the televised Challenger hearings was just a public relations gambit to engage the American people. After all, that is what TV is used for. I had not expected Feynman to actually put his finger on the problem, and do it with all the drama of courtroom TV. What an irony--truth outwriting screen writers! I roared with laughter at the simultaneous ludicrousness and seriousness of this event. But when it was over and Dick returned to his heroic battle against cancer, I knew I needed a break. GRO would not be going up for a long, long time while NASA sought the corrective measures required. I even contemplated withdrawal from my position as Co-Investigator with the *OSSE* instrument to return to my first scientific love, the synthesis of the chemical elements. I discussed this possibility with Jim Kurfess, the Principal Investigator (Chief) of the *OSSE* experiment. I am glad he talked me out of it, for I was still to have a thrilling relationship to *GRO*.

A fresh start on fatherhood and on England

On New Year's day 1987, Nancy gave birth to our son Andrew in Hermann Hospital in Houston. When we married in 1983 I had not planned to become a father again, and Nancy had doubted that she wanted motherhood. But love changes a great many things. Nancy's mother had broken down in joyful tears when we had told her of the pregnancy. It offered Nancy's parents the unexpected joy of having their first and only grandchild. We trained for natural childbirth, and I helped with the delivery. My son Don, in *terra incognita* by suddenly having a 27-year younger brother, dropped by Nancy's hospital room in afternoon and I took him out to *Angelo's* for a celebratory New Year's dinner. We had been drawing continuously closer in our slow healing of the hurt of my first divorce, when he had been nine. I called Devon in Los Angeles to share the good news with him. He responded with customary happy sarcasm. I had learned several years earlier that Devon was gay, and our closeness increased markedly with that open awareness. Nancy and I had twice visited him with his cohort of Hollywood friends and Devon and I had our best times together. And my daughter Alia, now nine years old, was genuinely excited to have a baby brother, even though she would not be able to see him because her mother had taken her away to Germany's Black Forest to live with her grandmother. This had been done in misguided secrecy, but I had located her simply by asking and had then visited her once in Wutöschingen. I had created a savings account for her there and deposited $2000, giving her grandmother power of withdrawal to prevent Annette from spending the child support while Alia lived free with her grandmother. For Alia on January 1, 1987, her joy was to hear her father's voice and to know that Nancy and I remained totally happy in the bungalow that she had helped us decorate prior to her removal to Germany. So Andrew's birth had galvanizing impact on the emotions of my entire extended family.

In June, 1987, Nancy and I returned to England, itself an unrequited love of our lives. Little Andrew, our son, was six months old and ready for anything. Nancy and I agreed that the north of England would be our choice. We had no heart for returning to Cambridge, with Fred Hoyle gone and his Institute reborn under new directorship and following new directions in research. Nor did I feel that I could avoid ghosts everywhere I looked in Cambridge. But our love of the north, especially of Scotland was strong, so north we went. My friend Arnold Wolfendale, who was nearing his eventual appointment as *Astronomer Royal*, welcomed us to Durham. Our photo of us can be seen at <1987 Wolfendale and Clayton in office>. Wolfendale maintained an active research

Andrew at seven months age with me age 52, and Durham Cathedral distant.

439

group in gamma-ray and cosmic-ray astrophysics, so their physics department provided a stimulating scientific home for me.

Durham University is an old English university, and one of the better ones. It has an impressively medieval feeling. Durham's great cathedral overlooks the River Wear, which winds around the pear-shaped old town, protecting it on three sides from Danes, Scots, or from whomever was in any given century abroad upon the land. The castle defended the fourth side unprotected by the river. Arnold proposed me for a visiting fellowship in St. Mary's College, and by good luck it was offered to us by the Principal, Joan Kenworthy. I was appointed St. Mary's College Fellow for Michaelmas Term 1987. St. Mary's is a women's college. This fellowship included a three-bedroom house next to the main building and dining rights in college. The generous terms included free use of the house, although we were asked to pay a modest rent for the time we lived in it outside of the Michaelmas term. The Fellowship included free lunch and dinner at High Table for Nancy and myself. It was a generous opportunity in the finest of academic traditions.

The Fellowship carried with it membership of *Durham University Society of Fellows*. I was expected to deliver a public lecture to be arranged jointly by the Society, the College and the Department of Physics. I was expected to participate to an unstated degree in "the life of the Senior Common Room of St. Mary's College." As in Cambridge, this latter duty, mysterious sounding to the uninitiated, actually meant drinking a glass of port with other Fellows after dinner and engaging in intellectual debate with them and invitees, often student officers— a pleasant enough duty.

My science ambitions had encountered a several-year delay in the launch of *Gamma Ray Observatory* at a cosmic moment steeped in irony. The first naked-eye supernova since Kepler last saw one in 1604 exploded on February 24, 1987. It was a stunning display, observable without instruments from the southern hemisphere following its discovery from Chile. A burst of neutrinos from it were recorded during the previous day, measuring precisely when the core had collapsed. *Gamma Ray Observatory (GRO)* was in storage at Cape Kennedy, waiting the go ahead for a shuttle. I therefore had to reorient my science time during this Durham sabbatical leave.

After settling in, I delivered *The Fellowship Lecture* on Monday, 26 October, in The James Duff Lecture Theatre, just prior to the Fellowship Dinner in St. Mary's Hall. The audience was quite respectably large for a lecture so ominously entitled: "Thermonuclear Origin of the Chemical Elements- are we all cosmic fallout?" The Principal wrote me a gracious letter after its delivery, exaggerating profusely my virtues:

It was a joy to hear a scientist at the forefront of human enquiry explain, in terms that laymen could follow, what it means to be involved in such research. For the first time I began to understand how work on the chemistry of the universe can be undertaken. It was a privilege to hear you.

I had not the slightest doubt that she said something similar to all of her Fellows; however, she did put her finger on one of the challenges of public discourse that I have always respected. I want to reach people, to touch them with the torch of science.

Nancy and I spent considerable time taking walks during this period. Two themes in particular were repeated. From St. Mary's College we would push Andrew's stroller down to the 1778 Prebends' Bridge, called "The New Bridge" in the 1824 Plan of Durham City, today a pedestrian bridge that spans the Wear near the southern tip of the central peninsula. Crossing it we would choose one of two walks up. The within-town route passed through the Water Gate of the walled medieval town, erected after the Prebends' Bridge made carriage traffic possible upward into the city from that point. The city wall is visible to its full height and thickness at Water Gate. It is one of the loveliest architectural settings in England. From there we walked up its lone cobbled street past the oldest colleges to the Cathedral and the Palace Green to its north, or to the castle and its stately keep standing guard near the destroyed North Gate. Alternatively we often took the rougher footpath outside the city wall, beneath its exterior west side, underneath the towering and stunning west facade of Durham Cathedral, until we passed through a passageway named Windy Gap to the

Palace Green and Cathedral. We were repeatedly rewarded by the pleasantness of strolling in an historic town without the constant curse of automobiles. Time stands still in such a setting, as it seemed to in my childhood. Auto traffic destroys that experience, making us rush rather than linger.

But our favorite walking was on the moors. For this we had a short auto drive in our used Ford, purchased from a previous Fellow, to spots above either Weardale or Teasdale. Across the moorland footpaths the spaces seemed to unfold, just a short step from there to infinity. Walking here resurrected our bond with the Scotland Highlands, which we were unable to do during this sabbatical stay. A half-year old is about nine years too young for the Highlands; so the York Moors sufficed. In all of these many walks our aims were simple: firstly to be together with our baby and each other; secondly, and what either of us would emphasize if alone, to yield to the senses, the quiet, the smells, the simple but stunning views, the breezes, the aloneness that graces such places, even in a land having considerably higher population density than the United States. Somehow, it just does not seem like so many people.

At work in Durham City

At work I turned immediately to several science issues at my core. I first dived back into meteorites and the isotopically anomalous minerals that were being found within them. It was always a little frustrating that gamma ray astronomy, in the doing of it rather than merely the theorizing over it, took so much time away from the other frontier areas of research. But that is the frustration of scientists like me who find it difficult to choose which question to tackle next. In a light hearted way I compare us to the three-letter high-school athletes who cannot decide which athletic scholarship to accept for college. A not insignificant virtue of Durham in 1987, and one that had figured from the outset in my planning, was the annual meeting of *The Meteoritical Society*. It was to meet in July in Newcastle, only twenty miles north from Durham! One of the ceremonial sessions was planned for Durham itself, a commemorative session for Professor Fritz Paneth, whose early research, conducted partly in Durham, on noble gases trapped within meteorites had pioneered a new appreciation for those stones that fall to the Earth. That our home during that meeting happened to be in nearby St. Mary's College produced some incredulous wisecracks from my colleagues at this meeting. They chided that this was the fifth time that I *just happened to be residing in Europe near the site of the annual meeting of our society.* This ribbing seemed to endow me with cachet, however, some recognition of a special calling, so I encouraged it. It never hurts in science to catch the attention of one's colleagues, no matter what the trick. But I was earnest over the information content of the meteorites, and my paper presented new results on a problem that had consumed me for over a decade, namely, the cause and meaning of the unusual isotopic patterns found in some minerals from meteorites.

My contributed paper at 8:40 A. M. on July 20, 1987, was the first paper of the entire conference! Such distinction bears no honor, only the likelihood of many sleepy colleagues missing it. Its title, "SUNOCON Oxidation and Sulfidation of the Isotopes of Ca, Ti, Cr and Fe", gave only a hint of the magnitude of the issue that it addressed. It continued a theme I had introduced eleven years earlier in a short publication[2] with S. Ramadurai. Our single point had been that different isotopes of the same element condense into differing chemical forms. This separates them at birth, so to speak. Our paper had addressed supernova condensation of isotopes into dust grains in sulfurous environments. Thinking its consequences to be more profound than evidenced by the scant scientific interest that it had elicited, I decided to hit it again at the Durham meeting. Chemical separation sprang from the first chemical environment that newly created isotopes find themselves in following their creation in supernova interiors. Although the separate isotopes of any element behave in the same manner chemically, the universe remembers distinct chemical histories for them owing to differences in the initial chemical environments in which they first appeared following their stellar nucleosynthesis. I called this "cosmic chemical memory". The separate isotopes of the elements do, at least in part, reside initially in separate chemical forms. This causes them to remain separated to some degree throughout their history.

A simple example is found in the element neodymium, whose *s*-process isotopes are created in red-giant stars, therefore condensing within silicon carbide grains, and whose *r*-process isotopes are created in supernovae regions that may not condense dust at all. During ejection, the *s*-process isotopes can condense much more efficiently from red-giant atmospheres than *r*-process isotopes can from hot, low-density *r*-process gas. For that reason dust and gas in the interstellar matter carry different Nd isotope abundance ratios from the beginning, so that differing proportions of collected interstellar gas and dust will form solids in the solar system having differing Nd isotopic ratios. It is a purely chemical difference in history. I called this aspect of cosmic chemical memory *polarization.* To me it was a very beautiful theory, having the attributes of a good theory. It was soundly based, it had specific calculable consequences and predictions, it could be confronted by observations, it could be built upon by the work of others, and it offered hope to explain one of the greatest surprises of the 1970s.

My presentation at the conference[2] detailing predictions for separation of the isotopes of the elements calcium, titanium, chromium and iron aroused little interest. I was by this time accustomed to such disappointments. The early audience was sparse and sleepy eyed, although most of the leaders of the society were there, they being more resolute at making early talks than the younger generation. But my thesis was almost too much to believe, too good to be true. It gave to some the impression of wild speculation, which was always entertaining but not something one easily saw how to base research upon. An example from that talk and its published table[2] in *Meteoritics* compares the two most abundant of calcium isotopes, ^{40}Ca and ^{42}Ca, and may help my point be understood. My table showed that in the initial SUNOCON (supernova condensate) component of interstellar dust the mass-42 isotope would be characterized by a greater percentage in oxidized SUNOCONs than would the mass-40 isotope, which has a larger portion appear within sulfidizing environments. The cosmic chemical memory theory predicted that after subsequent incorporation of those dust particles into second generation solids, some secondary minerals generated will remember the initial separation of oxidized ^{42}Ca from sulfidized ^{40}Ca, whereas some minerals cannot. I could not explain what minerals would be formed, however, because I am not a mineralogist and the chemical paths to mineral formation in the early solar system are not known. I was never able to detail the subsequent chemistry and the degree to which it would mix the isotopes back together again. Many such chemical-memory distinctions were in my table. Seeing little interest in such ideas led me to conclude that it was either too early in the growing understanding of cosmic chemistry or too early in the first morning.

My calculations could be nothing more than best estimates that I could make at that time. I always tried to draw the distinction between the theory itself, which should have legs of its own, and my poor ability to calculate its predictions. I argued that the theory had to be understood even though I was myself not yet able to use it with predictive confidence. This attitude did not play well with meteoriticists, who are primarily chemists by training and by and large practical and empirical in their philosophical outlooks. This is no criticism. It is not easy to appreciate new theory that can not be easily used. More of a theoretical physicist in my inclination, I was aware that, in the history of physical theory, the correct idea, or sometimes even merely the correct question, was equally important, sometimes more so, than the embellishments and applications that later secure for the theory a dependable status. Physical theory often struggled bravely to simply ask the correct questions, and those that were able to do so were pioneers of physics.

Despite going over like a lead balloon for me, the 1987 Newcastle meeting was a good meeting for the *Meteoritical Society*. Heinrich Wänke from Mainz was eloquent and gracious in his review of the work of Fritz Paneth in a ceremonial plenary session held in Durham in Paneth's honor. I enjoyed that history a great deal, along with the historical public remembrance of Thomas Wright, one of the first great cosmologists, who lived and worked in Durham. That Plenary Session took place from 6:30 to 8:00, and afterwards Nancy and I went out for a festive time with colleagues to the "3

Tuns" restaurant on Church Street. In old Durham one walked everywhere.

Several papers presented were to influence my activities during the years ahead. Colin Pillinger and his colleagues presented the work from their laboratory at *The Open University* in Milton Keynes showing that carbon and nitrogen isotopes could be partially separated by stepped partial combustion of the diamond residues extracted from meteorites. These diamonds had been present in the interstellar gas when the solar system formed. This impressed me enough that a few years later I would spend part of the spring in Milton Keynes (about 60 miles north of London) working on a paper with them concerning the origin of those interstellar diamonds. Edward Anders and colleagues from Chicago presented data of their chemical separation of distinct carbon dust from meteorites, a separation that was to revolutionize human knowledge of interstellar dust. I regard this as some of the most important experimental work ever done in meteoritic science, and I admire Edward Anders for his ability to lead this team through their long search of repeatedly dissolving in different acids. His research group extracted the silicon-carbide particles[2] that carried the *s*-process xenon gas. It was thrilling to see that one single kind of interstellar particle, the SiC particle, carried this almost pure *s*-process xenon. My prediction[2] of the effect had lacked conviction that the particles could actually be found in the meteorites. But here they were! It saddened me that Edward Anders remained fiercely critical of me.

The challenge for Nancy in Durham was of yet a different kind. An architect trained at the renowned architecture school at Rice University, an earnest artist who had not included motherhood in her goals, who had been bonded to me and me to her in a mutual storm of friendship and understanding, Nancy became a mother at age 33. With her abundance of creative gifts, she had turned them temporarily to creative writing. She had written several imaginative and stylized short stories and had won a *Pen Writing Prize* for Houston in 1986. I had enjoyed enormously her public reading of her story at the Houston Museum of Fine Arts. Now in Durham, separated from her books and her typewriter, she was thrust into the boa-like coils of a tight relationship with a six-month old son who was, by innate temperament, demanding of her attention. She tried to set aside some writing time, but you can already guess what happened. Andrew won all of her time with his own relentless energy. We underestimated the psychic grip of a six-month old. Writing creatively was not possible. We laughed at this peculiarly human problem.

The urge to relocate

Our 1984 decision to relocate from Houston, given the right opportunity, was always in our thoughts during the 1980s. So, loving England as we both do, Nancy and I could not but consider what it would be like living here again, although we had no inkling of a plan that might enable such a thing. But as we strolled through Durham and the towns of the dales we cast quizzical eyes at houses, markets, pubs and churches, wondering what kind of neighbors and life that might be. It did not appeal so strongly to us during this time. Earlier residences in Cambridge and in London, prior to our marriage, did not extrapolate accurately to the north of England, so we were curious. Dining with friends we would usually ask them how they found living in Durham. Those conversations were mental exercises for us, nothing more.

One concrete action did result, however. I applied for the Headship of the Astronomy Department at *Penn State University* when it was advertised, as well as for the Eberly Professorship of their Department of Physics. That Pennsylvania setting has much in common with the setting in Yorkshire and in County Durham, and we are fond of those similarities. Disappointed at the cross purposes that had frustrated our previous interview at Penn State in 1984, we decided, after much thought, that we would rather head the astronomy department and reside in State College than reside in Houston for the remainder of my life. Bear in mind that headship would require a very important career reemphasis in any American research university. Head is a full time job, requiring constant attention to political and administrative needs of a large and active department. I had, heretofore, worked more as a lone wolf, concentrating on my own scientific insights, pursuing them for the sheer joy of that chase. I was very aware of those career differences, which gave considerable emotional punch to my decision to apply. Many friends and colleagues could not understand why we would move at all, considering that Rice University is so excellent. An especially important example to me was the reply from my friend and prior mentor, Willy Fowler, when I asked him to again support me with a letter to Gerald Smith, Chair of their Search Committee for physics. He wrote, in a letter that he regarded sufficiently significant to be typed by his secretary:

Professor Donald D. Clayton Kellogg Lab 106-38, Caltech
#3 West Court
St. Mary's College
Durham DH1 3LR

   Dear Don,

      It was great to learn there is still an opportunity for you at Penn State although why you would wish to leave Rice is beyond me.

      I'll be happy to tell Gerald Smith what great contributions you have made. However, I cannot mention Academy prospects. As you must know the Academy elects only those whose work is generally recognized as first rate <u>and uncontroversial!</u> There have been exceptions but they are very few indeed.

      Anyhow, I will do all I can. With all the best to you and Nancy and to Andrew Adair.

   Sincerely,

   Willy (signed)

The underline is his. And that vexed me-- and not for the first time. My vexation was not with Willy Fowler himself but that hearsay remarks by a few were able to color a man's reputation.

I had watched this happen in spades in the life of Willy's friend Fred Hoyle, whose reputation as a wild man was created by a few biologists, abetted by Fred's stiff-backed reactions. Those critics had not the slightest comprehension of the power of Hoyle's thinking and the depth of his contributions to astrophysics, but they were able to color Fred "wild". I was vexed by similar attributions. My work is not and never has been controversial, although many who did not understand it thought it so, perhaps because two famous scientists railed openly against one specific piece of it, often rudely, often in public. However, a contemporary reading today of all of my papers in the period 1975-1985, the period that so offended some, reveals, I believe, the cautious skepticism and solid thinking that are the essence of science. Science *is* skepticism, not certainty. It was my critics who dishonored skepticism, not me. And as Fred Hoyle put it to me in words I will never forget, "The offender never forgives."

The trouble in this Fowler letter is that he had been frequently told by two very powerful members of the *National Academy of Science* that my work was too controversial for them to ever support (in effect halting my chance of election), when what was really the truth was that they very much *disliked* my work. They disliked that I was enlarging scenarios of cosmochemistry. They disliked my raising challenges to their published conclusions. Their frequent criticisms were disingenuous to an astonishing degree. They were incensed that I suggested skepticism about beliefs that they had held dear, primarily because they had invested so much in those beliefs. They maintained this stance even though my papers read as being very sound. Willy also reminded me in that letter of what I already knew well, that he could not discuss Academy election procedures or chances with anyone. His unexplained remarks about Anders in 1982 had been the limit of his willingness to talk. Two other Academy members also chewed me out for asking them the same question, although those two have been more supportive of my work without necessarily believing that the new ideas for which I was under fire would prove correct. That idea, to remind the technical reader, was my skepticism that extinct radioactivity within meteorites, especially million-year ^{26}Al and initially seventeen-million-year ^{129}I, had actually been alive when the meteorite formed. Because of well founded skepticism, *I was the radical*! I have described this conflict earlier, but it was forever rearing its head in negative remarks about me. Allegations have persistence, even when lacking substance. People remember allegations, and tend to think that where there is smoke there will be fire.

Nor did I hunger for Academy membership. I had seldom given it a thought until I began the search for a different university for our family life. When one seeks a prestigious position in a strong university, nothing makes administrators sit up and take notice like membership in the *National Academy of Science*, which is often (though wrongly) stated as the highest honor one can receive outside the Nobel Prize, or perhaps the *National Medal of Science* (both of which Fowler had won, in 1974 and 1983 respectively). Because extremely attractive professorships were, and are, very difficult to win, I realized that my chances for the change we sought would be helped enormously by election to the Academy. Then virtually *every* university would want me. I must underline this poorly understood influence by relating how firmly that it had been impressed on me by this very case. During my first interviews with Penn State in 1984 it was evident that many of their professors, including their Head, Satoshi Matsushima, were eager to make me a Penn State Professor. I and Nancy were treated very well there. And indeed it would have been a successful marriage, although, in retrospect, not as ideal as the solution that we were on the verge of discovering. The Penn State department had a well endowed Professorship to be offered, *The Eberly Professorship*, suitable for me and I for it, except that it carried the stipulation that it was to be so prestigious that only Members of the *National Academy of Science* could have it offered to them. This stipulation was not one required by the professional astronomers, but by administrators and the benefactor. The astronomers realized that it tied their hands very severely. When they told me of that restriction, I quite naturally became more interested in the foot dragging and opposition that had already prevented my election.

Two other reasons also activated my concern at that time over

National Academy of Science election. The first is an emotional reaction, an annoyance that I was not being elected, especially in comparison with the careers of those who were. Election falls primarily to those who are simply well situated in major science research programs at major universities already having many members. The Academy is more a prestigious club than prestigious honor, and is also a chartered organization for providing scientific advice to the government. As in all clubs, the club members elect their own. And they elect people they know well and with whom they are sure they can work effectively. Annually additional academy fellows are elected from Caltech, from Chicago, from Princeton, from Berkeley, from Cornell and from MIT, those universities most endowed with Academy members. Scientists outside the power base have little chance. Scientists in the south have little chance. One friend joked to me that a Caltech professor must have a mediocre career to escape election, so strong is Caltech's political base in the Academy. This is quite all right of course, except for the implication that those who are not elected (almost everyone) are regarded as scientifically weaker than those who are. Despite clear demographic evidence to the contrary, the Academy straight facedly maintains that it is a society based solely on science merit. It is indeed so in the sense that all Academy members do have very high scientific merit. But it is not so in the sense that the members have more scientific merit than the best of those who are not elected. In 1980 Rice University had no political strength in the Academy whatsoever, no one with whom members could jockey over election strategies. Rice also had no faculty who were elected to membership while they were Rice faculty, although Rice did begin to hire members from elsewhere, and it also hired new faculty of such talent that by the turn of the century Rice University was an institutional clubmember. After my later move from Rice to South Carolina's Clemson University my political base was even weaker, virtually zero. South Carolina had no members.

I lived with this dual inappropriateness, firstly that insecure scientists could spread the stigma "controversy" over my work, secondly that administrators would narrow my options because I had not been elected to the Academy. To finish this story, which

I am sorry will be offensive to some Academy members, two sequels pertain. Penn State was actually able to lure an Academy member, Icko Iben, Jr., to their Eberly Professorship. But his heart was not in it, and he returned to University of Illinois within a year. After our return to Rice from Durham my friend Gordon Garmire, distinguished x-ray astronomer at Penn State, wrote for the search committee that my lack of management experience made me an unconvincing candidate for the Head. I suppose that is correct, although many distinguished scientists become effective department heads. This annoyance all became mute when, within a year, we found the perfect opportunity at Clemson University in South Carolina, reaffirming the code of life inherited from my father-- that things turn out for the best, given sincere pursuit of high principles.

The age of the chemical elements

Isolated in Durham from the grounded *Gamma Ray Observatory*, I returned to other scientific goals that suited me while in Durham. One that has gripped me throughout my life is the scientific content in the natural abundances of radioactive atoms. *Where had they come from, and why were they not all gone? When were the elements created? What is the meaning of the now extinct radioactive abundances that were apparently present at the birth of our solar system?* These were questions that had won my early fame at the press conference in Washington D.C. in 1963. They are questions that have since then never been far from my thoughts. Now in Durham on what would be my last sabbatical leave seemed to me a good time to pick up these questions once again with an active new study. I had just recently discovered a profound new tool for their study. In my struggles over the meaning of detectable gamma rays from radioactive ^{26}Al in the Galaxy I had discovered that the abundances of radioactive nuclei were intimately tied the physical evolution of our Milky Way. I decided to amplify this new insight and present it to the astronomical world from Durham.

I was determined to discard artificial parameterizations of the history of the production of matter that had always been popular in nuclear cosmochronology, and to formulate that theoretical problem within a realistic context of the abundance evolution of the elements in the Milky Way Galaxy. Astronomers call this *galactic chemical evolution.* I wanted to create a treatment that could express realistic understanding in terms of time-dependent formulae rather than a table of numbers generated by a computer. My goal would be to present general solutions of the differential equations that govern this question. Differential equations are mathematical expressions that describe how basic features of our Galaxy change as time passes. It is easy to list those features. What is the mass of the interstellar gas in our Galaxy, and how does that mass change while new pristine material falls onto the disk of our Galaxy and formation of new stars consumed that gas? How does the rate of infall of metal-poor gas onto our galactic disk change with time? What is the total mass of all of the stars and the mass of interstellar gas, and how do they change with time? And how do the abundances of the elements, both stable and radioactive, change as time passes? These are the question of chemical evolution of our galaxy when it is approximated by a single zone of variables.

One time-dependent differential equation describes the answer to each of the above questions; but those equations are coupled together. The time-dependent value of each quantity appears in the equations for the rate of change of each other quantity. The quantities are linked inseparably together, so that one of them can not be found without simultaneously finding all of them. Not only did I discover how to do this but also that the radioactive age of the elements also depends on the time-dependent history of each quantity. The bottom line for chronology was that the apparent age of the elements, as inferred from the abundances of radioactive nuclei, depends intimately on how the structure of the Galaxy has evolved. Previously used expressions for answering the age question were oblivious to that dependence and were incorrect.

I found that I could obtain time-dependent mathematical functions for each quantity if the rate of infall of element-poor matter onto the Milky Way disk was approximated by a member of well defined families of functions. My procedure was to define yet another time-dependent function, specifically the rate of infall per unit mass of interstellar gas, and to generate all of the needed functions from that one. I showed how the mathematics works. I needed only the additional assumption that the rate at which new stars form from the interstellar gas is proportional to how much gas exists. Both assumptions are physically plausible, so that my analytic solutions characterize the real Galaxy. My Durham program was to present the families of solutions based on this approach and, more to the point, to try the hard task of applying the solutions to the growing body of data on the abundances of both stable and radioactive nuclei at the time of the birth of our solar system. This was my goal, and I worked on it tirelessly, joyously, during the entire seven months in Durham. Eventually I published this study[3] in the first 1988 issue of *Monthly Notices of the Royal Astronomical Society*, England's journal for astronomy research-- but not before a lot of action!

While settling in I was immediately stunned by the appearance in the 9 July issue of *Nature* of a remarkable new research paper by Harvey Butcher, an astronomer in the Netherlands. This came just ten days before the *Meteoritical Society* meeting began in Newcastle, and it settled any lingering doubt over the subject of my research in Durham. *The Times* of London described the impact of Butcher's paper as follows in their July 17, 1987 edition:

Universe gets its youth back

By Pearce Wright, Science Editor

A new method of analyzing the evolution of stars from their chemical composition has led scientists to conclude that the universe still carries the bloom of youth-in astronomical terms.

The results show that the universe is less than 12 billion years old, rather than the previous estimates of 16-18 billion years. Dr. Butcher of the Kapteyn Astronomical Institute in the Netherlands studied 20 sun-like stars to conclude that our galaxy, and by extension the universe, was between 10-12 billion years old. He has overturned previous estimates by devising what is in effect a third clock to tell the minimum age of the universe. His findings are published in a paper in Nature entitled "Thorium in G-dwarf stars as a chronometer for the galaxy".

The previous results that Butcher "had overturned" included my own. My rhenium-osmium galactic age had appeared to be at least 18 billion years when I discovered it in 1963. So although an enhanced decay rate in stars now lowered that result by several billion years, I was more than a little interested in Butcher's result. What he observed in the stellar spectra was that in the oldest stars, the ratio of the abundance of Thorium to the abundance of Neodymium was about the same as it is in the sun. That was certainly a surprise inasmuch as those elements have been created over galactic time at differing rates. Thorium is radioactive with halflife 14 billion years. This means that in a star of 14 billion years age, 50 % of the Th in that star should have decayed during the star's lifetime, whereas only about 20% of solar Th should have decayed during the sun's known age of only 4.6 billion years. The complementary effect was the gradual decrease with time of the Th/Nd ratio in the interstellar gases from which stars are born. Butcher therefore could conclude that his old star, one of the galxy's oldest, was no more than about twice the sun's age, in which case the galaxy is no older than 10 billion years.

Somewhat astonished at this result, it took me only a day of worry before I found a first theoretical flaw in Butcher's argument; namely, he assumed that the nucleosynthesis production ratio of Th to that of Nd should have been the same when the oldest stars formed as it was later when the sun formed. Only given that assumption does the observed equality in the Th/Nd abundance ratio between sun and stars imply much younger ages for the stars than previously believed. Underpinnings in nucleosynthesis argued to me that the assumption that Butcher had used was wrong. So I would contest his conclusion. That is, by the way, what science does.

The theory of the *s* process in stars demonstrated that half of the abundance of Nd in the sun was due to the past operation of the *s* process, whereas the other half has been the result of the *r* process. But the *r* process is responsible for all of the thorium. Furthermore, the *s* process could not happen until after many supernovae had increased the abundance of iron from which the *s*-process Nd is made by a series of neutron-capture reactions. The dependence of Nd production on the prior history of supernova iron made the *s* process a "secondary process". If stars contained no iron, they would make no neodymium. A star containing twice the iron of another will create twice the neodymium. Neodymium is a secondary element.

The *r* process elements, on the other hand, could be made even in the first supernovae, before any iron existed. Astrophysicists therefore call the *r* process *primary*. A star destined to become a supernova need not contain iron in order to make the *r*-process Nd. Given that, the oldest stars would have formed prior to significant occurrence of the *s* process, so the initial stellar ratio of Th/Nd should have been the ratio of those elements as produced in the *r* process alone. That ratio is twice the solar abundance ratio. This means that if Th/Nd has the same value in the oldest star as it has in the sun, about half of the Th would have decayed in the time between the formation of those stars and the formation of the sun. Thus the oldest stars would be about 14 billion years older than the sun, or about 18 billion years old. This was more in line with my early result with the Re/Os chronometer than it was with Butcher's new conclusion.

I was initially excited that this *seemed* to have solved the puzzle. Butcher's conclusion was premature because it had not taken into account the slower growth of the Nd abundance during the evolution of the galaxy. My calculation of the theoretical expectation for the old galaxy based on my changing production ratio exactly agreed with Butcher's data! The fit was perfect. So I immediately wrote up scientific correspondence to submit[4] on 22 July to *Nature*.

Then things began to heat up scientifically. First Harvey Butcher wrote a rebuttal in the pages of *Nature* pointing out that, as interesting as my theoretical perspective was, it flew in the face of astronomical observations. Observations of spectral lines of the elements in stars reveal that the ratio of s-process abundances to stable r-process abundances has been constant over the lifetime of the disk of our galaxy. Therefore, my recalculation was, he observed, no better than a theoretical exercise, inconsistent with facts. His method remained, in his opinion, secure. *Nature* provided me with his reply, the normal procedure, prior to publishing either. I made a few wording changes, as did he, and so our arguments appeared in print.

What most fascinated me now about Butcher's rebuttal had been missing from his original paper. He now claimed that the ratio of the abundances of s-process nuclei to the abundances of those nuclei created in the r process has not changed with time, even as both grew more abundant during the aging of the Galaxy. This did not agree with theory at all! That disagreement rang my bell. I had my conviction that the r nuclei should have increased in abundance much more quickly than had the s nuclei at early galactic times. Butcher's paper had been submitted without seeming awareness of the importance of his assumption and that it violated standard theory; so I thought that he had simply overlooked it. My argument called attention explicitly that key assumption; but Butcher's reply convincingly showed his assumption could be justified. His evidence was astronomical data that show clearly that the element europium (an element synthesized by the r process) and barium (an element synthesized by the s process) increased in abundance with time at the same rate, notwithstanding my theoretical reason that they should not. This conflict puzzled me, and also excited me. I had found the elephant in the room! I decided that the lock-step growth of s- process and r- process, not the age estimates for stars, was the premier astronomical puzzle underlying the entire question. Something was wrong with standard theory. As the reader now knows, I loved those paradoxes.

For several weeks I puzzled over the rapid growth of s-process abundances while the Galaxy aged. Their growth should have been delayed by the need to wait for the growth of the galactic iron abundance from supernovae, because it was from iron seed nuclei in later stars that the s-process abundances had to be synthesized. This quandary would not release me. I returned to it daily awaiting the idea that I knew would soon hatch. If I had any gift as a scientist, it lay in my recognizing among simple facts the existence of an important puzzle and wrestling with it daily until I grasped its solution. That personality trait had led to almost all of the great successes of my science life, and it served me well here. Thinking over the findings of my Ph.D. thesis, I zeroed in on its demonstration that a very modest increase in the number of liberated neutrons per iron nucleus during the *s* process would produce a large increase in the barium abundance produced. I saw that the puzzle could be resolved if the supply of neutrons in stars was almost unchanging within new generations of stars despite their inheritance of an ever increasing iron content from the enriched interstellar medium. The natural way for this to happen would be for red giant stars to *manufacture*, rather than to *inherit*, the ^{13}C abundance that is the neutron supply. The ^{13}C abundance serves as reservoir of liberation-prone neutrons. In AGB red giants it is created in the star by helium-fusion production of ^{12}C followed by some subsequent mixing into the ^{12}C-rich matter of protons from the star's envelope. Then hot protons can turn the ^{12}C into ^{13}C. If this were the correct neutron source for the s process, the strength of the source of neutrons would probably not depend on the amount of the heavy elements contained initially within each star.

That was surely it! I had discovered that ^{13}C nuclei made in this context from primordial helium and hydrogen constitute the true source of neutrons for the *s* process, not the smaller amount of ^{13}C left over from routine hydrogen burning, nor the ^{22}Ne left after routine He burning in stars. If this is even approximately true, early stars having low initial metallicity would experience a larger number of free neutrons per initial iron nucleus. Problem solved! Excitedly I wrote this analysis into my impending cosmochronology paper[3] so that I could intelligently analyze Butcher's Th/Nd method as well as its other goals. This idea, which simulated a *primary s process* despite requiring a secondary seed nucleus to capture the neutrons was so powerful that the referee of my paper, who identified himself as Bernard Pagel, recommended that I factor that pregnant topic out to be a separate short paper to be more visible. I declined to do this, however, because I saw the lock-step growth of the *s*-and-*r* process abundances as a key feature in the nuclear dating of our Galaxy. I now envisioned for my paper an original and powerful evaluation of radiogenic cosmochronology. The issues for my new paper were, taken together, so large that I saw that it could become a handbook of new arguments for the age of the elements.

On August 12 Willy Fowler sent me, in reply from Caltech to his receipt of my counterargument to Butcher's interpretation, a copy of a letter that he had written to Grant Mathews and David Schramm, two well known younger nuclear astrophysicists who had also addressed Butcher's initial paper with an analysis of their own. They had sent their short paper, which I had not even seen, to Fowler, and Fowler did not like it! He wrote to tell them so and sent me a xerox copy of his understated letter. Upon that copy (typed by his secretary) of his letter he wrote to me in hand:

Dear Don, August 12, 1987

Everyone wants to get into the act. The model of my Milne Lecture (published before Butcher)! gives his results if I exclude p=process production in the 17% early spike!

Your (published) note disturbs me. You should not call the process you have invented the r process. The B²FH r-and-s processes both built on iron seed nuclei (and are therefore secondary)

Cheers!
Willy

Willy's comment was, alas, very wrong, and in many ways. That disturbed, pointing to the decline of Fowler's thinking. His concern that his Milne Lecture predated Butcher's paper was irrelevant and beyond comment. His whole tone was uncharacteristically self serving, placing emphasis on the numerical value of the answer that each had obtained for the age of the disk and overlooking that Butcher was presenting a new scientific method, not a mere estimate of the answer. Secondly, the buildup of the *r* process, even if upon iron nuclei, was now unanimously regarded to be from iron atoms that are produced by nuclear evolution within that same presupernova star, not upon the initial iron atoms inherited by the star when it formed. The *r* process is for that reason *primary*. Thus his negative remark looked backward, not forward, which was not characteristic of the Willy Fowler I knew.

I wrote back to Willy, as I always immediately did, always glad to hear from the man that had been the major influence in my scientific life:

Dear Willy, 18th August, 1987

> *I think you should not be disturbed, Willy, that I mean by r process the rapid, photoresisted, capture of neutrons. B²FH should not live or die on the question whether iron is a required seed for that process. There is plenty of heavy element buildup (of iron) in the inner zones. Initial iron is forgotten. But the s process <u>does</u> require iron seed. Thus, primary versus secondary. What in that disturbs?*

> *Yours truly,*
> *Don*

Two more agitated letters from Fowler revealed a major falling out between us. That saddened me and worried me. I loved Willy. Not only had he been my thesis advisor, the man who gave me the *s*-process problem, but we had been close daily friends for three decades, especially during our unforgettable seven years together in Cambridge. My science disagreement lay with his naïve hope to pin down the age of the Galaxy with his hopelessly oversimplified analysis of radioactive abundances, as described in his Milne Lecture. The problem lay with his claim that it is possible to give the age of the disk from the abundances of the two isotopes of uranium and of thorium. I decided that I would have to demonstrate openly that galactic age could not be pinned down with that skimpy evidence. Not only were the relative production rates for those two nuclei by the *r* process too uncertain, but unknown factors in the growth and evolution of our galaxy played an overwhelming role in determining their initial value in the sun. Makiing that issue clear to the world of astrophysics, including Fowler, now became my Durham agenda. The theory I was now describing for my large paper showed that the ratio of the two uranium isotopes was almost *insensitive* to the time of the beginning of galactic nucleosynthesis. It gave no decisive information at all! Fowler had become convinced that he could see the correct answer from data that could never yield the answer. He might just as well have been reading tea leaves.

I regretted this tension between Fowler and myself so acutely that I considered just dropping my whole plan. *Was the most important friendship of my scientific life worth more than scientific truth?* Scientific truth trumps all else that we do, and I knew that Willy Fowler had always maintained that position. That speaks for carrying on. What's more, too many people quote Nobel-Prize winners as the last word, and Willy's proclamations on this topic were publicized and well known by many who were unable to judge his arguments. So I inherited a second challenge for the main work of my sabbatical year, the challenge of clarifying why Fowler's simple model calculation could not be relied upon to give the correct answer, despite Fowler's inordinate faith in it. Both challenges fit the project I had underway in Durham, to construct physically reasonable models for the history of the galaxy and for its abundances of chemical elements. I had found the arguments, the mathematical functions, and the opportunity in my life to provide this; and now I was being forced by Willy to explain it even more clearly for public scutiny. I realized that this was sure to be to his detriment. But had not Willy himself always told me, by his life and by his actions, that science truth was irrelevant to issues of friendship or loyalty? I had found what I had wanted when we arrived in Durham; namely a set of important issues and strong personal motivation to examine a problem close to the core of my scientific life. The problem is nothing less than the history of our universe, and of the time and place of the chemical elements' coming into being. My agenda for the final four months in Durham was set.

After one more written exchange with Willy I decided to let it lie. It would do no good to try to explain by letter why he was wrong; *i.e.* why his numerical parameterization was not a true astrophysical model. I just let it lie, knowing that further exchange would disappoint him as much as it hurt me to press on. In my 22 September letter of transmittal of my paper[3], I included the following:

> *I regard this as a work of extraordinary significance, even after discounting the usual author's tendency to overvalue his work. It addresses in fresh ways the three-decade-old effort to determine the age of the galaxy from natural radioactivity....I provide a new framework that will draw this specialist's topic more broadly into the main stream of astronomical research. For that reason I believe that it should not be reviewed by a nuclear astrophysicist , but rather by a broadly competent astrophysicist (and certainly not please by my old friend W. A. Fowler, whose influential Milne Lecture I find necessary to assault[3] rather thoroughly).*

There. I had said it. I felt I had to say it again when I wrote to my Karlsruhe colleagues on the *s* process, Hermann Beer and Franz Käppeler, who had recently attempted to reevaluate the cosmoradiogenic chronologies advanced in my 1964 paper. They were sure to be interested in my new argument for the 13C neutron source for the *s*-process, allowing its abundances to grow as rapidly as those from the *r* process. But on the issue at hand I wanted to calm their uncertainty over the cogency of Fowler's analysis of the galactic age. So in my 27 September letter to Beer and Käppeler I wrote:

> *Note that 'Fowler's exponential model', which you have used, Hermann, in your cosmochronology papers, is not a model at all. It is just an arbitrary parameterization, and*

*one that causes considerable confusion to most nuclear
astrophysicists. Fowler himself is no exception, as his
Milne Lecture shows very clearly. We must be careful to
not mislead cosmologists into thinking that the nuclear
evidence requires a younger galaxy than it really does.
The issue is too important. This new paper of mine should
clarify the astrophysical dependence much better.*

On the issue of cosmoradiogenic chronologies, I decided to
write to another old friend in Brussels, Marcel Arnould, who had
written a wonderful paper in 1983 on the application of my Re-
Os cosmoradiogenic chronology. With Yokoi and Takahashi he
had studied the extent to which the ^{187}Re beta decay is speeded
up in the high temperature conditions within stars. Normally a
radioactive halflife does not depend on temperature, but in this
special case it does. Their numerical model, that is, their computer
printouts, of the chemical evolution of the galaxy, including the
remixing of ^{187}Re into and out of stars, showed that my uncorrected
Re-Os clock did make the galaxy look older than it truly was. The
^{187}Re beta decay had to be faster in stars, as I had predicted from the
beginning it would be. While writing to him with my admiration
of his work and how I was utilizing it in my own paper, however, I
added a word about my fascination with new data from supernova
1987A. I explained why in my 8 October letter:

*Dr. M. Arnould 8 October, 1987
Institut d'Astronomie, BRUSSELS*

*Dear Marcel,
..............[first, about the ^{187}Re chronometer and my new paper,
then]...
The most exciting thing just now is that both GINGA and MIR
spacecraft have picked up the hard x rays from supernova 1987A. If
these are the Compton tails of the ^{56}Co radioactivity, we will easily
see it when we fly our telescope in Australia in March 1988. Of
course, others will look sooner, and a few upper limits have been
established already, less than 2×10^{-3} gamma rays per square*

centimeter per second. But our NRL instrument for Gamma Ray Observatory (OSSE) should attempt to get the best spectrum for future studies. It is especially exciting to me, because of my lucky prediction of this detectability 19 years ago; but I never thought I would see it in my lifetime!

Sincerely,
Don

A 1977 photo of Marcel Arnould and Koji Takahashi with this writer is <1977 European Renaissance in Darmstadt>.

Here is how those hard scattered x rays had come to my attention. On Wednesday 16 September 1987 I had taken the day to attend the *Anglo Australian Telescope Symposium* in Edinburgh. They planned a special session on the great new supernova of February 23, 1987. I rode up with Ray Sharples, a Durham astronomer, in his car. I presented a short talk at the Edinburgh Symposium entitled "Looking for gamma rays from supernova 1987A" in the special session on SN 1987A. Gerry Skinner from Birmingham was in the audience. He came to me immediately and told me that Rashid Sunyaev had reported the hard-x-ray detection of SN1987A from x-ray counters on the Russian *MIR* spacecraft, and had interpreted them as scattered ^{56}Co gamma rays emerging from the surface. That was very exciting, but I had still not been sure if anyone could actually detect the *unscattered* gammas. If they could, their origin from ^{56}Co could be unambiguously demonstrated by their energy. The editors of *Nature* asked if I would write for them a *News & Views* column describing my conclusion that, if hard Compton-scattered gammas were indeed already emerging with the reported observed flux, the gamma-ray lines themselves could not be far behind. Headlined "Supernova 1987A: hard X rays imply more to come", my essay appeared in the December 3 issue of *Nature*. The actual date of publication is memorable to me owing to a significant life experience that would happen on that very date.

Hill walking one last time with Fred

After all this conflict and excitement I needed to see England, to walk its hills and streams, and to talk to Fred Hoyle. The intensity (which was by no means finished) of my exchanges with Willy Fowler and my repudiation of his model-dependent conclusion[3] had left me agitated. This always happened to me when I disagreed with Willy; he was so much a father to me that an emotional equation always welled up within me at such conflicts. So at 9:00 am on Wednesday, 7 October, Nancy, Andrew and I climbed into the used Ford sedan that I had purchased for our stay, and departed for the Lake District.

I had loved the Lake District since my first visit of it with Fred two decades earlier, and even more since his move there in 1972. So in high spirits we headed west, up little traveled Weardale and its desolate wet hills and moors, and made lunch in Alston, a memorably pretty English town in hilly setting. By mid afternoon we had passed Penrith and made the *Ullswater Hotel*, a Victorian Gothic stone hotel on the shores of Ullswater. After the drive, and still in fever over my correspondences, I left Nancy and a napping Andrew in our very nice room and took a long hike up the crags behind Glenridding, past the Greenside Mine, and out onto overlooks that set me at peace. Sitting there, I looked, thought, and remembered the many times in such places with Willy Fowler and Fred Hoyle, then plodded back to dinner. I was freshly happy to see Nancy and our dear strong-willed eight-month old, Andrew. And now I was relaxed, confident I had done the right thing. We loved the hotel, sheep grazing on its lawn! Like most parents we called sheep "Ba, Ba's" to Andrew. After singing him to sleep that evening, we summoned the courage to go downstairs without him to dine. Nancy had to screw up her courage to leave him sleeping in our room, even though we could return to it often to check on his slumber. Life never seems richer to me than it does when dining in an English country hotel after a day out.

After a morning drive during Fred's guarded hours of work time, we came from the north into the village of Dockray, nestled on the slopes above Ullswater, from where the small road turns up the hill by Cockley Moor to their home[5]. Fred and Barbara met our son Andrew for the first time and we caught up on the gossip of parenthood. Then with a glass of sherry we settled down to chat, delaying lunch to enjoy the sherry and to talk. I told Fred straight away about the difficulty I was having with Willy Fowler over radioactive cosmochronology. Hoyle's 1960 paper with Fowler in *Annals of Physics* was a notable early work in the topic. There they had compared abundances of ^{235}U, ^{238}U and ^{232}Th by regarding their relative ages to be describable by an arbitrarily parameterized history of continuous nucleosynthesis. They had then stressed the importance of thinking of nucleosynthesis not as a single event but rather as a lengthy and continuing series of events in stars. Those stellar events distributed, little by little, the birth of the radioactive nuclei that the solar system would eventually inherit from the interstellar medium of the Galaxy. The birthdates of the radioactive nuclei ranged from the earliest stars formed in the galaxy to a few that exploded just prior to the solar system's formation. In such a picture some of our chemical elements were very old when the solar system gathered them in and some of them were quite young. Owing to the nature of radioactive decay, more of the old nuclei had decayed than of the young nuclei. This is the model that we all still use today.

The question that I had placed under the microscope in my new paper was how those birthdates for uranium and thorium nuclei were distributed in our solar system nuclei. Fowler and Hoyle in 1960 had simply parameterized their production curve—something quite different. They used an exponential time dependence for that production rate, letting the rate of decline of their age spectrum be a free parameter. It was not an astrophysical model. They did not realize that the *production rate* in the Galaxy differs greatly from the *age spectrum* of solar nuclei. So they simply ascertained for each value of that parameter what the model age of the galaxy would have to be to produce our known ratios of those three radioactive nuclei. It was a sensible survey, but nuclear astrophysicists had hardly progressed beyond it in their thinking during the subsequent three decades.. Astrophysicists needed to be shown that such parameterization was no substitute for a physical model and was in fact misleading.

Fred and I talked about that a good bit. He confessed having suffered some confusion over the difference between the *decrease of the galactic rate of nucleosynthesis* and the *decrease of the number of solar nuclei of decreasing presolar age.* I contrast those concepts to emphasize the heart of the matter. As a concrete example, consider the relative numbers of thorium nuclei that were created 5 billion years prior to the sun's birth and the number of them created during the last 0.1 billion years before solar birth. Their production ratio in the Galaxy differs greatly from the same ratio in those nuclei that we inherited on earth. Most of that Th created 5 billion years ago becomes buried inside new stars, whereas its recent production remains in the ISM. I explained to Fred that my formulation shifted focus away from the vague "rate of nucleosynthesis" to the spectrum of ages of nuclei in the Earth. Fred agreed that the flow of matter into the disk of the galaxy and the burying of interstellar nuclei within stars combine to make a knotty relationship between the nucleosynthesis rate and the age spectrum. And he agreed that it is the age distribution of terrestrial nuclei that matters, not the temporal distribution of galactic nucleosynthesis. This, I explained to him, was just what I had now solved. My new formulation[3] made the meaning of such quantities and concepts explicitly clear for the first time. When I commented that Fowler had a hard time with those conceptual differences, Fred said, "Willy is a nuclear physicist. He is not so strong in other areas of physics." This caused a pause. Fred said this simple truth without one ounce of malice, simply as an observation about a person who had been important to him as well as to me. Fred expressed his admiration for what Willy had achieved, but doubted that I should be upset over my divergence from him over the current issues. I was relieved that Hoyle saw the situation rather clearly.

Hoyle's attitude toward Fowler showed no sign of grievance over Fowler's Nobel Prize of 1983. When he evaluated Willy's abilities in general astrophysics, it was without envy. The specter of that Nobel Prize had lurked throughout our discussion, however, so I brought it up explicitly. Fred admitted surprise that the Nobel Prize Committee had worded their 1983 award to Fowler with such care that it excluded him. He had rather expected that he and Fowler would share it. Many of Fred's supporters were extremely unhappy over that omission. I was myself. But Fred did not seem to be. He was quite sincerely magnanimous, exactly what I would expect from one devoted to the principles of science. With Barbara Hoyle, however, that was not the case.

Barbara Hoyle felt instinctively that Willy could have found a way to let it be known that the only Nobel Prize he wanted was one shared with Hoyle. Hoyle had created the theory of nucleosynthesis in stars, and in the process made a believer out of Fowler. Fowler had inspired an army of enthusiastic researchers to measure the rates at which the key nuclear reactions actually occur in the stars. Both were efforts of great vitality and profound impact on the evolution of science. They were inextricably intertwined, however, just as the careers of Fowler and Hoyle had been for more than two decades. It certainly was a curious thing that they had been separated like Siamese twins by the Prize. But Barbara sensed even more. She suspected some kind of collusion between Fowler and a Nobel Prize Committee that had had its fill of Fred Hoyle for other reasons, perhaps both for his criticism of the prize to Anthony Hewish and for Hoyle's unprofessional appeal to the public over the interstellar-origins-of-life issue. These are the two considerations that I suspect caused Hoyle's excision from the prize. Hoyle himself has remained above such speculations in conversation with me. But Barbara Hoyle went further than I could go, suggesting that Willy had somehow encouraged this solution by the Nobel Prize Committee. It will be clear from my great admiration and affection for Fowler that I could not easily then or now believe that to be true, although I am all too aware of human weakness in the face of an ignoble path to a great gain. Barbara Hoyle's doubts over Willy in this regard were not simply her reaction to wounds to Fred. She recounted times when Willy, in his impish and playful way, would threaten to "take it away from Fred". She supposed that he would never have joked to her about that were there not some germ of truth behind it. Certainly I do not find myself persuaded by Willy's impish threats. The Willy I knew had delivered many an impish threat to attractive women, threats rather well received in fact.

Nonetheless, that story pains me, not only because it is historically meaningful but also because of my current struggle with Fowler. This was not an easy time for me. These two, Willy

Fowler and Fred Hoyle, were, beyond all comparison, the two most important influences on my scientific career and on my vision of man as scientist. Much of my mature personality I forged from my view of their greatness. Each had helped me more than I can easily relate, not only scientifically but also helped me to leave behind twisted preoccupations of self. They showed me that one must cast off arrogance, pride, greed, and selfishness in science. Indeed, those sins are built on the presumption of the self-made man, a type that can in honesty hardly exist in science. So I love and honor them both as the greatest men I have known closely. It renders it difficult to write of this period concerning them. I draw it to a close by relating that both Fred and Barbara Hoyle told me during this conversation that they had not heard from Willy once during the four years since 1983, since immediately after the Nobel Prize announcement. Just that fact carries with it a weight of great sadness. For what suspicions, what doubts, could so abruptly change the rapid and easy flow of two decades of close communication? Our conversation ended on this topic.

Then we had a terrific lunch, by then practically mid afternoon. After some strolling around in the gardens and the hills around the old 200-year-old Cumbrian farmhouse, we returned to one more serious topic. Fred was interested in Buckminsterfullerene. Not surprisingly so, for nothing could be more natural for a pioneer of interstellar carbon dust. He was aware of the discovery of this new form of solid carbon at Rice University, in the very building above the office that I had occupied for two decades. He asked me about Smalley and Kroto, and I told him all that I knew of their tremendous chemical discovery. We both felt there would be a Nobel Prize in it. We discussed two astrophysical questions concerning the possibility that "Buckyballs", as they are sometimes called, both names honoring the soccer-ball domes of Buckminster Fuller, could also be grown in stellar ejecta, especially within the carbon shells of supernovae.

I called that supernova-consensate particle by the name SUNOCON to distinguish it from other sources of interstellar dust. Fred Hoyle had, with a friend and colleague of us both, Chandra Wickramasinghe, advanced the SUNOCON scenario (without using that name) in a 1970 paper in *Nature*. There they argued that condensation within the supernova interior might be the main site of origin for interstellar silicate dust; and a few years later I discovered the fertile idea that brought SUNOCONs to life. I predicted isotopic tracers within such particles as a link to their origin. This gave to us both an intense interest in Buckminsterfullerene. So we spent some time speculating on two astronomical issues that they might, in principle, impact. First is the origin of the unidentified optical bands of absorption that had long puzzled astronomers. If foreign atoms can attach as imperfections to the Buckyball, either inside or out, it seemed possible that interstellar Buckyballs could be so abundant that this unidentified opacity could result. Second was the possibility that interior space could provide a trap for noble-gas atoms, especially xenon, that were so isotopically strange and known to be carried in some type of carbon within meteorites. Talking with Fred Hoyle about such things was the uplifting experience that I had found it to be for two decades. We forgot the troubles of being human in a world where success counts for too much. We turned from the past and asked questions that could only be answered in the future. This is the absolution of science. Unfortunately, we could not answer the questions we raised, knowing that more detailed knowledge would be needed even to present a credible astronomical argument.

This conversation occurred almost entirely in two large and comfortable reading chairs on the border between the 200-year and the 100-year old portions of the house, which their renovation and modernization had joined into a pleasing whole. The chairs sat in the younger part, opened up to, but two steps higher than, the lower dining room in the older part. Fred sat in these chairs every day, but to me it was a privileged chair. I had evolved from junior colleague at Fred's *Institute of Theoretical Astronomy* in the 1960s to friend in the 1980s. Fred's visits to Rice in the 1970s, and later to Clemson, and mine to England had offered me the chance to work with him and to know him deeply.

As Nancy and I drove with Andrew back to Durham, I knew it likely that I would not see Fred and Barbara Hoyle again in that priceless setting, *Cockley Moor*, "that heaven on earth", as Willy

Fowler had once described it. Fred and Barbara had spoken of their impending need to relocate nearer to people in their advancing age, nearer to help, further from the biting winter winds. We knew that they would probably be back in the south when we saw them next. Of their seventeen years on Cockley Moor, Fred wrote "..it was magical, not perfect.." Be assured that as we drove away down that winding road past Dockray, I stopped the car. Nancy and I savored a last lingering look and snapped a photograph. Even if we will go back to look again someday, it will not be the same. A little pensive, but relieved from my personal conflict, we drove back to Durham through incomparable Teasdale.

Baby Andrew and old Durham

From the remembered and forgotten imprints of a marvelous childhood and family, I carried within me an intense devotion to family. And my Durham work did not interfere with sharing life

with those I love most. Nancy and I devoted a lot of time to each other and to sharing Andrew's growth--his first steps, his first words, his first manifest stubborn-ness. We laughed ourselves through those hilarious adventures of life's begin-nings. On June 24 Andrew had cried tragically at having his first baby sitter so that we could attend the annual Senior Combination Room Dinner of St. Mary's College. We walked his baby pram over the streets of Durham, talking to him excitedly about every glorious sight. Sometimes I stayed after-noons with him

Andrew, now eleven months old, and Nancy, December at St. Mary's College.

471

in our college home so that Nancy could undertake the shopping, always more of a chore in England than in the USA. During these times I tried to be inventive with Andrew. By the beginning of September Andrew began taking his first steps. He was eight months old. When he made it to Nancy he would say "ma,ma, ma,ma...". On September 14 I asked, "Andrew, what does a doggie say?" He answered with enthusiastic guttural noises that we took to be a bark. On the 17th of September he stumbled to my chair and said "Da, da, da...". He certainly knew this to be correct. On November 12 he stooped to look at a book fallen under the secretary and said "Book!" Such monumental efforts of the budding intellect were enough for us to break out a bottle of wine for dinner. By the following week he walked two or three steps from one chair to another, at age 10 1/2 months. And every night when I walked him up to bed I counted aloud the sequence of seven steps to the landing as we took each: "one, two, three, four, five, six, *seven!*" Then I repeated this on the seven steps to the second floor.

By experimenting later I discovered that Andrew recognized an error if I intentionally miscounted. But above all, Andrew became ever cleverer at ways to hold our attention. We resigned ourselves slowly to a boy who would not easily relinquish center stage in our family life.

Nancy and I opened our modest doors in St. Mary's to as many kin as could come. Nancy's mother arrived and, in addition to sightseeing with us, kept Andrew while Nancy and I

Devon and I during his 1987 visit. Devon was here 26 years old.

took a trip to London. My second son, Devon, spent two weeks with us in October, in the middle of my science scuffles. Willy Fowler's younger daughter, Martha, had been Devon's baby sitter in Pasadena. He got me out of my worry a bit. We walked and talked all over Durham, Newcastle, York and London. We drove to the seashore for coffee. Devon is a breath of fresh air among people, among the easiest with whom to spend time without stress. Everyone loved him for that. This made hosting him very warm for two weeks in our small home. Our easy relationship was possible because of my acceptance of his homosexuality. I was fortunate to not have been put off by that.

Devon was in one of his seasons of high happiness, and in those moods he was very entertaining. But as we walked and drank coffee in cafes, our favorite vice together, I also heard rumblings of continued discontent with his life in Los Angeles. I rediscovered that a certain side of him found it difficult to be content. One side of him seemed depressed. Both characteristics reminded me of his mother. It worried me enough that after we later moved from Houston we encouraged him to leave Los Angeles in an attempt to help him find fresh perspective. But it was very hard to penetrate Devon's private nature, to identify his discontents. *How can such a wonderful person be discontent?* I have ideas about this, born of my own long suffering over Devon, and I broached them with him over those coffees, remembering for him their possible seeds from his earliest years. In this he was very interested, and yet remained inscrutable.

Nancy's knowledge of architecture and my fascination with it found happy union in Durham. There is so much to see of the architecture of this region. The colleges are neither so old nor so grand as Cambridge or Oxford, but more reflective of the past life of the city. We found a highly recommendable book, *The Book of Durham City*, by Peter Clack, which we read as fuel for our excursions. It provides considerable technical and scholarly background to the city, at depths exceeding that of tourist guides.

Near the end of our time there we received an invitation that was to have some considerable influence on our subsequent life, although it is admittedly impossible after the fact to ascertain the degree to which one's big life decisions are influenced by single events. Mary Hawgood, member of the Senior Combination Room at St. Mary's College, invited us to tea on December 3 at *Crook Hall.* Mary and her husband, who worked in London, had acquired ownership of a part of the remains of a series of manor houses built between 13th and 18th centuries. Situated on the north side of walled gardens, the hall is a linear progression of structures moving from 13th to 18th centuries. Mary and her family lived in the Georgian Hall, and it was there that we had tea and appreciated the massive and sound structure, so full of life and the graceful characteristics of fine Georgian buildings. We were told that its maintenance required constant work and funds. Nancy and I found ourselves identifying with people who would commit their extra resources to such purpose. Mary explained openly the commitment that one must have to do this, and also its rewards. One did not need to have the rewards explained, because you could feel the rewards just by being there. With Mary we toured the adjacent medieval hall, which had been filled in by centuries of silt and which they had partially restored. Nancy and I were inspired. Its massive fireplace, staircase, and medieval windows were simply beautiful. We wanted to have done it ourselves, to have owned it and cared for it.

Of course, we did not speak of such an impossible goal. But when we undertook similar objectives just two years later, this tea time with Mary Hawgood sprang to mind, giving us courage, determination, patience, and vision for an opportunity to commit ourselves to an ongoing renewal in late 19th-century Seneca, South Carolina. Now, twenty years into that adventure of renewal, I can attest to the pride of purpose and satisfaction that it has added to our lives together. My love of residing in substantial old houses is easily traced back through my history of lucky residences; *St. Clement's Gardens* on Thompson's Lane in the shadows of St. John's College in Cambridge, *The Grange* in Great Wilbraham, *White Cottage* on Storey's Way, far back to the renovated church on West Sixth Street in Davenport, and ultimately to the solid farmhouses of my family roots. Each reinforced my fascination with the unspoken remembrance of things past, things indelibly outlined by the architecture in which they occurred.

Gamma-ray lines from supernovae 1987A

Mark Leising telephoned me from *Naval Research Laboratory* on December 8, 1987.

"We've found it!" he exclaimed. I was silent, gathering my nerves for this long awaited moment.

"Found the gamma lines?" I asked, knowing that's what he meant.

"Yes. We see 847, 1238, and maybe 2559", Mark replied. He was referring to the energies in keV of the gamma ray lines emitted following ^{56}Co decay.

I was breathless with excitement, having waited eighteen years for such an event. I knew that at NRL they had been trying to extract the gamma-ray lines using *Solar Maximum Mission (SMM)* data from its sunward pointing Gamma Ray Spectrometer. Mark explained to me that the sum of a few months data showed clearly that supernova 1987A was emitting these gamma ray lines. What's more, they were emerging earlier than expected, just as the hard X rays had done. Russian scientists that detected those X rays, which could not exist in SN1987A except owing to the scattering of gamma-ray lines from electrons within the supernova, had concluded that the ^{56}Co radioactivity had, in part, been mixed outward, closer to the surface of the expanding supernova gases. From those shallower depths, some of the gamma rays could emerge without being deflected or absorbed. Mark Leising quickly published a short paper in *Nature* clarifying what the early arrival meant for the supernova structure. Stan Woosley energetically published a series of calculated models demonstrating fresh conclusions, as did also Ken Nomoto in Japan. .

Mark Leising had, since earning his Ph.D. from Rice University, been working with Gerry Share at *Naval Research Laboratory* (NRL) in an attempt to convert that *SMM* space satellite for observing the sun into an astronomical observatory. SMM pointed at the sun, so the technique for seeing elsewhere depended on the changing view of the galaxy behind the sun. It had not been easy, but they had succeeded dramatically. I declined Mark's invitation (from *NRL*) to travel to Washington for the impending *NASA Workshop on Gamma Ray Astronomy*. It would have been exciting to attend the delivery of the announcement papers; but I was content just to know that this goal that I had held so long had now been reached. Of course I had hoped that NASA's *Gamma Ray Observatory* could be the first to detect the gamma-ray lines, and that I might even have been the co-discoverer with the *OSSE* telescope. But *Gamma Ray Observatory* was not in space, and would not be for quite some time. So that hope was fruitless for me but exciting for my dreams. What satisfied so was having predicted this new gamma-ray-line astronomy and witnessing its confirmation. This historic detection of radioactivity in supernova 1987A was the first occasion on which man had been able to confirm newly born nuclei in any exploding astronomical object by this technique. Because the cobalt halflife was so short, the 20,000 earth masses of pure radioactive cobalt observed in SN1987A was comprised entirely of new nuclei! This too could have merited the headline "Happy Birthday".

It was a sublime day for me. Nancy and I opened a bottle of French wine to accompany our modest dinner, and sipped it enjoyably as we recounted the years of work and hope and the excitement of hearing of the final success, and of its coming while we were residing in St. Mary's College, Durham.

Help sprang to compensate for the grounding of *Gamma Ray Observatory*. Jerry Fishman had obtained financial support from NASA to pack his balloon-borne gamma ray spectrometer aboard a plane to Australia. At the *Alice Springs Balloon Launch Station* it floated in October 1987 to the top 4 grams of the Earth's atmosphere. The 847 and 1238 keV gamma rays can penetrate to that depth. Jerry's experiment recorded those gamma-ray energies with a germanium high-resolution spectrometer. Although they would not be first to publish this discovery, it was the earliest detection of those lines at a known time from supernova 1987A. The *Solar Maximum Mission*, about which Mark Leising had telephoned me with the initial discovery, had recorded some of its gamma rays prior to October, but their exact dates of arrival are less well documented than Fishman's were and included also later arrival times. Several more balloon observations were made by other scientists with high-resolution germanium detectors, but Fishman's had occurred earliest. These data were fruits of NASA's aggressive plan to support a flotilla of balloons observing SN1987A. This ranks as a romantic episode in the annals of supernova research and of NASA dedication to their mission. These historic detections were the first gamma ray lines from a single exploding star ever recorded by the human race. A new astronomy had been borne, one in which profound information about the exploding object could be obtained by the detection of its newly created radioactivity. Jerry Fishman's accomplishment was truly poetic. It had been far back in 1968 when we had first begun calculations[1] of the chance to detect these gamma rays, and two decades later he also became one of the first to find them. That beginning-to-end establishment of a new astronomy will go down in history. I wished I could have done this myself; but two of my students, Jerry Fishman and Mark Leising, had done it for me--for all mankind. This stands near the top of my scientific experiences.

Closing our Durham Year

Our sabbatical half year in Durham came quickly to a close. We sold our used Ford sedan for 750 pounds to a student in van Mildert College. It was a good buy for her and good for us to get back half the price we had paid. We simply advertised it on a university bulletin board and it went fast. It is common in university towns for a car to be cycled among faculty to visitors or post docs to students until its value is negligible. I took my camera for a final photo sessions, as I typically do, to the office of Joan Kenworthy, Principal of St. Mary's College, and to a Physics Department sherry party hosted by Arnold Wolfendale to bid us farewell. Arnold, soon to be appointed Astronomer Royal, had been a splendid host. That is needed to have a memorable year abroad.

On December 12 we bundled our luggage into a taxi to the spectacular Durham train station and caught the 3:05 train to

Manchester, our final England stop during this half year. Andrew was approaching his first birthday, and stood on the seat in the passenger compartment, bouncing up and down and burbling about the passing scenery. We had already encouraged his love of trains by repeated trips to the Durham station, high on a hill overlooking the old city, to watch them arrive and depart. So Andrew was quite keyed up to actually ride in one. How romantic it is, always has been, traveling by train.

Manchester was host for the international science conference *Dust in the Universe.* There we enjoyed arrangements by *Ellis Llwyd Jones Hall* of Manchester University, specifically staying in *Sunnyside,* a nice Victorian house in Manchester's Victoria Park, a once distinguished residential area threatened by decay and division. We loved the house, but were slightly saddened at the social evolution that leads to the dividing of a great family home into guest rooms for the university--a fine reutilization to be sure, certainly better than decay, but still a little sad when one listens for the ghosts of the turn-of-century children romping about their privileged home. That fate is one we would have the chance to prevent about a year and a half later, when we would move to South Carolina.

I presented two oral papers at that conference, both concerned with histories of isotopic memory within interstellar dust. One involved computational models[6] for the recycling of interstellar dust among the phases of the ISM. I had developed those models, and their underlying ideas, with Kurt Liffman, one of my last Ph. D. students at Rice University. Our calculations demonstrated that SUNOCON cores of interstellar grains can be expected to survive much longer against destruction by interstellar shock waves than current wisdom maintained. Their longevity arises from their cores being shielded from the destructive effects of sputtering collisions with fast interstellar ions. Those cores become shielded from the destructive ions by overlying cold mantles. Those mantles are accreted repeatedly during recycling of dust and gas atoms into cold molecular clouds. This was a truly pioneering study.

The other oral presentation in Manchester addressed the origin of the isotopic anomalies found within meteorites. Both papers may be found in the published proceedings. It was a satisfying conference for my own interests, considering that I had taken my scientific mission to be the fusion of interstellar-dust studies with those studies made by an entirely separate group, the meteoriticists. This conference was populated by astronomers, astrophysicists, chemists, and space scientists, whereas the one in July in Newcastle had been populated by meteoriticists. Those groups did not yet meet together. Today several scientists actively forge links between these communities. But within two years NASA would define a new program of research support entitled "Origins of Solar Systems" carrying an explicit purpose of bridging scientific disciplines related to that great question.

One interesting sidelight was a social hour after the conference banquet in Ellis Llwyd Jones Hall. I went to a bar with Harry Kroto to drink ale and discuss two problems. A delightful and handsome man[7], Kroto's eyes twinkle during conversations, marking excitement for science. *That reminds me of Dick Feynman,* I thought while we were talking, remembering Dick's electric eye excitement. On the scientific side Kroto fastened on my 1975 paper predicting almost pure ^{22}Ne trapped in carbonaceous stardust, and perhaps found within the meteorites. We were both excited to think that perhaps radioactive ^{22}Na atoms are caged within fullerenes that are part of the carbonaceous residue that remains after the meteorite is dissolved in acids. The fullerenes, larger cousins of Buckyballs, would have to have grown, and closed, in expanding gases faster than the 2.75 year halflife of ^{22}Na in order to enclose the ^{22}Na before it decayed to gaseous ^{22}Ne, leaving it trapped within. Kroto was emphatic that sodium would be trapped within fullerenes. Our conversation reminded me of the same topic hashed over with Fred Hoyle a couple of months previously.

But Kroto was also distressed about something. He was not slow to admit it. Harry was distressed at the feeling that Rick Smalley was freezing him out of Smalley's laboratory at Rice University, where they had worked together on the discovery of Buckminsterfullerene. That discovery had been made two years previously and was reported in the pages of *Nature*. I sympathized with his worry, but I also understood how Kroto's undeniable zeal might have caused him to overstep an appropriate role within Smalley's and Bob Curl's labs, where the discovery had been made. I urged him to be patient and to not press. The issues are big; there is sure fame awaiting the outcome of attribution for the discovery of Buckyballs. I expected it to reap a Nobel Prize, as it eventually did. It is not every day that a new allotropic form of carbon is discovered. Everyone knows of diamonds, of graphite, and of soot, although perhaps not everyone knows those are forms of carbon. Fewer know that the very stable soccer-ball arrangement of sixty carbon atoms is an entirely new form of solid carbon. It is more than just another molecule, much more. I am sure that issues of scientific credit were also on Kroto's mind, though he never said that. He talked only of his scientific frustration at not being easily able to contribute further to Smalley's program. But his agitation was understandable to me because the stakes were so high. I could not smooth his return to Rice University, and indeed I would not have wanted to insert myself into something that is no business of mine. I never told Rick Smalley of this conversation with Kroto, because that would only increase any misunderstanding that might actually exist. I certainly experienced a very engaging time with this spirited and highly motivated chemist. And my reason to tell of it now is simply that it is of historic interest in the wake of their great fame.

Only one thing really rubbed me wrong. Kroto spoke disparagingly of Fred Hoyle. Many biologists and chemists, ignoring the greatness of Hoyle as an astrophysical cosmologist, attacked his appeals to the public over the single issue of interstellar carriers of life. Hoyle's behavior as a scientist had indeed been wrong; but I did not like to hear my hero criticized. Hoyle had repeatedly, and perhaps brazenly, argued that life's chemical code did not originate in the "primeval soup of Earth", as all supposed, but was carried into Earth by interstellar dust; and he had done so in books for the public as well as in scientific publications. I think it unfair to great scientists to attack them, repeatedly, for their more questionable forays outside or inside the scientific methodology. I do not think it important to cling onto mistakes by scientists--a form of "three strikes and you're out" designed to judge them by their worst efforts rather than their best. More important is whether a scientist makes great contributions that move science ahead, as Hoyle did repeatedly. I also felt this same unfairness personally in my own career, especially the unwillingness of some leading meteoriticists to appreciate some of my most creative efforts because of their curious fixation on what they perceive to have been a mistake on my part.

What *is* Science?

History is piled high with the scrap heap of rejected ideas in science. The essence of science is skepticism rather than belief. Scientists do not prove ideas to be true, as many assume; we prove them to be wrong! Science is a process of successive approximation in which the best ideas are refined by physical measurements and retained within the framework of knowledge. When ideas stand the test of repeated attempts to disprove them, we develop increased confidence that they may be true. Incorrect ideas are quickly discarded to the rubble pile and forgotten. It is the true and lasting new ideas that enable this process of refinement to go further. Hoyle and I were luckier in that goal than very many who were never wrong.

CHAPTER 18. A MOVE TO CLEMSON UNIVERSITY

It is not easy to leave a professorship of twenty-five years at a distinguished university. We academics love our lives. It is a privilege to be able to spend decades prospecting along the boundaries of knowledge and striving to pass on that which we acquire. Rice University was a wonderful place for me to do this. It has become one of our universities of distinction, and it has the luxury of being able to support excellent teaching while restricting teaching load so that scholars have time for original scholarship. Rice does these things extremely well. Many professors, probably most, are unwilling to leave such an ideal setting. Our universities become part of our identities, and the most meaningful parts of our aspirations are linked to our intellectual home. Leaving can seem to be a loss of self. But in the 1980s that is what I was doing.

There comes a time when one wants a fresh start, and perhaps also a geographic change to a setting that tugs more at the heart. Since our 1983 wedding, Nancy and I sensed the weight of so much personal history in Houston. My failed marriages and their many artifacts lay all about us. All of my friends had come to know me with previous wives, and we shared a palpable wish for freedom from that. It was the desire to go forward in our lives in a totally new and secure relationship outside of the ruins of past

482

mistakes. To relocate seemed the way to focus on our lives without constant reminders. So during the1980s we agreed to check out other university opportunities that met our goals. Those goals were very clear: a comparable position in another fine university, one that needed me for its goals, but also one in which I could pursue my research ambitions; a suitable setting for a team working on data from *Compton Gamma Ray Observatory*, where I was Co-Investigator; a more rural eastern or Midwestern setting; near to natural outdoor spaces that alleviate the oppressions of unending city and unending summer. To seek such opportunity at the highest level we had to activate the plan long before I seemed near retirement, while the fire for research and leadership was still strong. Near age fifty is such a time. So in response to job advertisements, we visited several universities that had interest in me and that we felt might match our goals.

Emotional aspects of personality enter such decisions. I had clung to many things, fearful of their loss. One great early loss had been the farms of my families, my emotional bases in Iowa. And I feel sure that covered a deeper more infantile clinging. So I had clung to Grandma Kembery's farm, to my dying Grandma Clayton, to the elm trees on Fontanelle square, to our De Soto that brought us to Texas, to University Park Elementary School, to summers back on the farms, to the world of Sherry Lane, to my golf clubs and my stamp collection, to old family photographs and to the thousands of new ones that I took to remember my children and my scientific colleagues and mounted into picture books, and on and on. I am still surprised that I did not similarly cling to Rice University. But by then something had changed, some umbilical need had been cut, some childish personality traits discarded.

On the other hand I possess the personality of an adventurer. I was the first among our kin to attend university, to seek a graduate degree, to move away from Texas to California, to surf in the Pacific and to climb mountains in the Sierra Nevada, to sample Europe with seven-year periods of residence in England and then in Germany, to divorce, even to intellectually leave the earth for science in space. I was never bound down, restricted by my losses. I flew forward. If there is paradox in me, it was the competing tugs of parental influences. Surely adventurousness derived from my father's lead, who left farm life and thereby took us from the farm. It derived from a father who flew airplanes, and who challenged me, "OK then. You fly it."

The fourth of the universities to which we made recruiting visits was Clemson University in the northwestern foothills of South Carolina. From the moment we descended over snow-flocked farms and forests and landed in Greenville, the area captured our joint romantic sense of a simpler world. And Clemson reached out for us with a sincere hospitality for which the south is known. Clemson University wanted me to establish a leading astrophysics research group within their department of Physics and Astronomy. I outlined a program that I called *Nuclear and Gamma-Ray Astrophysics* that was unified by the discipline of nuclear physics in the universe and whose research would probe the nucleosynthesis of the elements, the gamma-ray-line astronomy of radioactivity, and the isotopes within stardust. Clemson University had rich century-old traditions as the land-grant university of the state of South Carolina. It began romantically on the plantation estate of John C. Calhoun (1782-1850), the celebrated US senator, Cabinet Minister, Vice-President and political philosopher who had led South Carolina during slavery and whose son-in-law Thomas Green Clemson had, after the devastation of the Civil War, donated the entire plantation to the state of South Carolina for a school for the scientific study of agriculture. The small town of Calhoun later changed its name to Clemson.

During that recruiting visit, Nancy and I had time free to explore the housing and the towns in the area. We did not find here the trendy affluence of the rural northeast, but rather simple small towns of late Victorian vintage facing troubling economies. They did remind me of the homes and businesses of boyhood Fontanelle. We had been sent a small tourist flyer about "Silk Stocking Hill" in Seneca, seven miles west of the university. We turned slowly onto West South First Street, creeping along in admiration of the solid old fashioned homes. After one block of what seemed to be a park, I pulled to a stop before the first house on the right.

"Wow. Look at that one!" I exclaimed.

"It's wonderful", Nancy said.

The large and stately Victorian home was not only grand but seemed to be a part of that park. Suddenly Nancy pointed to a small handwritten sign on a wooden stake by the front walk: *For Sale by Owner*. Amazed by that sign we stepped out of the car and looked around

G. W. Gignilliat House at Christmas 1989

briefly, two hearts beating hard for this solid and distinguished home that seemed with each glance to have our names written on it. Somewhat perplexed at the lack of an answer to the phone number, we entered the Lunney Museum across the street. Its curator, Betty Plisco, said,

"Oh yes, it's Thomas Gignilliat who has the house for sale. He lives just next door to the house". She gave us his phone number.

After arranging an interview, we visited Thomas Gignilliat in his lovely Georgian Revival home from the 1920s. Thomas, who was to become a great friend, confirmed that the house was for sale. It had been built in 1898 by his grandfather George Warren Gignilliat, the leading businessman of the town of Seneca following its establishment in 1873 after two railroad lines crossed here.

"The family calls it *the big* house", he commented, offering the first clue that this was a precious home to a large family of descendents. After explaining that I expected to be offered a professorship at Clemson University, and that we were very interested in buying the house, I asked the price, somewhat holding my breath.

"$150,000", Thomas replied.

We were stunned. Another grandson of the builder, Frank Adams, came over to guide us through the house. As we walked through we became increasingly certain. The floors were solid and a beautiful mix of narrow-gauge oak and "heart pine", the dark, hard center of the hardwood longleaf pine from which great southern homes were constructed. The twelve-foot ceilings and two sets of mahogany "pocket doors" took our breath away. Excitedly we noted that the house was in original structural condition. Because only the family had lived in it until now, no amateur architects had modified it at any point. The load-bearing walls of the walk-in basement were solid and plumb. Original glass-plate windows, some topped by stained-glass uppers, gave charmingly wavy views of the glorious grounds around.

"We want to buy it," I said to Thomas. Our minds were made up.

When he explained that the price did not include the park to its east, I immediately said that we wanted the entire land belonging to this historic house included in the sale. This was the surrounding four acres of what had originally been shown on the sale plat for the town as a small thirteen acre farm adjacent to the downtown square. Thomas's father, the builder's son, had acquired the western nine acres when he built his house next door. The Gignilliat family had built seven of the houses on Silk Stocking Hill.

"I have not yet officially been offered the professorship, but I expect it very soon", I warned.

"Well, I won't hold you to it if you don't get the position", Thomas offered. "And we can include the whole four acres", Thomas added, apparently having decided that we were what the family wanted for Gignilliat House

So our contract was sealed over a handshake, one which would be excused if we did not get the offer. In euphoria, we returned to our lodgings in historic Liberty Hall Inn in nearby Pendelton, the town having the most historic roots of the upstate towns, and prepared for dinner with Dean of Science Bobby Wixson. That occasion struck good rapport as I explained what my plan would be for Clemson astrophysics. His eyes really lit up on hearing that I would bring funds for my Co-Investigator position on the *Gamma Ray Observatory*. In return, mine lit up on hearing that what I wanted to do matched what Clemson wanted me to do for their renovation of astrophysics in the department. On the next day in his office he made an outstanding offer. I accepted. For Nancy and for me it had been one of the most exciting weeks of our lives. It set the course for the remainder of our lives, and we have never looked back.

Our minds were made up, but we had to think of others. Both Donald and Devon had some misgivings about our move. They expressed these and I sympathized with their sense of final loss of their boyhood family in Houston. Such loss had touched me deeply in my own life, so I was sensitive to their anxiety even though, at ages 23 and 22, they were well out of the nest. Nancy's parents had their misgivings. Her father was a Rice University graduate, as was Nancy, as was my son Donald, so that *Rice Blue* colored our family blood. But when they saw that we loved the idea of new life on the east coast, Nancy's mother caved in at once, admitting that she missed Philadelphia, where Nancy had been born. My daughter Alia, now age six, came to visit on a trip to Houston with her mother, and to her puzzlement I explained that since she no longer lived in Houston, our desire to remain there for her no longer was relevant.

"We will take along some of your stuffed animals for your room in our beautiful new house", Nancy wisely interjected.

Thus assured that she always had a home with us, Alia looked in awe at photographs of Gignilliat House. One by one, the misgivings of those dear to us had been allayed. Each was won over by the enthusiasm that Nancy and I evinced for our decision.

Rice University tried to induce me to stay. They offered a handsome salary increase to match my offer from Clemson. But it soon became clear that this was no power play on my part. There was one thing that Rice could not match, and it was no fault of theirs. The *Department of Space Physics and Astronomy* could not offer to hire new faculty members to build up those new areas of astrophysics that I was creating. Our department was now mature, which means that new hires had become rare. This era contrasted with that of the sixties, when we were constantly on the lookout for new faculty. Because of my extensive Heidelberg period, 1977-82, focusing there on cosmic chemical memory, moreover, I had let Rice's preeminent position in nucleosynthesis slip, a position that we had earned by 1975. So although I was wanted by Rice to continue my personal scholarship, it would have been a long uphill struggle to ask them to hire new young faculty in nucleosynthesis, gamma-ray astronomy, and stardust. Mature departments can undertake such an ambition only when it is clear to all that the architect of the plan is the leader for the future of the department. Our department had many leaders, so it would resemble a turf war for any of us to suggest that he holds that key. I certainly did not want to undertake that struggle. But large rebuilding is exactly the chance that Clemson University offered owing to their rebuilding of a retiring faculty and their large commitment made to me for leadership. These aspects of university life also need to be understood to comprehend why I would leave Rice. My research future was involved.

Before departing Rice I had research support tasks to get into place. I wrote a proposal to Jim Kurfess at NRL for the subcontract funds on the *OSSE* instrument for *Gamma Ray Observatory*. This was important research money for my Co-Investigator position, which would be the front line of my buildup of astrophysics at Clemson. The same was true for my proposals to NASA's *Cosmochemistry Program* and to the new *Origins of Solar Systems Program*, which supported my stardust research and which I asked to transfer to Clemson. The Welch Foundation did not transfer grants out of Texas, so I would, regrettably, have to do without that one.

I took two of my final three Rice research students, Kurt Liffman and Paul Scowen, to the annual *Lunar and Planetary Science Conference* sponsored by the *Lunar and Planetary Institute* adjacent to Johnson Space Center. There we presented coauthored oral papers[1] on new aspects of cosmic chemical memory. A conference banquet in Bay Shore Hotel featured a celebration of the twentieth anniversary of the *Apollo 11* landing on the moon. That lunar landing had changed the research lives of all in attendance, about a thousand scientists from around the world. Then in April I dared not miss attending the *Gamma Ray Observatory Workshop* that NASA was sponsoring at *Goddard Space Flight Center* in preparation for launch in 1991. There I presented my talk "The Astronomy of Radioactivity" and also met with OSSE team members and described my ambitions for Clemson University to Mark Leising and Jim Kurfess. And from my Rice office I mailed dozens of letters to leading scientists urging their support with their congressmen for making *Gamma Ray Observatory* a budget line item. Such tasks highlighted a whirlwind spring, but the impending adventure kept my energy levels high.

In early June, 1989 we departed Houston in my gold Cadillac Coupe de Ville for South Carolina, nursing five infant kittens and one six-month old son as we went. We had found the kittens underneath our house on Bartlett, which we had fortunately sold rapidly. The kittens survived the trip, and were greatly enjoyed at rest areas by fellow travelers who laughed at the sight of five little kittens on five leashes emanating from Nancy's hand. The motels were another matter.

One week later, in Clemson we closed on Gignilliat House. The remainder of the summer was spent on initial restorations and on getting settled into my Clemson office. We rewired the entire house, and later replumbed it as well. We replaced the floorboards in the kitchen owing to its century of hard knocks, assorted buildups, and generations of new holes into the basement below. Nancy took the lead on planning all renovations. Moreover, she took this on as her job in our partnership, an attitude that I love in her. These were the beginnings of a commitment that never ends. The preservation of historic homes is an ongoing task, a labor of love, and it is here that the inspiration of Mary Hawgood at *Crook Hall* in Durham fueled our resolve.

Comic battle over the Leonard Medal

Before the fall semester at Clemson had even begun there occurred an unprecedented professional event, one that demonstrates the almost unbelievable battle that I had faced since 1974 over stardust.

A friend and distinguished meteoriticist, Ahmed El Goresy, telephoned me from Heidelberg on July 27 with a confidential message.

"Don, you have just been selected by the Council of *The Meteoritical Society* to receive its Leonard Medal, the highest honor in meteoritic science".

I was overjoyed at this news! I felt sure that it must eventually come my way, because stardust had revolutionized both meteoritics and astrophysics, and I had interjected the concept into both disciplines. But its announcement just as I arrived to my new position in Clemson was timely and seemed politically useful. Nothing save a large research grant so validates a scientist to university administrators as the award of a top scientific prize. Fortunately I made the lucky decision to not announce this until the official letter arrived from the president of the society.

On August 8 I received a second phone call from El Goresy telling me that I was not the Leonard Medalist after all! Ahmed explained, naturally with some embarrassment, that the Council's selection had, in an unprecedented move, been vetoed by the Society president, Edward Anders. This left the Council members in anger and shock. To me it was predictable, as Anders's disparagement of my contributions had been occurring continuously since 1974. So to be consistent he must have felt that he must now reject them once again.

During the following year I heard the sequels. Council members threatened to resign and disclose Anders's action to the society unless he reconsidered. He had to accept that the president's duty was to act on the judgment of the distinguished council of the society, whose job it was to select the awardee. Anders focused on a technicality rather than ad hominem objection; namely, that I had not been nominated save by the committee itself. At the committee's request, therefore, a formal nomination was submitted from Heidelberg. This only took more time. Eventually Anders acquiesced. When, two years later, I did receive this society's award at the 1991 meeting in Monterrey CA, Nancy and I were by tradition seated during the society banquet at the table of the president, none other than Edward Anders. This irony was enjoyed by many at the banquet who knew what had happened; but the conversation at our table was very strained.

The reassuring aspect of science is that good ideas can not long be suppressed. There exist no czars who can dictate truth. So I received my medal at the same time that stardust and cosmic chemical memory were being validated as a revolution in science, despite seventeen years of attempts to discount it. I entitled my Leonard Medal Lecture "Meteoritics and the Origins of Atomic Nuclei"[2]. When I telephoned a weakening Willy Fowler to share this good news, his joy almost matched my own. In some exaggeration he said that he had thought all along that my ideas would prove to be important. The comic opera does not detract from the truth that I prize that 1991 Leonard Medal. I had worked two decades to develop my picture of stardust and cosmic chemical memory, and I had been selected for that honor by a committee of my peers in the society.

Defining astrophysics at Clemson University

My Co-Investigator status on *Compton Gamma Ray Observatory* was my leading program for building astrophysics research at Clemson University. Research costs money, and this gave me a running start. It included funds for a postdoctoral Research Associate, and as quickly as I could advertise the job and screen some applicants I selected Lih-Sin The, an Indonesian who had completed his PhD research with Adam Burrows at University of Arizona. I telephoned Lih-Sin on August 16, 1989, to tell him that he was my choice for the job. Lih-Sin was already familiar with the astronomy of radioactivity because his thesis had expertly computed the escape of gamma rays and x rays created by radioactivity in an expanding new supernova. His computer models described detailed evaluation of ideas that I had introduced in my 1974 paper. In a larger sense, Lih-Sin's training fit well into the intellectual umbrella that I was preparing to describe to the Department of Physics and Astronomy faculty. He arrived on October 20 to a department with very few Research Associates owing to its low research funding.

The OSSE team meetings in preparation for launch were continuing at *Naval Research Laboratory* on the Potomac in southern Washington D.C. For the first time I attended by train, catching the AMTRAK that passed just behind my home through Seneca and which made a passenger stop in Clemson. In those years I stayed in Washington at the *Cosmos Club*, not because it was especially convenient but because I loved being a club member. I loved the grand clubhouse on Massachusetts Avenue NW and enjoyed its distinguished members. During this lively meeting, run by our P.I. Jim Kurfess, I found social time to explain to Mark Leising that I wanted him to visit Clemson and present a department colloquium. I hoped that he would apply for a faculty position at Clemson University. I described my strategy for building gamma-ray astrophysics there. On September 20 Mark did arrive to spend two nights with us in Seneca, and his invited department presentation, "Discovery of Radioactivity in Supernova 1987A", impressed the faculty highly, both with the topic and with Mark himself. This set the stage for my proposal to the department faculty.

In a department faculty meeting called to discuss hiring strategy, I described my strategic proposal for a general research program that I called *Nuclear and Gamma-Ray Astrophysics*. It would involve a faculty of at least four eventually, and would focus on gamma-ray astronomy, on the theory of nucleosynthesis, and on isotopic patterns in stardust. I suggested that each hire should share the knowledge and culture of nuclear astrophysics to provide an intellectual umbrella that would unite us into a powerful group. My proposal created enthusiasm among the faculty, but also consternation when I advocated hiring all three during the very next year!

"We don't have three faculty positions to offer, do we Pete?" one professor asked of our Head, Peter McNulty.

"No. We might have two hires this year owing to retirements," McNulty replied, "and these have to be allocated among our target disciplines: solid-matter physics, theoretical physics, atmospheric physics and astrophysics".

Then after expressions of doubt that my aggressive plan

would be possible using the routine application prodecure of advertising generally and evaluating applications, I stunned the faculty by replying that I had already spoken to all three! This was presumptious of me, and somewhat risky, because it is the faculty that votes up or down on hiring suggestions.

"I have had conversations with each of my nominees: Mark Leising at NRL, who already presented that good research colloquium here, Dieter Hartmann from University of California, and Bradley Meyer from University of Chicago." They were impressed by the group's credentials, agreeing that Mark Leising, who had already visited, looks like a winner.

"Would guys like that come to Clemson?" one professor asked.

"Probably not on our past reputation", I replied. "That's why I propose we hire all three! They know and respect one another, and I think they might agree that if the others will do it, they will too. All three might come to join my plan when no single one of them would do so alone--sort of a bandwagon effect. Promising physicists want promising colleagues."

We would need to induce all three to apply to our search advertisement. I was confident that they would apply to the ad.

"I think I could get the Provost to approve three hires next year by borrowing from future hires", Pete McNulty interjected. That set a very helpful tone.

Huge discussions followed. I feared that they could turn territorial by being seen as a threat to the other disciplines of their own growth. But to my relief, that did not evolve. Pete McNulty had made the right suggestion at the right time. That is what good department heads do. So in the end the faculty agreed to mortgage the next three hires, provided that the Dean of Science would approve, in return for three immediate hires. This came about when in 1991 our three new hires were three young stars of nuclear astrophysics: Mark D. Leising, Dieter Hartmann, and Bradley S. Meyer. This was widely regarded on campus and around the country as a recruiting coup.

Dieter Hartmann, Bradley S. Meyer, the author, and Mark D. Leising in Clemson House, November 1990

All were aware that they were trusting in an investment in me, whom they had hired to point the way. Our group would be close relatives to me, using the family metaphor. Leising and Meyer had once been my students at Rice, though at the very different levels of PhD student and senior thesis student, respectively; and Hartmann was the PhD student of Stan Woosley, a world-leading former student of my Rice program. We three held much in common. The inbreeding assured that we would speak the same language and, hopefully, stimulate the ideas of each. Along with Lih-Sin The, my Research Associate, we four were the backbone of nuclear astrophysics at Clemson University for the next decade.

A major discovery with *Gamma Ray Observatory* was not long in coming. Its 17.5 tons of detectors and electronics roared upward from Cape Kennedy aboard shuttle *Atlantis* on April 5,1991 into its orbit around the earth. Imagine our excitement after a dozen years of waiting. The *OSSE* team first needed to demonstrate that our instrument was functioning correctly in orbit, so we performed initial observations of the *Crab Nebula*, the famous supernova of

1054 that had been observed by Chinese astronomers. We could not hope to find radioactivity there, because all that was present 937 years ago was long ago dead or of far too small abundance to provide enough gamma rays. Our purpose in observing the Crab was only to calibrate the performance in orbit using an astronomical object that was already known to be a continuous-energy source of energetic x rays. The flux and energy of x rays from the Crab had been measured several times, so there was little doubt what we should see. These x rays today are produced by the pulsar that the supernova left behind. The team was relieved to find that the OSSE instrument confirmed the earlier data.

Our Clemson team goal was to next turn the telescope toward supernova 1987A in search of the remaining amount of live radioactive ^{57}Co. The problem is that most of that radioactivity had already decayed! The halflife of ^{57}Co is 272 days, but SN1987A had exploded four years ago. More than five halflives had passed since the ^{57}Co was synthesized by nuclear fusion within the explosion, meaning that less than three percent of the initial ^{57}Co would remain alive. But ^{57}Co was nonetheless expected to be the strongest remaining source of gamma-ray lines, which we hoped might still be detectable. Model calculations based on theoretical expectations suggested that about four of the ^{57}Co gamma rays, each having the measurable energy 122 keV, would enter our detector each minute. The experimental challenge for gamma-ray astronomy is the considerably larger rate of counts that are produced by the routine background of accidental cosmic rays. Someone likened it to trying to see stars in the daytime. To seek this "needle in a haystack" required a strategy of taking a long observation in which two of the four *OSSE* detectors point at SN1987A and the other two point to a different nearby direction to obtain a background of the sky. Our observing technique was to search for a small excess near122 keV energy in the two counters pointed toward SN1987A that was not present in the two looking another direction at the same time.

This worked! By staring for two different two-week-long periods, the excess counts from the SN1987A direction became evident[3]. Considerations of statistical significance are paramount in such studies. We observed about eight of the 122-keV gamma rays entering our detector each minute, but we had to stare for weeks to make those stand out above the background of random counts. The measured significance sufficed for us to publish a secure discovery. Our detection of radioactive ^{57}Co in SN1987A was only the second case in human history of a radioactive nucleus discovered within the exploding object that created it. More abundantly created ^{56}Co nuclei had been the first, having been detected prior to our launch by the balloon-borne gamma-ray telescopes. Those once abundant ^{56}Co nuclei had essentially vanished now, however, when we got our chance to look. Both isotopes of cobalt had been synthesized in the thermonuclear explosion of 1987 as radioactive isotopes of nickel that transmute quickly to cobalt as a result of their first decays. Our measurement provided direct unambiguous evidence that nucleosynthesis was happening in supernovae in very much the way that I and my colleagues had predicted two decades earlier.

Author, James D. Kurfess, OSSE PI, and Mark D. Leising at Clemson University writing the 57Co discovery

Personal satisfactions attended this discovery. Foremost is simply contributing to new knowledge. Doing so is the basic scientific satisfaction that rewards as none other can. Usually there is no financial reward, perhaps no fame if the contemporary world does not perceive what has happened, and sometimes even misunderstood scorn from competitors, Despite all that, creating new knowledge is its own reward. On top of that in this case was the sense of closure to a three-decade effort. Predicting the ^{56}Co and ^{57}Co radioactivities and their gamma-ray lines had triggered decades of service as spokesman for the future of this astronomy. I had fully embraced its hope. I had worked to discover scientific payoffs for pursuing its goals with gamma-ray telescopes. Measurement of the ^{57}Co radioactivity was my personal observational consummation of all that effort. It was a unique beginning-to-end satisfaction that likewise rewarded Jerry Fishman when he had observed the ^{56}Co radioactivity three years earlier from a balloon above Australia. Another satisfaction was making the discovery from Clemson University with the *OSSE* team. It validated my move to Clemson. Jim Kurfess, the *OSSE* Principal Investigator, and Mark Leising, who had begun their efforts at Rice University, had played leading roles in the observations of ^{57}Co radioactivity and the data reduction. My contribution was more to make sure that we drew pregnant inferences from measuring a smaller amount of ^{57}Co radioactivity than contemporary science was glibly inferring from other considerations. A provincial pride is that each of us-- Jerry Fishman, Jim Kurfess, Mark Leising and myself, as well as Neil Johnson, the *OSSE* Chief Scientist-- had begun this field together at Rice University in the 1970s. We were a family inspired by Bob Haymes and his pioneering balloon-borne gamma-ray telescope. This was a great achievement of Rice University during the 1970s. Each of these connotations of our 1992 discovery are woven together in my mind like a fine lacework.

Yet another satisfaction is that any great discovery, be it either theoretical or experimental, brings unforeseen profound consequences. This is the magical, expansive property of science; that new facts cause new ideas to be born. In this case, our discovery produced an unrecognized consequence of gamma-ray astronomy. That story begins with the very old idea from the 1950s and 1960s that supernovae would quickly become dark were it not for the presumed existence of radioactivity to heat the expanding gas so that it can shine. Expanding gas otherwise cools suddenly. Following our 1968 discovery that the most abundant of isotopes of iron (^{56}Fe) was created in the form of radioactive nickel (^{56}Ni), which then decays to radioactive ^{56}Co, which in turn decays to ^{56}Fe, we recognized that supernovae shine owing to the reheating of the expanding gas caused by those radioactive decays. It is sobering to realize that when observing the brilliant light from these astronomical explosions we are seeing the afterglow of cosmic radioactive fallout. SN1987A itself confirmed this emphatically during the first year after the explosion, because the decline of its light output traced precisely the decline in abundance of the ^{56}Co nucleus as it decayed with 77-day halflife. The plot thickened in 1992 when astronomers observed that SN1987A had by that time become too bright for this explanation. The remnant's luminosity exceeded what could be supplied by ^{56}Co radioactivity. This prompted one influential team of astronomers that had been observing the light output of SN1987A from our national observatory in Chile to argue that the excess light was powered after three years by the decay of the ^{57}Co isotope, which produces more decays than ^{56}Co after two years. This argument caught on in the astronomical world, even causing others to predict that after almost ten years the radioactive ^{44}Ti would be providing the optical power. An uncomfortable corollary of this explanation in 1991 was that it required that the number ratio of the two radioactive parent nuclei, namely ^{57}Ni/^{56}Ni at their time of nucleosynthesis in SN1987A, had been five times larger than the ratio of their daughters, ^{57}Fe/^{56}Fe, as measured in natural iron on the earth. Still, the world of astronomy was accepting this as the correct explanation of the excess light. Then our OSSE paper[3] ruled that out. Our data showed that the production ratio, ^{57}Ni/^{56}Ni, had been only 1.5 times greater than the ratio in terrestrial iron. The abundance of ^{57}Co failed by a wide margin, at least a factor three, to explain the excess light from SN1987A.

Surprising contradictions in physics often reveal underlying new phenomena; so I immediately brought that attitude to this apparent discrepancy. Within a week I had glimpsed the new underlying phenomenon, and advanced it in our 1992 paper. It amounts to a new aspect of the astronomy of radioactivity, by which I mean observable phenomena that are manifestations of radioactivity. The excess light is actually powered by the recombination of excess numbers of ions and electrons left over from earlier times when more intense radioactivity had ionized the gas more fully. The mechanism works as follows. When free electrons combine with positive atomic ions, light quanta, carry away the excess energy. Recombination in space is a slow process, requiring a year or more for an ion to capture an electron because of the low density of the supernova nebula. So the much more intense ^{56}Co radioactivity during year two was not being radiated by recombination until year three or four. We called this *delayed power* in our paper because it was based on the time delay required to release the energy stored by the ionization caused by the ^{56}Co gamma rays. In effect, we were saying that the apparent excess of light was caused by earlier ^{56}Co radioactivity rather than by current ^{57}Co radioactivity. This new explanation testifies to the power of new facts. If we had not measured the actual amount of ^{57}Co in SN1987A by recording the rate of arrival of its 122 keV gamma-ray line, astronomers would still be talking contentedly of ^{57}Co decay powering the light curve. Facts often destroy seemingly lovely explanations, but thereby point us in the right direction. That corrective process epitomizes science. It is the brilliant discovery of the scientific method. When men say we know such and such to be true, we are relying on the scientific method. If that seems self evident, consider all of the truth that men espouse in the world today based on myth, on religion, on superstition, or on simple hope.

Life in the 1898 George Warren Gignilliat House

Although my scientific life continued at Clemson University with the same pace and intensity that had characterized my career, it also assumed a day-to-day tranquility that I had not known since childhood. I liken this to the feelings of those on the great sailing ships when they emerged from dangerous storms to plain sailing on a calm sea. Residing in our gracious historic property set a tone that matched our emotional dedication to doing all things well for each other. Nancy and I were partners in the simplicity and goodness of living. She became an expert health-conscious cook, and our daily dinners with Andrew became a highpoint of each day. She also took responsibility for our financial affairs, which I gratefully relinquished with a trust and assurance that I had not experienced in marriage. I was liberated to work daily on astrophysics, a passion that for me always required a lot of concentrated thought for every intellectual product that it produced. Nancy understood that about me and helped me to work in that mode. Calm sea unleashed within me a torrent of creative thinking that I had not expected when leaving Rice University.

A counterpoint to astrophysics was the inspiring opportunity to transform the old and overgrown gardens around Gignilliat House into simple spacious beauty. We cleared an overgrown forest that had marched to proximity of our rear veranda. I traded about fifty huge eighty-year-old pine trees, which had been planted by the Public Works Administration under President Roosevelt, to a local tree-harvesting company in return for their removing the tree stumps and clearing and grading of the two acres behind the house where they had stood. I removed unwanted plants, of which there are many, as South Carolina is a setting conducive to runaway growth, especially of non-native invasive plants, such as wisteria, bamboo and kudzu. I composted and double dug our vegetable garden toward the northwest corner of that cleared land. I purchased a John Deere tractor for mowing grass in spring and summer and for picking up huge numbers of fallen leaves each fall from the old oak trees. I threw all garden waste into our forest to the north of the house, including the fall leaves and the unwanted plants. I thinned those forests by cutting down small sickly trees, amputating the invasive wisteria that choked them, and leaving standing only the healthiest of hardwoods-- oaks, poplars, maples and elms, and of course the native dogwoods that thrive in the damp shadows of those forests and that return such resplendent beauty in springtime. I loved the resemblance of outdoor work to that on the farms of my boyhood. These were magical years.

With a hired crew of four we repainted Gignilliat House, twice in eighteen years, a huge and admittedly expensive labor of love. Through paint removal Nancy discovered the original 1898 colors that had adorned the molded contrasts of Queen Anne houses and that had long ago been painted over with a uniform sea of white paint. We rebuilt some damaged wraparound porches and installed new standing-seam copper roofs on them with the aid of a Charleston contractor for historic preservation. We invested $50,000 on that work alone. And Nancy designed a new carriage house-- this one for automobiles-- with a beautiful apartment above. We built it on the site of the original carriage house that had been devastated by termites and rot. That new carriage house fits and augments the main house so appropriately that later visitors assumed that it had always been there. Understand our effort as an expression of enthusiasm for the romantic goals and our confidence in each other. Our efforts seemed nothing more than the natural expression of a life of contentment and the associated will to create beauty.

My adult life had never been so free of conflict. It had become easy not only to work on programmatic growth for our department but also on original new thoughts about the origin of our material world. New ideas were coming easily to me here, in career twilight. Calm seas empower intellectual energy.

Fathering Andrew

Our son Andrew is quite a talented boy, and we invested heavily in his growth. Of my four children, Andrew was the only one with whom I was able to live every moment of his life from birth to college, and I grasped the opportunity that had earlier been denied me. Because the public schools are weighted down by the burden of trying to teach so many who are unprepared for learning, especially in South Carolina, we concluded, with substantial philosophical regret, that we would enroll him in private schools. When it came to my own gifted son, I departed from a lifetime of fierce democratic support of our public school system. From preschool through sixth grade I drove him daily to Clemson with me, where he attended Clemson Montessori School. Our intellectual activities during those drives say much about my urge to lead with fatherly support and of Andrew's subsequent comfort with mathematics. From third grade onward I set him problems in algebra. A typical question was,

"There are twelve children in the classroom. There are two more girls than boys. How many girls and how many boys are in the classroom?"

Andrew's good natured responses not only led to tougher problems, but allowed me to explain to him that the question is one of *algebra,* because there are two unknowns (the numbers of girls and and the number of boys) and there are two constraints (total number of twelve, and two more girls than boys). I spent a few minutes daily preparing intriguing questions for him, and our drives passed in swift fun. I need not recount the reward that I got for this. And in December 1994 I invited the Montessori kids to Clemson University for my demonstration of the effects of motion in magnetic fields. They were wildly interested, thinking me a kind of magician. But before long they learned that they could do the same thing with my arrangement of magnets, electric currents, and moving objects. Forces that act at a distance through invisible fields fascinate even old children.

Following Montessori School tears we enrolled Andrew in *Christ Church Episcopal School* in Greenville, one of the many excellent private schools supported by the Episcopal Church. In order that we not burden our time and energy by driving the forty miles between Seneca and Christ Church, we purchased a condominium across the street from the school. There we lived most weekdays for seven years while school was in term until Andrew graduated in 2005. Then we sold that condominium, which remains dear in our memories because it was central to a plan of high purpose. During these years we were typically busy parents. The most remarkable activity was Andrew announcing at age three, after attending a concert at Clemson University, that he wanted to play the violin. We obtained private lessons and nurtured his skill and his practice schedule until, at age eleven, he was a fine child violinist. His highest achievement was probably playing second violin in a performance of J.S. Bach's double concerto for two violins. But playing and practicing had sneaked ahead of our motivation, even as Andrew's waned, and at age twelve he wanted to stop. This was a mild heartbreak for us except for our reminders to ourselves that our goal had been for him to be able to pursue *his* goal. He had achieved it and enriched himself and us. But playing the violin was not it.

Andrew was a very talented soccer player, and for ten years we moved heaven and earth to get his to his classic-soccer team practices, to arrange our lives to accompany the team to frequent weekend tournaments, and to enroll him in summer soccer academies run at the universities. Before he became too good, I took him frequently to a nearby sports field where we could practice many soccer skills. This too came to an end as he decided that he would not move to the next level, his high school soccer team. This again engendered disappointment in us; or would have but for our focus on the goal being for Andrew. On the positive side, the blessed relaxation of our released weekends was a relief! Such activities required years of dedication on our part, and that dedication was freely given for the fun and reward of helping a son discover who he is and what he wants to be. Andrew's develop-ment to a high skill level before dis-continuing a talent was a priceless lesson to me, because I struggled during

With son Andrew and Nancy at his 2009 commencement ceremony at Clemson University. I wear my Caltech Ph.D. academic regalia.

my own lifetime with regrets over having given up my own special skills. I had in part missed such fathering experiences with my other three children, each living with their mothers as a result of divorce.

Andrew has just graduated in 2009 from Clemson University, majoring in economics, with a second major in Spanish. Today Andrew plays the guitar very well, has resumed soccer, and plays volleyball with the Clemson Volleyball Club.

The isotopes within STARDUST

At Clemson University I found myself seeking a strategy for making the astrophysics of stardust an integral part of our group effort. By asking a leading colleague who was utilizing the main technique for studying these particles in the laboratory, I found that the startup costs required to have our own laboratory would be too steep for Clemson University to underwrite. The only way forward would be to encourage members of our group in nuclear astrophysics to develop their knowledge of this field so that we could together develop its ideas in much the same way that I had been doing for fifteen years. Several good things had happened that would place Clemson University in a strong position to compete for NASA research support. The first is that the award of the Leonard Medal of the *Meteoritical Society* validated me as a leader in this field rather than a controversial upstart. Rather than patting myself on the back here, this was a necessary validation for a large number of uncertain practitioners of meteoritics and for NASA program managers who would consider our proposals.

A new science landmark had just occurred at this time, namely the isolation and identification of gemlike interstellar grains within meteorites. These stardust grains were already being shown without doubt to be solid chunks of individual stars that had died before our sun was born. Their stardust nature was demonstrated by the silicon-carbide (SiC) particles that could be isolated from the meteorite host and demonstrated to be of such unusual isotopic composition that they could not have formed any place other than in a single star having that unique isotopic composition. Individual stars generate very unusual isotopic compositions owing to the new nuclei that are synthesized by nuclear reactions within each star during its lifetime. Subsequently those nuclei are dredged up to the surfaces of the stars where they enrich the escaping winds of material. Within those cooling winds the stardust condenses. But after the star's material is ejected into the interstellar gas and mixed with the older ejecta from other stars, the total interstellar gas maintains a very normal isotopic composition, much like the composition of our sun. Only solids that condensed from a star's gas before it mixed with the interstellar matter could preserve the bizarre isotopic ratios that SiC grains were being demonstrated to possess. To further ice the argument, the SiC grains are hard, refractory minerals that can grow within a slowly cooling hot gas, a gas still having temperatures near 1800°C, a heat that would evaporate less refractory minerals.

What do I mean by strange isotopic composition of the SiC particles? The abundance ratio of the two carbon isotopes, $^{13}C/^{12}C$, is in each SiC grain roughly half of its value on earth. This would by itself demand explanation. The silicon is typically more abundant in both of the rarer isotopes of that element, ^{29}Si and ^{30}S, relative to ^{28}S than is the silicon on earth. Nitrogen is chemically solid, bound quite strongly within the SiC mineral structure, even though we experience nitrogen as gaseous in the earth's atmosphere. The ratio of abundances of the two nitrogen isotopes, $^{14}N/^{15}N$, is typically fifty times greater in the SiC minerals than in the earth's atmosphere. These great discoveries rocked the science world. The only scientists not amazed at these isotopic compositions were those who did not know of them. They meant that these SiC grains were not formed in our solar system, but rather long ago and far away. And many more dramatic discoveries of this kind were breaking as the 1990s began. These discoveries validated the stardust concept and undoubtedly advanced my name for the Leonard Medal.

But how could the good reputation of my medal be converted to a research program at Clemson University? I spoke about this with my friend Ernst Zinner, a research professor at Washington University in St. Louis. Ernst was leading the way in this flood of new experimental documentations of stardust. He was doing so with a secondary-ion mass spectrometer, commonly shortened to SIMS. This device controls the paths in space of charged particles (ions) with the aid of electric and magnetic fields; moreover it creates those ions by sputtering them away from the SiC target grains. In essence a beam of oxygen ions is accelerated to high speed while focused by the fields in such a way as to bombard the SiC grain. Each oxygen ion penetrates within the SiC grain and deposits its large kinetic energy as an explosive burst of energy within the grain. That energy explosion ejects ions from the SiC grain, and these are in turn focused in a way to strike a detector after their energy and momentum have been measured. That procedure suffices to identify the mass of each ion, so that the relative numbers of sputtered ions having differing masses can be counted. This procedure gives very normal results when applied to SiC grains made on earth, but when it is applied to an SiC grain from the meteorites there results the zoo of unusual isotopes recounted above. For his leadership in these discoveries Zinner would be awarded the 1997 Leonard Medal at the Maui meeting of the *Meteoritical Society*. A photograph of us with our wives, Brigitte and Nancy, at the Maui banquet is <1997 Ernst Zinner, Donald Clayton and wives on Maui>.

Ernst Zinner discussed with me a plan that had first arisen from Ernst's colleague Robert (Bob) Walker at Washington University, who was the leader of its McDonnell Center for Space Sciences. At the 1984 meeting of the *Meteoritical Society* in Albuquerque, Bob had proposed to me annual meetings between my Rice research group and his laboratory to further the science, then only hoped for, of presolar stardust grains. I explained to Bob that this attractive idea was untimely for me, as I was competing for a professorship at Penn State University in the hopes of establishing my group in nuclear and gamma-ray astrophysics there. Bob expressed surprise and considerable interest in this, even proposing that I come instead to Washington University for a year as *McDonnell Distinguished Professor*. I also had to decline that offer while focusing on my future. But now, after I had made my move to Clemson University, the idea became workable. For Washington University an annual meeting would bring an influx of theoretical ideas about the isotopes and their origins; and for Clemson it allowed my research group of five to hear annual reports of new measurements and to participate in discussing their meanings.

At the *Lunar and Planetary Science Conference* in Houston in spring 1990 I spoke of this idea to Joseph Nuth, who was serving as the first program manager for NASA's new research program, *Origins of Solar Systems.* Joe expressed interest because this new NASA program intended not only to award research grants but to sponsor workshops that advance to goals of the program. We quickly decided that Clemson University would host the first such workshop sponsored by that NASA program, to occur in November, 1990. Thereupon Ernst Zinner and I agreed that it would also be the first workshop in an annual ongoing relationship between Clemson University and Washington University. A group photo on Clemson House Penthouse of the attendees at that *Workshop on Isotopic Anomalies,* as well as many other candid photos of attendees, can be viewed at <1990 Workshop Isotopic Anomalies (Group)>.

This annual workshop has now been held for two decades, alternating between Clemson and St. Louis, later enlarged to also include *Carnegie Institution of Washington* and the University of Chicago's *Enrico Fermi Institute.* This became a world famous collaborative meeting, and because our mode of operation was to help each other with unpublished ideas and data we resisted inviting others to participate. In this same year I obtained a new research grant from NASA's *Origins of Solar Systems* program and held it for fifteen years. And one of the new Clemson group of assistant professors, Bradley S. Meyer, took enthusiastically to stardust research and became a leader in theoretical explanations of isotopic effects. Details of these meetings since 1999 can be seen at this website maintained by Washington University:

http://presolar.wustl.edu/events/workshop.html

While preparing to leave for Houston for the 1992 *Lunar and Planetary Science Conference* I received good news from NASA Headquarters. This news came by telephone from Al Bunner, NASA program manager for space astronomy:

"Don, we have selected you to receive the NASA Headquarters *Exceptional Scientific Achievement Medal.* This is the highest award that NASA bestows for science research."

The award cited my origination of two new fields of astronomy based on newly created nuclei in supernovae. The citation read:

> *for outstanding contributions to theoretical astrophysics relating to the formation of elements in the explosions of stars and to the observable products of these explosions* ·

News of this created excitement at Clemson University and approbation from its administrators, who were relieved to see that their investment in our new research group was paying off with academic prestige. Attending the awards ceremony involved awkward effort, however, because I had to first fly to Houston for a portion of the *Lunar and Planetary Science Conference,* leave in the middle of it to fly to Washington for the ceremony, then return to Houston for the big Thursday session on isotopic anomalies in stardust, in which I presented my oral paper with my new postdoc[4] on possible interpretation of the silicon-carbide presolar grains. A photograph of NASA Administrator Richard Truly and Associate Administrator Aaron Cohen presenting the medal to is viewable at <1992 Clayton NASA Medal>. This award was almost Richard Truly's last act as NASA Administrator, causing me to joke that apparently the award to me had been the last straw. Richard Truly had been a distinguished astronaut and test pilot, but the wave of the future for NASA was proven administrators who could deal with its especially knotty budget problems.

Another strategy for the science productivity of our Clemson research team was achieved through ongoing relationships with international science leaders. Setting these up could effectively increase the size of our small faculty. The first was Mounib El Eid, a Lebanese astrophysicist in Göttingen who had by that time returned to Lebanon the become chairman of the Physics Department of the *American University* in Beirut. I had first met Mounib during my visit to Darmstadt <1977 European Renaissance in Darmstadt>, a date prior to his development in Göttingen of his excellent computer code in stellar evolution. Mounib spent many research summers with us in Clemson, working primarily with Lih-Sin The and Brad Meyer on nucleosynthesis aspects of the evolution and explosion of massive stars. They demonstrated in their publications many new aspects of *s*-process neutron addition as it occurs in the helium-burning and carbon-burning thermonuclear phases of those stars. These happen, according to their computer models of stellar evolution, within stars that are 15-20 times more massive than the sun. It occurs in spherical thermonuclear shells surrounding the hotter and denser cores of these stars. They computed that within such stars the weak neutron exposure of a significant fraction of all galactic iron occurs, locating what I had named "the weak s process" two decades earlier[5] without knowing where it happened. I always got a kick out of any new understanding of the *s* process, pulling me back three decades to my PhD thesis. Science never truly completes the understanding of the *s* process. It is a beautiful and intricate aspect of nucleosynthesis that we understand more and more of as years pass. Our financial support of Mounib's visits was minimal, and I wished it could have been more; but it was the beauty of South Carolina, and our extraordinary personal hospitality, and the research devotion of our team that provided Mounib with his own peaceful pursuit of science, which was so difficult in Lebanon. The depth of our research team was aided markedly by his participation. A photo from 1992 of Mounib El Eid with the author and the Clemson group can be viewed at <1992 Nuclear Astrophysics at Clemson>.

A second ongoig European collaboration was with gamma-ray astronomer Roland Diehl, from *Max Planck Institute for Extraterrestrial Physics* in Garching, just north of Munich. It was here that the compton-scattering gamma-ray telescope had been designed and constructed for inclusion as one of the four instruments on NASA's *Gamma-Ray Observatory*. That instrument, called *COMPTEL* (COMpton TELescope), was a cousin to our *OSSE* experiment because both were designed to be able to detect radioactivity, but using differing detection technique. The great advantage that *COMPTEL* enjoyed was being able to observe much of the galaxy at all times, without having to point in any special direction. This constant viewing of many directions simultaneously was instrumental in its many discoveries, especially in its mapping of the radioactive ^{26}Al that is distributed around the disk of our Milky Way. Stars do not eject enough ^{26}Al to be detected emerging from a single star. The observed ^{26}Al distribution has emerged from many dead stars during the past million years, which is the lifetime of ^{26}Al. I was at first surprised by these gamma rays, each having energy 1.81 MeV, being much more numerous from specific regions of the Milky Way, showing higher ^{26}Al concentration there than the average. I had calculated that several million supernovae have exploded during the last million years, and it seemed to me that such a great number would be more or less randomly distributed in our galaxy, more or less like the stars themselves. But in fact, much star formation concentrates at any time in specific select regions of our spiral arms, so that those regions had experienced many more massive-star births in the past million years than other quiescent regions of the galaxy. This was stimulating new truth from COMPTEL, not only proving that new nucleosynthesis is occurring today in our galaxy, but that it does so in galactic hotspots. A corollary is that those same regions are temporarily enriched in all of the nucleosynthesis products of supernovae, at least until their gas is slowly mixed into the bulk of the galactic interstellar gas.

Roland Diehl's first visit occurred in 1994. His second in 1995 was the occasion on which he and my Clemson colleagues planned a special conference to be held at Clemson in March 1996 in honor of my sixtieth birthday. Roland had a sabbatical leave of five months from Max Planck Institute to spend with us at Clemson University in 1996. A photograph of the attendees at that conference can be seen at <1996 Workshop Group, the Radioactive Galaxy>. That conference, entitled "The Radioactive Galaxy", was the first of a series of cutting-edge workshops that we bundled under the rubric "The Astronomy of Radioactivity". These took place at Clemson and at the *Ringberg Castle*, a Max Planck conference site on a mountain overlooking the Tegernsee south of Munich. The castle is a twentieth-century monument commemorating the idiosyncrasy, originality, and remarkable single-mindedness of two men: Duke Luitpold of Bavaria, a member of the Wittelsbach family who ruled Bavaria over 800 years, and his friend, the all-round artist, architect and interior decorator, Friedrich Attenhuber. The castle is entirely their creation, from the massive Renaissance-inspired exterior, right down to the fittings and furniture which, in every detail, were designed by Attenhuber himself and executed by native craftsmen. Details of the entire series, including photographs, can be seen at these web site:

http://www.mpe.mpg.de/gamma/science/lines/workshops/
radioactivity.htm

During the second of these workshops, held at Ringberg Castle in fall of 1999, that I heard of a thrilling discovery by the *COMPTEL* team. Attending was also an excuse for a romantic trip for Nancy and myself. After two nights in historic Regensburg on the Donau (Danube), including two seats to *La Traviata* purchased for a mere 47 Deutschmark each, we met Roland Diehl in Freising, his home, to ride together to Ringberg Castle, where Nancy and I were housed in the Duke's own bedroom for the meeting. Anatoly Iyudin presented the *COMPTEL* team's evidence for a 1.16 MeV gamma-ray line emitted by radioactive ^{44}Ti nuclei in a 340 year old supernova remnant called *Cassiopeia A* (or *Cas A*). Light from this historical supernova was apparently observed in 1680 by the Astronomer Royal, Flamsteed. Detection of live ^{44}Ti within that young remnant confirmed our prediction made in 1969 after discovering that most of the mass-44 isotope of calcium was first created in the form of ^{44}Ti, which decays with 78-year halflife to ^{44}Ca. This detection was the third radioactive nucleus to be measured within supernova remnants. The *Cas A* supernova exploded about four halflives ago, so that the present number of remaining ^{44}Ti nuclei in that remnant is but about 1/16th of the initial number; however, the gamma rays from that depleted population remained detectable with statistical significance.

Our immediate worry was why our *OSSE* instrument did not confirm the COMPTEL detection. Confirmation is very important in science because first observations of any discovery may be flawed in unknown ways. Within a few years, further analysis of the data showed that the initial *COMPTEL* report of the gamma-ray flux had been somewhat overestimated, and the correct value lay just below the threshhold value that *OSSE* would have been able to confirm. But the discovery remained secure, because the much greater total observing time by *COMPTEL* enabled it to see a fainter 1.16 MeV line than *OSSE* could extract. This discovery raised another astrophysical problem, however, whose answer is still not known. Where are all of the other supernova remnants that should have occurred in our galaxy during the past three centuries? Astronomers estimate that three supernovae happen every century, at least *on average*. For that very reason a half a dozen or so supernova remnants should be detectable within our Galaxy at all times. The 340-year old *Cas A* remnant is very old for ^{44}Ti's halflife but close enough that its gamma-ray line flux was strong enough to detect. But that was three centuries ago; where are the six supernova that should have occurred during the past two centuries? Any such would be bright enough to detect anywhere in our galaxy? It so often happens in science that a new detection raises puzzles that are deeper than the initial question. That is the fascinating thing about science.

Another of my first predictions surfaced at that 1999 Ringberg meeting, but not at all in the form that I had proposed. Guenter Korschinek from Munich presented his group's astonishing discovery that live radioactive ^{60}Fe nuclei existed on earth in deep sea sediments. I had expected[6] that the ejection of this radioactive isotope from supernova would be detectable through the gamma ray lines that accompany its radioactive decays. I showed it would be too weak for detection from individual supernovae but that, like ^{26}Al, it would sustain a galactic glow of its gamma-ray line from millions of past supernovae. Korschinek's presentation convinced us that we have such radioactive atoms on earth today, still alive, deposited by radioactivity-bearing supernova winds that impacted the earth only a few million years ago. What an astonishing and unexpected delight! Apparently a single nearby supernova occurred about three million years ago, and some of its ejected atoms penetrated the earth's atmosphere, where access is difficult owing to the earth's magnetosphere and ionosphere. Those atoms succeeded to fall onto the oceans. There the atoms accumulated in a two-million year old sedimentary layer of ocean bed. Truth again shows itself stranger than fiction. That explosion, by the way, would definitely have been a stressfully bright sight to nocturnal mammals and to precursors of the human species. Only in the last few years have has the detection of the gamma rays from our galaxy been achieved. The final word on these sediments containing radioactive ^{60}Fe is yet to be written.

First love is forever

My first science love, the wide-eyed passion of my scientific youth, was the theory of nucleosynthesis. Such work consists of formulating mathematical descriptions of the nuclear transmutations within a system of nuclei in astronomical settings. In the mid 1990s I was able to return to this love in collaboration with Brad Meyer. Working with Brad has comfortable roots that had begun at Rice University when he worked on a senior physics thesis with my guidance. Since that humble beginning Brad has moved into the forefront of nucleosynthesis theory.

With a Ph.D. student, Tracy Krishnan, we were using Brad's computer codes to investigate what happens when a very dense, very hot gas of neutrons and protons is expanded and cooled. Such events occur in supernovae because the heavy nuclei are broken into their elementary nucleons by the intense heat. With about 15% more neutrons than protons the nuclei bond again into nuclei as they cool during rapid expansion. We were exploring their attempt to reassemble into iron nuclei. My attention was drawn to one nucleus, namely ^{48}Ca, which was conspicuously maintaining nearly constant abundance while most others were either increasing or declining rapidly as time passed. The large abundance of ^{48}Ca renders it an important isotope whose natural abundance had not been explained by theory. It is an isotope of puzzling superlatives. The ^{48}Ca nucleus is "doubly magic" in that its proton number (20 for element calcium) and its neutron number (28) make excessively stable clusters. Both 20 and 28 are in the set of magic numbers for nuclear structure. The ^{48}Ca nucleus has a greater percentage of excess neutrons[7] than any other nucleus that is created explosively. It is of surprisingly high abundance in natural calcium, being almost fifty times more abundant than ^{46}Ca, which has 26 neutrons and is the next most neutron-rich isotope of calcium. It is even more abundant than ^{43}Ca, which is synthesized in the main line of explosive oxygen and silicon burning in supernova mantles, wherein no ^{48}Ca at all is produced. Such arcane facts are precious clues to the nuclear theorist. I seized on these puzzles and would not let go until we had solved them.

To make a long story short, the ^{48}Ca nucleus was shown by us to be locked into a quasiequilibrium pool of surrounding nuclei that, because of their buffering effect, would not allow its abundance to change rapidly. The ^{48}Ca abundance became a structural pillar to which other nuclei attached in a large, systematic and sluggish group. Furthermore, ^{48}Ca became hugely abundant if the density was very high, as at the center of a Type Ia supernova. At high density, the nuclei reassembled into *too many nuclei* to achieve statistical equilibrium, with the resulting huge abundance of ^{48}Ca. At lower density, as in the center of a Type II supernova, the nuclei reassemble into *too few nuclei* to achieve equilibrium, with the resulting small ^{48}Ca abundance. We moved nucleosynthesis theory forward by recognizing that the number of nuclei assembled was a measure of the distance of the system from the demands of statistical equilibrium[8]. Perhaps it is something that only a nucleosynthesis theorist can love, but love it we did. Collaborating with Brad Meyer made this possible for me. It lifted my heart to publish at age sixty the natural origin of this special isotope.

Coping with dreadful loss

On one sunny and blue-skied afternoon in April 1996, while our azaleas bloomed exuberantly, Nancy met my car as I drove in and told me to follow her. She appeared unusually determined and businesslike, even grim. She led me to the lawn overlooking our rear garden and asked me to sit on the grass. Then she told me that my son Devon had committed suicide in his small Atlanta apartment last night and had just been found.

Words will not convey my shock, my disbelief, my grief, my anger, my guilt, each replayed over and over in my broken heart. This was the moment that I had long cast from my mind in fear. I had worried about Devon for all of his thirty five years, not expecting this awful end, but sensitive to his depression and to his stunted sense of joy when alone. From his final letter to me and from his papers I learned that he had long anticipated that he would go that way. I learned that his depression was deeper and darker than I knew. I argued with God that, no, this can not be. *I will do more to prevent Devon from being backed into that dark corner. I just need another chance.* I grieved unceasingly. I cried often, certainly every time that I spoke with someone about Devon.

Life brings moments of great grief, but there is none in my experience to compare with the loss of one's child. It is not right that a son whom I had loved every minute of his life should die while I lived on. There is no grief to compare with my repeated recalling of his steps of childhood, every sweet aspect of that life, picturing his many childish expressions and childish insecurities, and to learn after the fact of his planned departure from a life with which he could not cope. Even today, after a subsequent decade of joyful, loving and successful living, tears overcome me trying to write of this. It will never go away. I will never get over it. I just forget about it for longer periods of time.

I had tried to help Devon in the best ways that I could. But in my tormented mind I accused myself of not helping enough. Nancy and I had persuaded Devon to leave the gay Hollywood life on which he had centered his being, and to move to Seneca to live with us in Gignilliat House. We did not reject his homosexual nature, and he knew that, but rather urged him to admit the dead-end patterns that the Hollywood social scene imposed upon him, keeping him from broader growth. Devon accepted our offer gratefully, saying to friends about living once again with his father,

"I am going to get it right this time."

Devon lived with us for a year. Daily he played nostalgic old romance tunes on the piano in our music room, and he also took some hours daily for creative writing. These skills and his friendships, of which he always had very many owing to his charm, wit and openness, were Devon's ambitions. Devon played *Lego* building blocks with Andrew by the hour in a generous effort to know his little brother. I had become optimistic and supportive of his decision a year later to get an apartment and a job in Atlanta. He became a writer and editor for *Southern Voice*, an Atlanta newspaper devoted to gays. I remained optimistic that he might be depressed but stable, coping in his way with his gloomy side. Then one awful event pushed him over the edge. In a recognized bleak mood he checked himself into an Atlanta hospital, where a psychiatric physician recommended that he begin treatment with *Prozac*. This medication has, the world now knows, the potential for suicidal initial reactions in some patients, and, in a careless act that anguished me, Devon was allowed to check himself out of the hospital on the grounds that he had checked himself into it, whereas the doctor had intended him to remain for initial observations. I knew nothing of this until after he went straight home and hanged himself.

We were able to purchase an unused family plot in Seneca's historic *Mountain View Cemetery*, a place of tranquility and beauty. After Devon's burial there a philosophical acceptance seemed to slowly grow within me. If I needed any more proof that I can not control everything, this tragic episode provided it. Finally I admitted that fact. I reached a new peace with accepting the bad news of the world. It seems providential that my impassioned theory for the origin of ^{48}Ca was happening during this time of great personal grief. Explaining the origin of ^{48}Ca with new theory was emotionally like creating a new child, and I am thankful that it was able in some way to provide solace for the loss of my second son. This could sound strange, even heartless, to some; but beautiful scientific explanations have always had the power to make life seem meaningful to those of us who struggle to create that understanding.

CHAPTER 19. ON CALM SEA

My unanticipated delight as I passed age 65 has been that creativity did not desert me. Throughout my sixties, a renewed sense of scientific vision inspired me. This unexpected blessing came as to a Captain looking outward at calm sea from his deck, no enemy ships to attack, no pirates to repel, no storms to threaten the ship. Calm sea allows a man's mind to drift to inner fascinations with puzzling known facts that will not release their grip, but call him back to the puzzle day and night. That intense concentration is, at any age, how creativity in physics happens. That is what Einstein meant when he admitted that working in the Bern Patent Office had been the happiest years of his life. In that boring bureau he could repeatedly pull his notes on relativity from his desk drawer and think about them.

My work on stardust after age sixty-five is still not nearly as well known as the main line of my career, but I personally hold it equally high. I think for this that I owe great debt to the Clemson University/Washington University annual workshops on stardust, wherein Ernst Zinner, Sachiko Amari and Larry Nittler in particular kept pounding their data into me. That forced me to consider how, within an astrophysical picture, such data could be possible. Here are the two puzzles that fueled my final creative thrusts. I close my career with these two profound puzzles by trying to point the way for the future of stardust studies.

The mainstream silicon-carbide stardust particles, which have been shown to have condensed from hot gases within the cooling atmospheres of giant red AGB stars, appeared from their silicon isotopes to have arisen in stars younger than the sun—a manifest impossibility. Such stars must be born, age and die before the sun was formed in order that their stardust could be present in the material from which the sun formed.

The second puzzle concerned *SUNOCON*s, those stardust particles that condensed within the expanding and cooling interiors of supernova explosions. Many aspects of *SUNOCON* isotopic compositions suggested that those isotopes had been formed by thermonuclear reactions within matter containing more oxygen than carbon, a situation which is in fact true of almost the entirety of the supernova structure. But chemists insisted that in order to condense these SiC and graphite *SUNOCON*s, the hot gaseous vapor must contain more carbon than oxygen. They were adamant that chemical laws required this awkward conclusion. To the contrary, I describe their future as seeking the true picture of how these particles condense from hot gas.

Those two puzzles really grabbed me, and in fact became almost the only science that occupied me during the decade 1996-2006, during which time I published ten papers on these two subjects. Because these two issues are among the greatest in chemical astrophysics today, I will close my scientific life by describing for both puzzles their essence and their future. I hope to take you along with me in this description.

The Future of Mainstream Silicon-Carbide

All silicon carbide crystals (SiC) extracted from meteorites are stardust grains. They are tiny gems about $1/10,000^{th}$ of a centimeter in size. What is it that makes the mainstream SiC grains appear to have come from stars younger than the sun, even though that is impossible? It is the composition of the silicon isotopes in these crystals. To understand this requires detour into abundance ratios among isotopes measured in these grains. If one constructs a graph letting the y and x axes represent numerically the isotopic ratios $^{30}Si/^{28}Si$ and $^{29}Si/^{28}Si$, the values of those ratios measured within each mainstream grain can be plotted as a point on that graph. Each grain is but a single point because the entire grain seems to have but a single composition. When about a thousand known mainstream grains are placed in that diagram, they align themselves along an almost diagonal line. That line is called "the mainstream". So where does time come in?

In 1988 I had published a paper[1] showing that such an alignment, which had not yet been discovered, might be expected by the evolution of interstellar silicon isotopic abundance ratios as the Galaxy ages. Both ratios, $^{30}Si/^{28}Si$ and $^{29}Si/^{28}Si$, should increase with time in the interstellar gas. In 1996 I followed with more detailed demonstration[1] of this with my postdoc Frank Timmes. No question that stars being born at later times from the interstellar gas should contain silicon having larger values for $^{30}Si/^{28}Si$ and $^{29}Si/^{28}Si$ than those born earlier. Not only that, theory predicted that they should align in the isotopic-ratio diagram described above. Initially this was pretty well accepted as the reason for the alignment of stardust SiC along the mainstream line.

Frank Timmes was one of the postdocs supported by the Gamma Ray Observatory science office. He had been the Ph. D. student of Stan Woosley, who had been my Ph.D. student at Rice University. I joked with Frank Timmes that he was, therefore, my grandson. Frank was a great colleague to work with, coming back every day with the answer to yesterday's question. Just like his old man.

Coming back to the mainstream line, the sun's location on it, as measured by the $^{30}Si/^{28}Si$ and $^{29}Si/^{28}Si$ ratios of the solar system, falls near the bottom of that mainstream line of stardust grains, just the opposite of what I expected! The sun's silicon suggests that the sun formed earlier than the AGB stars that contributed their stardust near the end of their own lives. The galactic-chemical-evolution picture would seem to require a manifest impossibility. It was the pre-sun AGB stars that, in their dying years, condense SiC stardust grains and scatter them into interstellar gas. When the sun formed later from interstellar gas, it contained those stardust mainstream grains that are now routinely studied in terrestrial laboratories.

They are extracted from meteorites that originated during the infancy of the new sun, when meteorites were being assembled. The manifest contradiction befuddled us all, like some Escher waterwheel, or the old hillbilly favorite "I'm my own grandpa". If those AGB stars formed after the solar birth, they could not possibly distribute their stardust in time for its inclusion within meteorites that formed contemporaneously with our sun. This major contradiction jeopardized our enthusiasm for seeing new astronomical science in the mainstream grains. Something was amiss, and that worried me incessantly.

The problem of the slope of the mainstream line was not as severe. In another publication[1] with Frank Timmes, we examined ways in which the unit slope of the theoretical mainstream line could actually be near 4/3, as observed, owing to a somewhat unusual silicon isotopic composition in the sun. But the value of the slope was the minor problem. The sun's composition falling near the bottom of the mainstream line was the elephant in the room. I lived day and night with this mainstream line. While other scientists just shrugged and went about their business, I became a loan voice worrying about the elephant in the room.

My preoccupation produced two novel resolutions. I will describe how each suggests connections with details of the movement of galactic matter and the buildup in time of the galactic mass. In the first[2], published in 1997, I argued that the same two ratios, $^{30}Si/^{28}Si$ and $^{29}Si/^{28}Si$, should increase, at any fixed time, in our galaxy as a function of distance from the center of our Milky Way disk. The sun's orbit lies about 25,000 light years from the galactic center, which lies in the direction of the Sagittarius constellation. Stardust condensed from AGB stars whose births occurred closer to that galactic center than the sun should possess higher values for $^{30}Si/^{28}Si$ and $^{29}Si/^{28}Si$ than does the sun, whereas those from stars at greater distances sfrom the galactic center hould have smaller values. I argued in my paper that our mainstream stardust had come preferentially from stars that had been born *interior* to the sun's orbit. During their stellar lifetimes, their gravitational scatterings from galactic masses had propelled them outward, so that in their dying breaths they delivered their stardust near the sun's birth location. The mechanical mechanism is similar to the slingshot effect by which NASA sends space probes to the outer planets. A spaceprobe's orbit is set initially so that it scatters first from near encounter with the earth, picking up extra velocity that impels it toward Mars, where another scatter slings it further outward toward Uranus. Such are the standard orbital tricks in NASA to move something near the orbit of earth outward to the orbit of Uranus without having to rely on fuel for that uphill journey. What I loved about this explanation is that it had required every bit of my physics training, from nucleosynthesis to stellar evolution to galactic evolution to the orbital mechanics of many-body systems. As much as I love it, however, I doubt that it is the correct explanation. I think my second one is better.

My 2003 paper[3] proposed that the mainstream line was established when our Milky Way galaxy absorbed a smaller less evolved satellite galaxy. Two galaxies actually merge. This audacious idea is not as implausible as it sounds. Astronomical research has established that massive galaxies, those that dominate the cosmic scene, grew so over time by gathering into themselves a large number of smaller galaxies that initially orbited the large galaxy. Many call this *galactic cannibalism*. Twenty one such small galaxies still orbit our Milky Way and will, in time, be drawn into it. Probably a dozen more have already been consumed. Two of the brightest remaining satellites are the famous Magellanic Clouds, within which supernova 1987A occurred. Their interstellar gas and their stars are observed to have compositions of lower metallicity than the Milky Way, as do other small galaxies, implying that their silicon is also characterized by a lower values of both $^{30}Si/^{28}Si$ and $^{29}Si/^{28}Si$ isotopic abundance ratios than is our Milky Way. Chemical evolution in small galaxies has not progressed as far as in the very massive ones.

My proposal utilized that expectation in the following way. When the merging satellite galaxy collided with the Milky Way, the stars pass right through with negligible impacts with other stars, but their gaseous components collide and begin to mix. Pressure forces establish compressed regions where a burst of new stars form from the mixing gases at a stage when they were dominated by the Milky

Way gas. AGB stars born from differing admixtures of the satellite-galaxy gas established the mainstream silicon isotope correlation line. It is a mixing line between the gases of these two galaxies. The sun, on the other hand, formed later from gas that had a higher fraction of the satellite galaxy's gas. More detailed justification of this bizarre scheme is too complicated to share here, but it renders the entire scheme plausible. This explanation, which may or may not stand the scientific test of time, also required a lifetime of experiences with many different astrophysical disciplines. Indeed, in retrospect it seems to me that it was I who made this step in 2003 precisely because I was the one focused on all of its issues. I retain high hopes that this explanation provides a thrilling future for the astrophysical meaning of the SiC mainstream line.

Incombustible soot, the future of SUNOCONs

Scientific inspiration is a consequence of relentless motivation. My muse swept me latterly into the field of condensation chemistry. Turbulent weather roared all around me because of this, but I had been through that before and now remained on calm sea. The condensation of carbon solids from hot supernova gas had interdisciplinary roots that spanned much of my career. That story began with a bright Rice University physics major, Lucy Ziurys, who came to me in 1977 to seek a senior thesis problem. This was during the years when formulating cosmic chemical memory dominated my thoughts, so I gave Lucy a problem concerning the carbon-monoxide molecule (CO) in interstellar gas. Stars should be prolific CO machines because when they eject their carbon atoms, bathed as they are in abundant hot oxygen, the carbon becomes oxidized. That is, carbon combusts: $C + O \rightarrow CO$. Very few bare C atoms should get out of stars, I had reasoned. Today I take pride in Lucy Ziurys's brilliant subsequent career in radio astronomy in which she discovered for the first time a large number of interstellar molecules.

In December, 1994, Ernst Zinner emailed to me the Washington University discovery[4] of the initial presence of radioactive ^{44}Ti isotope in a certain type of SiC meteorite grains, proving beyond

doubt that those specimens were supernova condensates--
*SUNOCON*s as I named them. I had awaited this moment since
predicting it in 1975[4]. Their discovery proved that the condensed
solids from supernovae really exist, making this one of the most
exciting days of my life. I had taken much abuse in the 1970s
over that idea, whose correctness I never doubted, even though I
could not be confident that these interstellar *SUNOCON*s would
be found within meteorites. Ernst's group had now demonstrated
their existence. My excitement was damped, or maybe I should say
inflamed, by a nagging question, however; namely, how can the
SiC grains grow in a supernova core that has more oxygen atoms
than carbon atoms? As hot supernova gas expended and cooled,
the carbon atoms should have, according to conventional wisdom,
combusted to CO molecules long before any SiC solid grains could
begin to grow. This was a problem that chemists found insoluble.
My attitude was that because the particles exist there must be an
error in chemical thinking.

The breakthrough occurred in 1995 when I noticed an exciting
paper by Harvard astrophysicists Weihong Liu and Alex Dalgarno
about the observation of CO molecules in supernova 1987A. The
CO molecules were undeniably present, but there were *too few of
them*. Their total number amounted to only 0.005 solar masses
of CO molecules, whereas it should have been a hundred-fold
more if all of the supernova carbon had been combusted to CO
molecules. Liu and Dalgarno presented in their paper a solution to
this problem that sped me along a well-worn track of my career.
Supernovae are profoundly radioactive. Their radioactivity keeps
them shining for years, and it produces the gamma rays and x rays
that had been so excitingly observed in SN1987A. Liu and Dalgarno
computed that the radioactivity also broke CO molecules apart
into free atoms of C and O. *Of course!* That is the answer. Because
most of the carbon then emerges as free atoms, it has the capability
of condensing carbon solids even in the presence of oxygen.
Admittedly, the growing carbon solids would eventually combust
with hot oxygen if the environment were to remain hot and dense,
but the supernova expands and cools so rapidly that there exists
insufficient time to combust carbon grains. I presented oral papers

on these ideas at the 1998 *Lunar and Planetary Science Conference* in Houston and at the 1998 *Meteoritical Society* meeting in Dublin, where Nancy and I enjoyed a week in the Georgian residence hall of *Trinity College* with our son Andrew and my first son Donald, who joined us for the associated holiday. Both presentations met very skeptical, even hostile, audiences. Even Robert Clayton, who had become a friend and supporter, said to me, "You have to have overlooked something".

Moving beyond chemical skepticisn would require presenting an explicit calculation that demonstrated the growth of carbon solids in radioactive oxidizing supernova gas. I spoke with both Weihong Liu and Alex Dalgarno about their paper on CO in supernovae and about this problem; and I mailed them my abstracts that I had submitted to the two meetings. I suggested that we together construct a computer model that calculated growth of solids in an expanding environment in which C was being oxidized while CO was being dissociated by the radioactivity. They agreed to think about such a model, and within a month Dalgarno's vast knowledge and experience with atomic astrophysics had helped us come up with a detailed chemical model that we could program onto a computer. They estimated atomic reaction rates for a huge number of molecule-building reactions and destructions. Our paper[5] published in the 26 February, 1999, issue of *Science* demonstrated that the ideas were sound and that, using a limited set of atomic reactions that nonetheless emulated a more general and more complex system, carbon solids did grow amid the chaos of CO formation and dissociation, even when bathed in more abundant oxygen atoms. The hot oxygen tries to combust the carbon solids, but they grow even faster owing to the free C atoms. Our calculation had been both kinetic and time dependent, eschewing entirely the misleading chemical notion of equilibrium. I regarded this as a triumph of scientific reasoning.

I pressed on with related investigations. During the next five years I published three *Astrophysical Journal* papers[6] extending those results with Ethan Deneault, my final Clemson University Ph. D. student, with Brad Meyer, my Clemson colleague, and with

Alexander Heger. I believe that these map the course of *SUNOCON* research for the next decade.

I had one more task to perform before thinking of my retirement. Geoffrey Burbidge had put it to me in the form of a request for a review paper on stardust for *Annual Reviews of Astronomy and Astrophysics*. I definitely wanted to write that review paper. It afforded an influential chance to lay before the astronomical community a summary of how new astrophysical science can be done with stardust. I loved that discipline so much, and it was still not well known to astronomers generally. Perhaps this review paper could leave a lasting testament to three decades of my research as well. But I needed a young and vigorous collaborator who was totally familiar with the state of the data, preferably a contributor to drawing astronomical inference from the data. I did not think long before Larry Nittler came to mind. He had entered the field as a Ph. D. student with Ernst Zinner, my incomparable colleague on stardust. I had spoken with both every year at our annual joint workshops with Washington University, and over a decade I had witnessed his progress from beginning student to world research leader. His was the same wondrous process that had happened to me forty-five years earlier at Caltech, and that I had witnessed in each of my own successful Ph. D. students. It represents the human side of our lives in graduate education by research. Fortunately Larry accepted my invitation to coauthor our publication[7], "Astrophysics with Presolar Stardust".

Final thoughts on myself in the 21st century

A consideration for scientists that is not widely relevant for people at large is that the works of senior scientists seldom attain the brilliance and creativeness of the works of their youth. We scientists may remain very productive owing to our experience at leading research teams, but our own conceptual ideas usually pale. Much data demonstrate this property of science creativity. My 21st century work necessarily draws more and more on experience and less on the fresh insights that come easily to youth. More and more, I began to question where this great quest of my life should halt. I unquestionably could continue to publish research papers. I could continue to lead students and teams along fertile avenues of research. I could continue to attend conferences and be recognized and respected. But I knew that my most inspired works, those for which I will be remembered, were now behind me. Creative works are not done in spare time, but spring from an impassioned perseverance that brushes lesser issues and day-to-day concerns aside. I became less and less confident that such single mindedness was what I desired for my last decades.

My scientific life has been an adventure of unique opportunities. By beginning in 1957 to develop new ideas swirling about among a few Caltech cognoscenti, I had become one of the pioneers of the picture that our chemical elements have been created within thermonuclear cores of the vast numbers of stars that had been born and died prior to the existence of our solar system. We have been assembled by chemistry from their debris. Our atoms are fallout. This has been a rare privilege for me, to have lived in the only time in the long history of life when mankind could solve that eternal mystery. In the twinkling of an eye, mankind had emerged from mechanical tools of agriculture to walk the moon and understand the origin of the atoms of our bodies as the result of cosmic evolution. My own life had repeated that evolution. I had been born at precisely the right instant of our species, allowed by fortune to hike that path a few steps behind its formative genius, Fred Hoyle. I had the good luck to know him well personally, along with other giants of science history. Another of these, William A. Fowler, had sponsored my Ph. D. thesis and my beginning studies toward elaboration of Hoyle's picture. His support and sympathetic guidance gave me my opportunity. Through exciting scientific decades of discovery I had discovered how those nuclear and stellar details worked, how mankind can understand the cosmic memory that has preceded our earth. I stumbled across new ideas for confirming the truth of that fantastic picture. I was *homo sapiens* carrying a flashlight in search of keys to truth through a dark room of human ignorance.

I had conceived two new branches of astronomy, each of which orbits the central idea of nucleosynthesis in stars. With colleagues I had introduced the science of testing nucleosynthesis via astronomical observations of gamma-ray lines from fresh radioactivity. These demonstrate that new nuclei are created in those explosions. With other colleagues I had introduced nucleosynthesis studies based on stardust samples from individual stars. And now, as I contemplated when I should retire from this incomparable adventure, I was aware that I would not likely climb such heights again. Once I had glimpsed that truth, once the motivating hopes for single-minded concentration seemed unlikely to produce comparable innovation, I realized that I should retire.

I can share the secret to my immense good luck down those decades. It had been simply that *I thought longer than most.* I am stubborn. I carried my deepest questions in my consciousness all day, every day. Whatever my current question, it refined under the assault of daily thought and fresh information. I forbad distractions. Even on rare vacations, mostly taken in conjunction with international scientific conferences, even through my sequence of personal upsets and tragedies, I had slipped away daily into my private mental realm, puzzling over my key questions about the known universe and how its matter originated. Wrestling with those questions made all of life seem meaningful. If I have an intellectual strength it probably lies in my ability to ask the key questions, to identify those puzzles that demanded answers having high payoff. I thought about those at some level for hours daily. So when I retired in 2005, I did so with a happy new purposefulness. For the first time in five decades, I liberated myself from my own bonds and made new dedications. One of these is the desire to write this book, to share the complete life of a scientist at the front in hope that it can enrich others, especially those who share deep love of science without having had chance to walk it byways.

Nancy McBride had appeared in my forty-fifth year, at first glimpse an apparition of mystical empathy. Despite her beauty, Nancy became friend before lover, helper before wife, stepmother before mother, guardian of contentment rather than agitator. Our marriage is totally unlike my first two attempts. Nancy recognized and preserved my thinking time for creative work, and she came more and more to protect me from the daily demands of an increasingly complicated world. She took over our financial bookkeeping. She handled family affairs with sensitivity and generosity. But more than that, she loved me. We have loved one another unabated for twenty seven years while remaining best friends to one another. As simple as this description sounds, it had taken me decades to glimpse it. She taught me life's lesson as no other had. I had wasted decades pursuing sex first and asking questions later. Such is surely the young man's folly. I had been unable to handle my life with the same sure footedness with which I handled my career. Nancy enabled me to see my problems differently, and so viewed, they tended to vanish. We have sailed that calm sea for twenty-seven years.

One commonly speaks of love as if it is well defined, as if we know it when we see it. My experience with Nancy enlarged my emotional relationships with everyone. I became more fond of all people and more interested in them. The large gap between men and women and between young and old melted away. Emotional confidence in my mate and the lack of tension between us enabled this. More than this, our love opened me to reconsideration of past regrets. Eyes opening widely, I saw that a long list of regrets that punctuated my life in adulthood concerned sexual love or the lack of it. It is a litany of sexual disappointments and frustrations, leading to a sense of regret that love had not functioned as I had wanted. Suddenly every liason, whether years of marriage or a season's dalliance, came back to me weighted with regret, even remorse. I wished that I could speak once more to Mary Lou to share this understanding with her. I regretted undeniable shortcomings in my capability to love. Unable to say this before the confidence of my love with Nancy, I preferred to blame others. Seeing truth did not dispel those regrets; to the contrary, they became more poignant with each recollection and each admission. So much of sexual love was wasted. Admitting and coming to terms with those regrets from as long as fifty-five years ago has been a twenty-five year mission. Gradually those regrets were replaced with increased love for others. Love between Nancy and myself emancipated me, my surprise in life. I rejoice in the depth and scope of love. I always had loving capability, but frustrated it by my wrongheadedness and infantile fear. The great capacity that I do have was inherited from a happy infancy in which I was loved tenderly by both of my parents, and it is to them that I give thanks for that gift.

When Andrew reached high school Nancy was able to resume her passion for painting. Life's forces had required her to set watercolor painting aside, but with Andrew's life centering on school, she began again. This accelerated when Andrew entered Clemson University and lived on campus rather than at home. Nancy had formally studied artistic anatomy and drawing in New York in the mid 1970s with a famous teacher of drawing, Robert Beverly Hale, at the Art Students League. Her work features a wonderful ability to draw, with an eye toward its architectural or abstract construction. Her resumption of her own high goal has added immensely to my life by adding to her own. I understand passion for achievement, for constructing something beautiful. I resonate with it. So talented is she that for the past four consecutive years (2005-2008) her newest painting has been juried into the annual show of the South Carolina Watermedia Society. This run has earned for her the accolade *Member with Distinction.* It is such fun to attend the annual meetings with her as husband of the artist, just as she attended so many science conferences with me. I am no art critic, but I anticipate high success for her. Nancy's studio occupies the upper floor of the garage that she designed and built when it was not possible to rescue the original carriage house. Her building added very sympathetically to our historic property.

Back to the mountains

Fulfillment of a major dream for Nancy and me was made possible by my retirement from active teaching. We signed up for a Sierra Club backpacking outing during the last week of August 2006 in Wyoming's Wind River Range. Nancy made no apology for yearning to make a trip on foot across significant and beautiful terrain. It was a lifelong dream for her. I longed to again look at remote and special places on our earth, places where the beauty and the history of our planet spoke eloquently. The Wind River Range along the Continental Divide in central Wyoming is such a place. At age seventy-one in 2006, would I still be able to trek with loaded backpack to high altitude and continue for the planned week? Our group departed the Scab Creek Campground with llamas to carry part of the gear. I was glad of that because I was unsure how well I would hold up carrying much of our gear. I had always been a strong hiker, but two very busy decades had passed since I last took vigorous climbs. My only recent strenuous uphill climbing had occurred in the Shining Rock wilderness in North Carolina.

Two nights later we pitched tents beside Valley Lake on a plateau beneath the peak of 12,254-ft Mount Victor. It looked inspiring, but our trip leader, Paul Minkus, thought it might be too difficult for our mixed group of eight. With such a group, his plans must be formulated beforehand. On the next day we hiked strongly, carrying only daypacks, up Europe Canyon to the Continental Divide at Mount Europe. From there we looked east on a magnificent day into the Wind River Indian Reservation, from which water drains to the Gulf of Mexico, and turned back west to view the drainage path toward the Pacific Ocean. This range contains seven of the ten largest glaciers in the continental United States, all shrinking in response to global warming. An essential message resonated in my thoughts: that our earthly structures are born and fade away, just as the family farming life of Adair County had faded away. These mountains are also part of my earth, very different from the fertile topsoil of the Iowa prairie. The vigorous upward thrust of these sharp granite crags spoke of their geologic youth, only about fifty million years old, which is much older than our species but only 1% of the age of the earth.

The Wind River Range is one of the undeniable beauties of the Rocky Mountains. This majestic high craggy range aligns southwesterly from the Grand Tetons. The Rocky Mountains took shape during a period of intense plate tectonic activity that formed much of the rugged landscape of the western United States. Three major mountain-building episodes reshaped the west from about 170 to 40 million years ago (Jurassic to Cenozoic Periods). The last mountain building event (about 70-40 million years ago) is responsible for raising the Rocky Mountains. This included the

Wind River Range of central Wyoming, where large blocks of granite dominate. Water and evidence of water is everywhere, from the eroding peaks to high lake-containing corries to gathering streams that at bottom rush outward in one of our great western watersheds. Watching one such waterfall during a peaceful stop, I marveled inwardly at the water molecule, H_2O, with its strong surface tension that produces such beautiful sculptured flow patterns. The two hydrogen atoms of each molecule had been created from pure energy during a split second almost fourteen billion years ago during the Big Bang. The oxygen atom of each molecule was produced later by nuclear fusion within the explosions of stars some twenty times more massive than our sun during a lengthy period between twelve and five billion years ago. And now their molecular combination waters our land, our plants, and us.

Nearing Mt. Washington with Nancy
on the Appalachian Trail in July 2008

Lunching daily with views of this spectacle my mind flashed back thirty five years to my first climbs with Willy Fowler and

Fred Hoyle across the much more ancient Scottish mountains, Liatach and Beinn Eighe, in northern Wester Ross. Those ancient and heavily eroded sandstones have Precambrian quartzite along their tops, giving to those heights an otherworldly white shimmer. Lower layers contain no fossils, a fact speaking dramatically to the origin of life on earth. How different from place to place is our world. The natural forces of evolution had produced this variety from an initially uninteresting, unlivable earth. Natural forces of evolution had also produced the atoms of this earth. The atoms of my body had been created within the fiery interiors of untold numbers of stars that lived their lives and evolved before this earth had even been born. In that sense I too am stardust. The Fontanelle farms suddenly seemed far away, long ago, but that distance in time and space is nothing compared with the remoteness of the origin of our atoms. A lifetime of pursuit of their origin has made them intellectually and emotionally part of my life. They became dear to me. Each isotope has its own personality, each a distinct chapter[8] of the story they have to tell. They are my atoms, and this is my earth.

END NOTES

Chapter 1

[1] Thomas Franklin Kembery and Martha Rosena (Keisel) Kembery were my mother's parents. They met and married after their parents immigrated near 1850. Frank Kembery was the son of Thomas Kembery, who arrived from western England (Somerset), whereas Martha Rosena Kembery was the daughter of German immigrants from about the same time. A portrait of Thomas Kembery can be found on p. 79 of the Biographical Volume II of [2]. Their families migrated west, to farming land in Iowa. I have written a separate document concerning my known ancestors, where they came from and highlights of their lives. But it is too long and too geneological for inclusion in this book.

[2] *History of Adair County, Iowa*, by Lucien M. Kilburn (Pioneer Publishing Company, Chicago: 1915).

[3] Mention of James Campbell is prominent in Chapter 1 of [2], entitled *First Settlers*.

Chapter 2

[1] In these tales I speak of my parents as if they are alive. My mother was alive when my first draft was composed, and we spoke often

of her vivid memories in order that I could capture them correctly. My father died in 1983, before I began writing.

[2] In 2009 my father was inducted into the Iowa Aviation Hall of Fame.

[3] On the western banks of the Mississippi River, Davenport was already far from the Adair County farms. My feeling about the three years living there is that they cut my parents' umbilical cords to the farms. This experience made them ready for the big move to Texas. To me, these were momentous changes. Much of who I became has roots in those three years.

Chapter 3

[1] On one occasion my mother described a milky white fluid; but in later years she said that she did not know what the vaccine looked like.

[2] My mother typed her memoir considerably later, using her diaries and saved memorabilia. This is unpublished. It was a gift to her descendents. Its style is refreshing, and it captures some early times much better than I could do myself, even were I given her knowledge of events. I here and elsewhere quote from her memoir or otherwise make use of it.

Chapter 4

[1] A nostalgic book of full-page photographs, *THE PARK CITIES*, by Diane Galloway (Mercury Printing Co; Dallas 1989) is very useful to one wanting to experience the history of this famous Dallas community of separate cities, Highland Park and University Park, together known as *The Park Cities*. A half-page photo from 1936 of University Park Elementary School can be found on p. 159.

[2] My mother, Avis Irene Kembery Clayton, wrote late in life her own autobiography. She typed it out on a manual typewriter, and inserted Xerox copies of news stories and of photographs that detailed her family history. These have sympathetic descriptions

of her life on the Kembery Iowa farm and of her Fontanelle schooling. But her great love, her chief interest, was our life in Texas with my father and three Iowa children. I am indebted to her for many insights into events that I lived and remember but did not understand.

[3] The east-west streets of University Park were named for American Universities. Their names were, moving south from Vincent Street to Lovers Lane: Colgate, Caruth, Greenbrier, Southwestern, Bryn Mawr, Hanover, Purdue, Stanford and Amherst. An aerial photograph of our neighborhood can be found on page 142 of *THE PARK CITIES*[1]. Taken in 1943, two years after our move there, from an airplane flying low above Southwestern Boulevard, it looks northwest at the new University Park Methodist Church at Preston Road intersection with Greenbrier and Caruth Streets. Northwest across Preston Road are Caruth and Colgate Streets, at the edge of University Park, then a two-block open field to their north, then Sherry Lane in the middle of nowhere. Our house is clearly visible, third from the left, west end. The view northward in 1946 along Douglas Avenue from the Lovers Lane corner is found on p. 131. Every detail is vivid in my memory from countless bicycle trips down that road. But it does not look at all special today to younger eyes trained in the modern images of affluence. Our home lay three houses to the right of Douglas Avenue, but down that well remembered road further than the photo can see. On p.130 is the view eastward on Lovers Lane from the same corner, toward the University Park Elementary School.

[4] Diane Caylor Galloway's articles on the Clayton family appeared in the *Dallas Morning News* on November 19, 1997 and on November 26, 1997. These newspaper stories also contain photographs of the University Park Elementary School (Nov. 19) and of the corner of Sherry Lane and Douglas Avenue (Nov. 26), looking west in winter's snow, which is being swept from the sidewalk on Sherry Lane by my grandfather Frank Kembery. My three-year old sister, Barbara, stands beside him in that 1947 photograph.

[5] What I did was to realize that any increase to the length of the base must be compensated by decrease of the height in order to maintain constant perimeter. I wrote down the relationship between height and base that would yield perimeter L. I then tried letting the height of the rectangle be L/4, L/5, L/6, L/7, L/8 and so on. This trial-and-error was possible because they set aside an extra 30 minutes for that final problem. Most simply gave up and turned in their test. But I persistently evaluated the area enclosed for several choices of the height. I discovered that the largest area in that sequence was the height L/7. Even smaller changes of the height showed it to be closer to L/7.1. Because π entered into the contribution of the semicircle to the perimeter, I guessed that the correct height was $L/(4+\pi)$. Because I had not studied differential calculus, I did not know that zero rate of change for the area gave the correct answer; but I reasoned it out with my sequence of trials. The school newspaper, *The Bagpipe*, asked to publish my photo holding my winner's cup. But I was most proud of solving that problem. It gave me a new view of myself.

Chapter 5

[1] During the five-year period 1959-63, Oswald Jacoby won the McKenney Trophy four times, awarded for being the player to win the greatest number of masterpoints during that calendar year. In 1963 he became the first player to earn more than 1000 master points in a given year. He was one of the most celebrated bridge players of all time. Dr. John Fisher was McKenney Trophy winner in 1972. He won his first big event, the Open Pairs at the national championship in 1958, just three years after I met him at the Dallas Downtown YMCA. Both of these greats were fixtures in our weekly club game. Today no top players are found at local club games.

[2] The motion of a pendulum on the rotating earth demonstrated the rotation of the earth. Perhaps more accurately, it demonstrated that a pendulum retains its inertial plane while the earth rotates daily. This famously demonstrated the power of physics, especially the relationship of experiments on earth to a cosmic

setting (the ponderous rotation of the earth about its polar axis). It is commonly called Foucault's pendulum, after the first experimenter, Jean Bernard Leon Foucault (1819-68). I loved that experiment, which is difficult to set up because it requires a very long pendulum and no friction. It demonstrated that some cosmic events can be registered by experiments on earth. Later examples, similar in that spirit, were experiments showing that the apparent speed of light is unaffected by the motion of the earth (Michelson and Morley), and that the expansion of the universe revealed itself through the red shift of light (Hubble) from distant galaxies. On a somewhat mystical note, I even compare the large-scale forces in my own life to my own swinging motion as a physics student.

Chapter 6
[1] Burbidge, Burbidge, Fowler & Hoyle, *Reviews of Modern Physics*, **29**, 547 (1957). This work soon became known throughout the world as B^2FH, last initials for its four authors. It was drafted in 1956 in Kellogg Radiation Laboratory in the same year that I entered Caltech as a graduate student of physics.

[2] *The Dark Night Sky*, Donald D. Clayton (Quadrangle, New York 1975) p. 8

[3] *Explorer 1* was launched on January 31, 1958. It was launched easterly from Cape Canaveral, FL to take advantage of the fact that Florida already has an easterly speed near 1000 mph owing to the rotation of the earth toward the east. Its launch utilized a German first stage that the army had captured after WWII and stored in Huntsville AL. The also captured Wernher von Braun was instrumental in cobbling the mission together as a launching rocket. *Explorer 1* was the first satellite to carry a scientific experiment, the Geiger counters developed by the principal scientist James A. Van Allen. It was these particle counters that turned a show for the American people into a scientific discovery of the first rank; namely, the high counting rates when the satellite passed through certain portions of the earth's magnetic field structure. The trapped charged particles on those magnetic

field lines are now known as lying within the *Van Allen radiation belts.*

[4] The lack of stable isotopes at masses 5 and 8 is a consequence of the nuclear force. The problem is not caused by the stars. But it was within the stars that Hoyle sought a situation that could surmount that problem. $^4He+p \rightarrow Li^5$, and $^4He + {}^4He \rightarrow {}^8Be$, neither of which is stable, so they quickly break apart again into their two constituents. It was to this impasse that Salpeter calculated that the ^8Be nucleus lives just long enough that a tiny fraction of those formed transiently would have sufficient time to capture a third alpha particle: $^4He + {}^8Be \rightarrow {}^{12}C^*$, where $^{12}C^*$ is an excited configuration of three alpha particles that might stabilize itself by emission of an energetic gamma ray. This whole sequence was called the triple-alpha reaction. It was dominating much of Kellogg activity[5] when I arrived at Caltech in 1956. E.E. Salpeter, "Nuclear Reactions in Stars without hydrogen", *Astrophysical Journal* **115**, 326 (1952)

[5] Hoyle's prediction of the existence of a particular spinless excited state was made in Fowler's office in 1953; but it was subsequently published: F. Hoyle, "The synthesis of the elements from carbon to nickel", *Monthly Notices of the RAS* **106**, 343 (1954). It may be the earliest prediction of a nuclear detail based on the "anthropic principle"; namely, if it were not true, we would not be here. Fowler loved to joke about this wacky prediction, pretending that Hoyle was bothering him; but he proceeded to measure it and found it to be true. C.W. Cook, W.A. Fowler, C.C. Lauritsen and T. Lauritsen, *Physical Review* **107**, 508 (1957). I arrived into the excitement of this discovery. It was the remainder of Hoyle's brilliant 1954 paper that I heard nothing of, and that I had to discover later on my own when *B²FH* did not review it prominently.

[6] Robert Jemison Van de Graaff was born in 1901 in Tuscaloosa, Alabama from Dutch descent. In Tuscaloosa he received his BS and Masters degrees from the University of Alabama before earning a Ph. D. degree from Oxford University. He invented a device

for charging up a spherical dome by depositing electric charge on it with the aid of an insulating belt that carried the charges upward to it. Later that device was used as an electrostatic particle accelerator that was instrumental in many of the experiments in low-energy nuclear phyiscs. Kellogg Lab had three Van de Graaff accelerators. Today the Van de Graaff home in Tuscaloosa is maintained as a museaum.

Chapter 7

[1] H.E. Suess and H.C. Urey, *Reviews of Modern Physics* **28**, 53 (1956). Hans Suess became my lifelong friend, and we met frequently in Heidelberg during 1977-84, years when we both resided during summers in Heidelberg. Suess's insights into the correlation between nuclear magic numbers and the natural abundances were prescient, and well noted by Fowler. Suess deserves regard in that sense as a pioneer of nucleosynthesis.

[2] Donald D. Clayton, *Physical Review* **128**, 2254 (1962). Tommy Lauritsen's drawings adorning the walls of the Kellogg hallway were submitted from time to time for publication in *Nuclear Physics* with his collaborator Fay Selove. In this way they were shared as published knowledge. These were an important aspect of the evolution of nuclear physics. What impressed me deeply was that drawings adorning the walls of a building would be the up-to-date knowledge in science, more up-to-date than the published literature. The difference from undergraduate education was glaring.

[3] Donald D. Clayton, "Hoyle's Equation", *Science* **318**, 1876 (2007). Fifty years after the publication of B^2FH I would attempt to redress the fame imbalance between Hoyle's 1954 paper and B^2FH. Not entirely coincidentally, my public foray into that arena began at Caltech in my presentation http://www.na2007.caltech.edu/program2.html during a conference called to celebrate the 50th anniversary of B^2FH. A third such foray was published one year later in *New Astronomy Reviews* **52**, 360 (2008). But during my graduate student years I was, like others, not aware of the sweep

of Hoyle's 1954 paper. That awareness grew during the next two decades, as did my sense that B^2FH was overcited as a primary source for nucleosynthesis in stars.

[4] Clayton, Donald D., Fowler, W.A., Hull, T.E. and Zimmerman, B.A. "Neutron capture chains in heavy element synthesis", *Annals of* Physics **12**, 331 (1961). Tom Hull was an applied mathematician on sabbatical leave at Caltech who suggested the technique for matching the Laplace Transforms of an approximate solution to the exact Laplace Transforms. Barbara Zimmerman was a technical member of the Kellogg staff who did many practical things for Fowler, including the computations for the figures of this work. A second paper providing the first quantitative decompositions of heavy isotope abundances into their *s*-process and *r*-procss components followed almost immediately: Clayton, D.D. and Fowler, W.A. *Annals of Physics* **16**, 51 (1961). Our technique, based on the new *s*-process theory, has been repeated many times down the decades, and has had large consequences for understanding the chemical evolution of the early galaxy.

[5] The rates of change of the abundances of the species in the *s*-process capture chain are expressed by a sequence of coupled differential equations. Those species form successive integer steps from ^{56}Fe to ^{209}Bi. Their identity is determined by the nuclei that are generated when neutron capture increases the atomic weight by one unit and time is allowed for beta decays that happen within a week or so. If that nucleus at any point on the chain is represented by the subscript *k*, the equation for the rate of change of the abundance of nucleus *k* is $dN_k/d\tau = \sigma_{k+1} N_{k+1} - \sigma_k N_k$, where τ is the flux of free neutrons in the stellar environment and σ_k is the neutron capture cross section for nucleus *k*. If one assumes, as B^2FH did for the sake of presentation, that the abundances are in steady state, unchanging with time, then $dN_k/d\tau = 0$, with the immediate consequence that the product $\sigma_k N_k$ has the same value for each nucleus on the chain. What my thesis did was to retain the time derivatives and seek solutions of the tine-dependent coupled equations. I sought those solutions that described the result of

neutron irradiations of iron for various total neutron exposures, the integrals of the neutron flux τ over time.

[6] Macklin, R.L. and Gibbons, J.H. , *Reviews of Modern Physics* **37**, 166 (1965). A photograph of us working together fifteen years later is < 1980 Clayton, Beer, Kaeppeler and Macklin >.

[7] Seeger, P.A., Fowler, W.A. and Clayton, Donald D. *Astrophys. J Suppl.* **11**, 121 (1965). Phil Seeger was already hard at work on this problem as Fowler's student when I joined that effort.

[8] Donald D. Clayton, "Cosmoradiogenic Chronologies of Nucleosythesis", *Astrophysical Journal,* **139**, 637 (1964). Previous methods had been based on a different principle, originally suggested by Rutherford in 1929. Fowler and Hoyle had mined that idea in their attempt to determine the age of the elements by comparing the surviving abundances of uranium and thorium, both long-lived radioactive elements.

Chapter 8
[1] "Chronology of the Galaxy", *Science* **143**, 1281 (1964). My credentials for writing such an article rested on my recent discovery of new treatments in radioactive chronometers for the age of the elements. My rhenium-osmium clock was showcased to a wider science audience than readers of *The Astrophysical Journal.* The globular clusters are spherical collections of perhaps 100,000 stars held together by their own self gravity. Their great ages were evidenced by the fact that their most massive stars, those more massive than about 95% of the mass of the sun, were missing from the cluster. The evolution time required for those to have aged and disappeared could be calculated, thereby giving the time since the cluster had been born. Ages so estimated for the famous globular clusters M2, M3 and M5 seemed to lie between 20-25 billion years. The nuclear chronometers can not accommodate such a great age; and today we know the age of the universe itself is but 13.7 billion years.

[2] "Stellar Winds and Ages of old Clusters", *Astrophysical Journal* **140,** 1604 (1964). It had been first submitted on March 25, 1964 and was revised on July 31. My calculation showed that a mass-loss rate of 0.01 solar masses per billion years would reduce the computed age from 22 billion years to 15 billion years. The paper carries no mention of the refereeing controversy.

[3] "Radioactivity in Supernova Remnants", *Astrophysical Journal* **142,** 189 (1965), by D. D. Clayton and W. L. Craddock. We focused on the Crab Nebula for our initial calculation. It had the advantage of having revealed the telltale 60-day decline of light output, and also had known age of 911 years and known distance of 3500 light-years. These made the flux at earth specific.

[4] "Neutron Capture Data at Stellar Temperatures", R. L. Macklin and J. H. Gibbons, *Reviews of Modern Physics* **37,** 166 (1965). My photograph of the three of us at Joe Fowler's house was unfortunately badly underexposed.

[5] Seeger, P.A., Fowler, W.A. and Clayton, Donald D. *Astrophys. J Suppl.* **11,** 121 (1965). Phil Seeger was already hard at work on the mass-formula problem as Fowler's student when I joined that effort. Our paper could say nothing, on the other hand about the hardest part of the r-process problem; namely, what caused the large density of free neutrons? Our speculation was a good one, that thermal decomposition of iron produces a gaseous bath of alpha particles and free neutrons and protons, and that reassembly of the alpha particles into heavier nuclei produces the heavy nuclei that were able to capture all of the remaining free neutrons. If so, our start of the process with iron was not correct, but was an idealization of a starting condition that enabled our calculation to proceed. Nor could we compute the change with time of the density of free neutrons, so we simply assumed it to be constant until it turned off. All of those harder problems became the focus of the subsequent five decades of r-process research. But I think it fair to say that our paper showed how the full process

would work, and did so more clearly than B^2FH had done. We also showed the superposition nature of the process.

[6] "Photon Induced Beta Decay", P. B. Shaw, D. D. Clayton and F. C. Michel, *Physical Review*, **140**, B1433 (1965). This may be the first calculation of a specific beta decay transition in a nucleus being speeded by the ambient environment, which at normal temperature has no effect on the rate of radioactive decay. The process calculated by us is quite different than a similar theme that was introduced by A. G. W. Cameron; namely that high temperature photons can cause the ground state of a nucleus to be excited to a configuration that decays faster than does the ground state. Although Cameron glibly called that "photobeta reactions" it was actually "excited-state beta decay". Ours was the true photobeta process, in which the photon is absorbed by the nucleus and emitted electron. It is the inverse of a process studied in the laboratory called "internal bremsstrahlung".

[7] "Particle induced electromagnetic deexcitation of nuclei in stellar matter", P. B. Shaw and Donald D. Clayton, *Physical Review*, **160**, 1193 (1967).

Chapter 9
[1] That 1966 letter can be found in my collected papers in the Clemson University Archives. An electronic scan is available from me. Because of our relationship, I never disclosed this letter until after Fowler's and Hoyle's deaths.

[2] I have published the history of radiogenic iron and the many incorrect conclusions that rested on Hoyle's 1946 picture synthesizing iron as itself rather than as radioactive progenitors. *Meteoritics and Planetary Science* **34**, A145 (1999). At the time, 1966, Fowler and Hoyle had maintained for twenty years the correctness of Hoyle's 1946 paper in which ^{56}Fe is synthesized as itself. This had caused them to advance several mistaken conclusions. These are detailed in my historical review.
[3] D. Bodansky, D.D. Clayton and W.A. Fowler, *Physical Review*

Letters **20**, 161 (1968) was submitted in January 1967. A second, longer version was submitted to *Astrophysical Journal Supplements* **16**, 299 (1968). Our discovery was, quite simply, the most convincing triumph of nucleosynthesis in stars that existed in 1967. It filled the only ambiguity of Hoyle's masterful 1954 paper on the nucleosynthesis of the elements between carbon and nickel. I felt at once that ours was the greatest paper yet written on nucleosynthesis save for Hoyle's. Fowler urged more modesty, considering that I was a coauthor; but privately he agreed.

[4] The PPI and PPII chains are explained in my textbook *Principles of Stellar Evolution and Nucleosynthesis* (McGraw-Hill 1968), as are the neutrino emitting processes. My paper with Bob May was not submitted until January 1968 owing to my departure for England in April 1967 and was published during 1968. Robert M. May and Donald D. Clayton, *Astrophysical Journal* **153**, 855 (1968).

Chapter 10

[1] *Home is where the wind blows*, Fred Hoyle, University Science Books (Mill Valley CA 1994). Hoyle discusses the founding of the institute rather briefly in Chapter 23.

Chapter 11

[1] A mass $1/10^{th}$ of the sun's mass is 2×10^{29} kilograms. That mass is about 30,000 times the mass of the earth, and in this case of almost pure ^{56}Ni. I enjoyed saying that such a large ejected mass of ^{56}Ni radioactivity ranks as the largest nuclear accident of all time! The energy release when that radioactivity decays is huge. When converted into light it creates the brightest stars seen in the sky.

[2] The first suggestion that radioactivity provided the power to keep supernovae hot during their expansions, which otherwise cools quickly, chose the ^{7}Be nucleus owing to its halflife being about right. L. B. Borst, *Physical Review* **78**, 807 (1950). The halflife coincidence was the same one that later inspired the hypothesis of californium radioactivity G.R.Burbidge, F. Hoyle, E.M. Burbidge,

R.F. Christy and W.A. Fowler, *Physical Review* **103**, 1145 (1956). Both ideas had the flaw that the supernova does not produce a sufficient quantity of such nuclei to provide the needed radioactive heating. The nucleus ^{56}Fe, on the other hand, is hugely abundant naturally, so its radioactive parent, ^{56}Ni, is amply abundant for the heating task.

[3] "Gamma-ray lines from young supernova remnants", Donald D. Clayton, Stirling A. Colgate and Gerald J. Fishman, *Astrophysical Journal* **155**, 75 (1969). This paper was published prior to Colgate's publication on interpretation of the supernova light curves in terms of ^{56}Co radioactivity. S.S. Colgate and C. McKee, *Astrophysical Journal* **157**, 623 (1969).

[4] A 1974 photo of Colgate and Clayton in Socorro NM can be viewed on my photo archive website <1974 Colgate and Clayton in Socorro>. Colgate was at that time President of New Mexico Institute of Mining and Technology. A photo of Haymes and Clayton with Haymes' gondola for balloon launch of the telescope is <1974 Haymes and Clayton with gamma telescope >. The balloon gondola is the metal frame that hangs from the balloon, whose launches were from Palestine, Texas. Haymes' gamma-ray telescope is the assembly seen between Haymes and Clayton, here mounted onto the gondola.

[5] "Explosive Nucleosynthesis in Stars", W. David Arnett and Donald D. Clayton, *Nature* **227** 780 (1970). Although we did not describe it in those words at that time, we followed precisely what I later named "Hoyle's Equation" in a publication in which I tried to restore scientific priority for Hoyle's leadership. "Hoyle's Equation", Donald D. Clayton, *Science* **318** 1876 (2007). But it was with this 1970 paper with Dave Arnett that I began my reemphasis of Hoyle's 1954 paper.

[6] Donald D. Clayton and Stanford E. Woosley, *Astrophysical Journal* **157**, 1381 (1969). Our calculations showed that the large ^{58}Ni abundance ruled out the chance that ^{56}Fe was synthesized

primarily as its stable self. The large ^{58}Ni abundance required ^{56}Ni parentage for ^{56}Fe.

[7] "The explosive burning of oxygen and silicon", Donald D. Clayton and Stanford E. Woosley, *Astrophysical Journal Supplement* **26**, 231 (1973). In my opinion, this work is one of the great papers of nucleosynthesis in supernovae. It launched Stan Woosley onto the career path at which he has excelled. He repeatedly has illuminated the nucleosynthesis of the elements inside the concentric shells of presupernova massive stars and their subsequent explosions, making him the next great disciple of Hoyle's equation.

[8] W. N. Johnson, F. R. Harnden and R. C. Haymes, *Astrophys. J. Letters* **172**, L1 (1972); "Positronium origin of 476 keV galactic feature", Donald D. Clayton, *Nature Physical Science* **244**, 137 (1973); "New prospect for gamma-ray line astronomy", Donald D. Clayton *Nature* **234**, 291 (1971). My mail address on that paper, as on many others, was Institute of Theoretical Astronomy, Cambridge UK.

[9] A clear description of the CN cycle of hydrogen fusion into helium can be found in Chapter 5, pages 390-400 of my textbook (*Principles of Stellar Evolution and Nucleosynthesis* (McGraw-Hill:New York, 1968)

[10] "Explosive Nucleosynthesis in Helium Zones", W. M. Howard, W. D. Arnett and Donald D. Clayton, *Astrophysical Journal* **165**, 495 (1971). The weakness of this calculation proved to lie in its initial conditions; namely, there may exist little ^{14}N in the shock wave caused by core collapse of the Type II supernovae. If the matter is sufficiently dense to be heated strongly by the supernova shock wave, it also was probably hot enough that the ^{14}N would have reacted with helium prior to the shock wave. In that case the nucleosynthesis products that we sought would not be produced, for they required ^{14}N seed for the required nuclear effect.
"Nucleosynthesis of rare nuclei from seed nuclei in explosive carbon burning", W. M. Howard, W. D. Arnett, Donald D. Clayton

and Stanford E. Woosley, *Astrophysical Journal* **175**, 201 (1972). This work utilized sudden neutron irradiation of solar-like initial abundances within presupernova stars to change those solar nuclei into rare unexplained stable isotopes. The free neutrons are liberated as a byproduct of carbon fusion with itself. Confirmation of this rapid neutron addition had to await the discovery a couple of decades later of its signature within dust formed in presolar supernovae.

Another photo of Mike Howard, Stan Woosley and Clayton with others of the Rice group in Cambridge is <1971 Rice Mafia at lunch >.

[11] Information on any of these collections can easily be found on the web. Search Google for example. They became significant jewels of Houston culture.

[12] A web color photo of Roberto Rosselini with the author is <1970 Rosselini and Clayton in Sardinia>.

[13] The important ^{57}Co opportunity for gamma-ray astronomy came in the form of another publication: "Line ^{57}Co gamma rays: new diagnostic of supernova structure", Donald D. Clayton, *Astrophysical Journal* 188, 155 (1974). This paper was submitted in July 1973, the first summer in six years that I did not reside in Cambridge. Hoyle's resignation had made it too hard to go back that soon. An opaque overlayer of 10 gm/cm^2 is an absorption column for gamma rays about equal to a four-inch depth of water, which, though quite transparent to light in human experience, cuts the gamma-ray escape roughly in half. The overlayer above any point in the interior decreases rapidly with time owing to the fast expansion of the explosion. It is when that thinning has fallen to about 10 gm/cm^2 that the gamma rays from the radioactivity begin to emerge from the surface and can be detected. In foresight I also noted that the overlayer can be mase smaller not only owing to general expansion but also owing to gaseous hot bubbles containing radioactive matter being stirred outward to

thinner overlayers. This effect would be detected about thirteen years later in SN 1987A.

[14] My s-process studies at Rice University are listed below:
"Termination of the s-process" Donald D. Clayton and M. E. Rassbach, *Astrophysical Journal* **148**, 69 (1967). We analyzed for the first time the effect on the s-process abundances of lead (Pb) of the radioactive recycling of nuclei at the end of the s-process chain. This was quite significant for the cosmoradiogenic chronology based on Pb.
"Thermally enhanced alpha decay and the s process" F. Perrone and Donald D. Clayton, *Astrophysics and Space Science* **11**, 451 (1971).
"Weak s-process irradiations", J.G. Peters, W.A. Fowler and Donald D. Clayton, *Astrophysical Journal* **173**, 637 (1972). We argued that the s-process abundances at atomic weights between iron and strontium (A=56 to A= 88) were the result of a different process than the main, heavier s-process abundances (A=88 to A=208). That different process was the burning of ^{22}Ne in the helium-burning shells of massive presupernova stars. As a result the distribution of neutron fluences to which iron has been exposed during galactic history is a superposition of two distinct sources of those neutron fluences in stars. That idea remains valid today.
"s-process studies: exact solution to a chain having two distinct cross-section values", Donald D. Clayton and M.J. Newman, *Astrophysical Journal* **192**, 501 (1974). This remarkable result demonstrated many significant properties of s-process chains by postulating a trial problem in which each nucleus could have only one of two distinct values for the neutron-capture cross section. We called them "big" and "small".
"s-process studies: exact evaluation of an exponential distribution of exposures", Donald D. Clayton and R. A. Ward, *Astrophysical Journal* **193**, 397 (1974). This calculated abundance ratios of s-process chains if the number of iron nuclei that have been exposed to neutron fluence τ declines exponentially as the size of τ increases. Certain mixing models within stars could be shown to have that property.

"s-process studies: branching and the time scale", R. A. Ward, M. J. Newman and Donald D. Clayton, *Astrophysical Journal Supplement* **31**, 35 (1976). We studied the mathematics of s-process chains taking into account the powerful idea presented by B^2FH; namely , that both the neutron flux and the ambient temperature could be studied by the competition between neutron capture and excited-state beta decay.

"s-process studies: xenon and krypton isotopic abundances", Donald D. Clayton and R. A. Ward, *Astrophysical Journal* **224**, 1000 (1978). This paper was first submitted to *Geochimica et Cosmochimica Acta* in 1975, intended as a diagnostic finder for my new idea that stardust could be recognized by the s-process isotopic ratios of its trapped xenon and krypton. It was rejected as too speculative, but later resubmitted in 1978 after that stardust was discovered within residual dust within meteorites. Chapter 14 describes more of this historic occurrence that changed the face of astrophysical science.

Chapter 12

[1] A space ship sent from Earth to the Moon cannot be captured by the Moon's gravity without significant preorbital burn. Unaided it would gain speed as it falls toward the Moon, pass by its distance of closest approach, and leave again. To be captured by Moon's weaker gravity, it must apply the brakes to slow down while approaching the Moon. That braking action is provided by firing of the onboard rocket engines.

Chapter 13

[1] Alastair G. W. Cameron was the third voice to which I refer. Both he and I, and our Ph. D. students to come, moved nucleosynthesis in stars past the point where Hoyle and Fowler were in 1969.

[2] The Munros are those Scotland mountains whose peak height exceeds 3000 ft above sea level. They are so named in honor of Sir Hugh T. Munro, who compiled the list in his historic book (first published in 1891), *Munro's Tables, and other Tables of lesser Heights* (Scottish Mountaineering Trust: Edinburgh)

[3] A photo of the Rice group leaving The Green Man in Grantchester can be seen at <1971 Rice Mafia at lunch>.

[4] "Measuring the rate of nucleosynthesis with gamma-ray detectors" Donald D. Clayton and Joseph Silk , *Astrophysical Journal* **158**, L43 (1969). We reasoned that the number of 1.24 MeV gamma rays per unit volume in the universe was about equal to the number density of ^{56}Fe nuclei in the universe. An analogous argument exists for the other gamma-ray lines from ^{56}Co decay. Each gamma-ray line is no longer a monoenergetic line, however, but is spread in energy by the historic expansion of the universe.

[5] A photo of Arnett and Clayton back at Rice can be seen at <1970 Arnett and Clayton at Rice >. Dave Arnette's initiated a series of calculations of explosive nucleosynthesis in the shells of supernovae was *Astrophys. J.* **157**, 1369 (1969). His initial foray focused on carbon shell burning overheated by the shock wave launched after interior matter fell onto and rebounded from the incompressible neutron-star core. This paper was followed promptly by one with his friend and colleague James Truran, also from Cameron's Yale group. They treated the explosive burning of the oxygen shell in *Astrophys. J.* **160**, 1369 (1970). Truran had been the one who had collaborated with me and David Bodansky at Caltech in 1968 to compare our new quasiequilibrium treatment of silicon burning with the dynamic results, using a nuclear network of reactions on a computer, of oxygen heated so highly by the shock that it was all consumed in nuclear burning. Our Rice group treated these explosive shells in more detail in Stan Woosley's Ph. D. thesis after Arnett assumed his Assistant Professorship at Rice. In my book *Handbook of Isotopes in the Cosmos* (Cambridge University Press, 2003), I summarized isotope by isotope the nucleosynthesis results of these and countless later calculations concerning the elements created in supernova shell burning. Arnett was the primary mover of this movement, and Woosley became its heaviest subsequent contributor. These collected works display dramatically the soundness of Hoyle's initial vision of the cause of the growth of primary heavy element abundances (24<A<62) during the aging

of our galaxy. I paid this tribute to Hoyle, long overdue, in a short published paper in *New Astronomy Reviews* **52**, 359 (2008; also available online at www.elsevier.com/locate/newastrev).

Chapter 14

[1] R. N. Clayton, L. R. Grossman and T.K. Mayeda, *Science* **182**, 485 (1973). The frustrating aspect of this eye-opening paper was trying to discern what the authors meant with their loose suggestions of how these ^{16}O-rich solids might have come about. They indirectly stimulated much effort on ideas that would not work quantitatively.

[2] H. E. Suess, *Annual Review of Astronomy and Astrophysics* **3**, 217 (1965). Hans Suess was one of the great pioneers of solar-system abundances of the elements and their relation to nucleosynthesis processes. His paper "Abundances of the Elements", written with Harold Urey (*Revs. Modern Phys.* **28**, 52 (1956)) had a strong facilitating impact on creation of theories of those abundances. Suess's coauthoring of the discovery paper for the magic numbers of nuclear structure had made him very sensitive for signs of the magic numbers in the solar abundances. But Suess had gone too far with the argument that all presolar dust had been evaporated by a hot solar nebula (or else isotopic anomalies in meteorites would be evident). Later, with increased experimental resolution they did become evident, but only in individual grains of dust that had to be extracted from the meteorites.

[3] L. R. Grossman, *Geochimica et Cosmochimica Acta* **36**, 597 (1972). About 8% of certain class of meteorites, the C3 chondrites, consisted of white assemblages of very refractory minerals rich in calcium, aluminum and titanium (their oxides and silicates). The fraction of Al_2O_3 structural units in those minerals was high. Grossman showed that this suite of minerals was characteristic of a residue of solid dust from which most other elements evaporate. It was in such minerals that the ^{16}O excess was found[1] to be greatest in value. Grossman and others took their existence to represent high-temperature condensation from hot solar gas;

but that picture is not necessary. They may also be the solid residues of interstellar dust that has been heated by the solar disk to temperatures that evaporate lower-temperature minerals. If solar-composition solids are heated sufficiently, the last thing to evaporate is Al_2O_3 and other high-T minerals.

[4]. The arguments to which I publicly objected repeatedly were of two types: (1) allegations that the CAI must have condensed from pure hot vapor because that is how our system began; or (2) discounting of physical ideas because their chemistry was inconsistent with an assumed initial hot gas. I routinely rose to object to both such assertions in an attempt to keep before the community that science must not rely on dogma. These public interchanges would occur mostly at the annual March meeting Lunar and Planetary Science and during the annual summer meeting of the Meteoritical Society between 1975 and 1985. Many meaningfull concepts were presented during the public discussion time following each spoken presentation of a research result. There is no record of these exchanges, so they are mostly lost to posterity. But I became known far and wide for my relentless assault on the paradigm of the hot gaseous beginning for our solar system. I need to be clear about my primary motivation at that time, which was my expectation that cosmic-chemical-memory isotopic anomalies owing to unvaporized STARDUST would be evident in meteorites. That became my contribution to meteoritic science. But the senior leaders of meteoritics were uniformly skeptical, and either objected or tolerated me in polite silence. Only after 1987, after STARDUST had been definitively found within meteorites, did it become manifest that solar material did not begin as a hot gas. Today the argument is less clear, less black and white. It appears that dust-rich regions of the disk were locally heated enough to evaporate most of the chemical elements locally, and that their thermal recondensation onto refractory dust that had not been totally evaporated resulted in the dominant condensation features of the CAI.

[5]. A.G.W. Cameron, in *Interstellar Dust and Related Topics*

(1973) made the argument that "carbonaceous chondrites are probably collections of interstellar grains which have been mildly transformed through exposure to higher temperature". This obscurely published suggestion carried little influence, in part because Cameron did not predict properties of those grains that would allow them to be identified. He seemed not to think of the vast possibilities for isotopic markers. That is what my papers on STARDUST achieved. My own papers were full of isotopic predictions, and could thereby be tested.

[6]. Donald D. Clayton and Fred Hoyle, "Gamma Ray Lines from Novae", *Astrophysical Journal Letters* **187**, L101 (1973). We predicted detectable positron-annihilation 511 keV radiation owing to ^{13}N, ^{14}O, ^{15}O and ^{22}Na as well as nuclear gamma-ray lines at 2.312 MeV and 1.274 MeV. Measurement of the flux of these would diagnose what occurred in the nova.

[7] The vulture in the road was photographed on my previous trip along the same route to Big Bend. It appears on p. 201 of *The Dark Night Sky* (Quadrangle: New York 1975)).

[8] *The Joshua Factor* was published by Texas Monthly Press (1985), and is now out of print. The novel is built upon my fascination with the solar-neutrino problem. The real rate of capture of ^{8}B neutrinos from the sun's center was observed by Raymond Davis Jr over a many year period, and it was consistently smaller than current models of the sun predicted. In the end this puzzle was solved by new properties of the neutrinos, but in 1982, when most of the novel was written during six weeks on the west coast of Scotland, there seemed a real conflict between the sun's calculated rate of production of neutrinos and their rate of detection on earth. My novel used a black hole at the solar center as its explanation of the neutrino puzzle and ended as a garden-of-Eden parable about the origin of man. My very exciting mechanism in that novel was stimulated by our pioneering work at Rice[18] on the existence of a solar black hole as the explanation of the neutrino deficit.

⁹ A photograph of Klaus Fricke and me at lunch with others can be seen in <1971 Clayton, Klaus Fricke and Woosley at lunch>.

¹⁰ Isotopes of the noble gas, xenon, have a very rich history, which Ed Anders had been reviewing precisely when the new idea arose within me. Xenon had been the element for which the first documented isotopic anomalies had been discovered by John Reynolds in Berkeley during the 1960s in samples of meteorites. The first was discovered during my graduate school years by Reynolds, and consisted of excess ^{129}Xe within unknown chemical sites rich in the element iodine. Reynolds inferred with a beautiful experiment that it exists in meteorites owing to the decay of radioactive ^{129}I, which, he reasoned, must still have been present, still alive, when the meteorites formed in our solar system. Reynolds' experiment consisted of irradiating the meteorite sample to free neutrons within a nuclear reactor. Neutron capture by stable ^{127}I, the only stable isotope of normal iodine, would convert a small part of the stable ^{127}I to stable ^{128}Xe. This ^{128}Xe would be in the minerls that had contained iodine when they formed. He then heated the meteorite sample and found that the ^{128}Xe was outgassed by the increasingly higher temperature in proportion to the amount of excess ^{129}Xe that had been there from the beginning. This correlation proved that excess ^{129}Xe was related to where iodine was in the meteorite.

The abundance of excess ^{129}Xe then was only 1/10,000th of that of stable ^{127}I because its 17 million-year halflife was much shorter than the time since the elements were first created. So most of it had decayed, accounting for its low abundance at the time it could be incorporated into the meteorites. Reynolds called this an "extinct radioactivity" because it is completely vanished today, having undergone radioactive decay for 4.6 billion years after the formation of the meteorites in our young solar system. Only its radiogenic daughter, stable ^{129}Xe, now exists, apparently parentless, in the meteorites.

J. H. Reynolds, *Physical Review Letters* **4**, 8 (1960).
     Reynolds immediately set forth incorrect conclusions from

his measurements, arguing that the elements are much younger than their actual age. He made this error by using an inappropriate astrophysical model for the interstellar abundance of ^{129}I. Thus even at the very beginning, astrophysics was the *bete noire* of xenology. Subsequently Reynolds laboratory discovered the other isotopic xenon patterns carried in the meteorites. A photo of John Reynolds having coffee with two of his famous disciples is <1976 Reynolds and disciples Lewis and Podosek>. A very good photo of him with Robert Clayton is <1976 John Reynolds and Robert Clayton>.

Later Reynolds and coworkers would experimentally discover anomalous Xe isotopes resulting from the spontaneous fission of ^{244}Pu within the meteorites. And most intriguing of all is the third anomaly, what Reynolds called "the general anomaly". Much more abundant than the other two, it seemed to be enriched in the heaviest of xenon's nine isotopes. Edward Anders had been describing it to be the fission-product daughters of a now-extinct superheavy nucleus that was supposed to have initially resided within the meteorites. This idea gave it its common name of the time, "carbonaceous-chondrite fission xenon", called "CCF" for short. What's in a name? Very, very much actually, because whenever a physical effect bears a name that derives from a picture of why it might exist, scientists tend to come to accept that picture as actually having occurred

[11] A photo of Hoyle opening the gate for our car can be seen at <1974 Hoyle at Cockley Moor>. The older part of the farmhouse is behind him.

[12] A photo of Hoyle and me in his office is < 1974 Hoyle and Clayton in Hoyle's Cockley Moor home>. The impending research paper that we were discussing was published *Astrophysical Journal*, **191**, 705 (1974). This paper was published a bit prematurely and seems now to not have scientific value. This was my fault, and happened in part because I was committed to trying to keep Hoyle before the astrophysics community. The only means at my disposal to

keep him active was to obtain his collaboration on nucleosynthesis issues, which were my strength.

[13] A photo of us on that ridge is < 1974 Hoyle and Clayton ascending Helvellyn>. Annette, who loved climbing with us, took this photograph for me.

[14] "Extinct Radioactivities: Trapped Residuals of Presolar Grains", Donald D. Clayton, *Astrophysical Journal*, **199**, 765 (1975). I had improved the wording in several ways and added a framework for how a chemical memory of presolar dust might appear in larger stones formed from that dust. The latter included my schematic Figure 1 showing the concept of the fraction of grains containing element A and how that fraction would decline with time in a cooling solar nebula.

[15] David Black's paper on Ne-E had been published three years earlier in *Geochimica et Cosmochimica Acta* **36**, 377 (1972). In his paper he made the profound suggestion that some type of presolar dust carried that component into the meteorites. I had already been in hot pursuit of evidence of nucleosynthesis of radioactive ^{22}Na by detection of its gamma-ray line emission: "Gamma_ray Lines: a ^{22}Ne Diagnostic of Young Supernovae", *Astrophysical Journal*, **198**, 151 (1975). It was at the same time that I decided to submit my supernova-dust paper "^{22}Na, Ne-E, extinct radioactive anomalies and unsupported ^{40}Ar", *Nature* **257**, 36 (1975), which David Black reviewed for *Nature*. In this same paper I presented arguments for several additional observable extinct radioactive nuclei.

[16] My paper with my student Richard Ward, "s-Process Studies: Xenon and Krypton Isotopic Abundances" *Astrophysical Journal*, **224**, 1000 (1978) was a resubmission of the same paper rejected in 1975 by *Geochimica et Cosmochimica Acta*. My correspondence on this is in my papers; but it can also be found in the *GCA* office. It had not been rejected as wrong or uninteresting, but rather as too speculative for a journal devoted to chemists. All doubted that the

s-process pattern could actually be found, undiluted by r-process nuclei. I resubmitted it after a phone call from Edward Anders, who received my 1975 preprint as one recipient of my blizzard of mail, disclosed that they had found the s-process pattern in a residue of meteorites dissolved in acid! My personal relationship to Edward Anders was very poor owing to his single minded promotion of his own ideas and his disparaging assessments of mine, but I have always been grateful for the scientific honesty that prompted him to telephone me about their discovery of s-process STARDUST in Chicago.

[17] "Solar Models of low Neutrino Counting Rate: the Depleted Maxwellian Tail", Donald D. Clayton, Eliahu Dwek, Michael J. Newman and Raymond J. Talbot, Jr, *Astrophysical Journal*, **199**, 494 (1975). It was my hypothesis that the frequency of collisions of two particles declined faster than the classical Maxwellian distribution. This was suggested to be the result of the many-body collisions attendant to the long-range Coulomb force. If the extra depletion in the two-particle relative energy is characterized by $\exp(-\delta(E/kT)^2)$, the low count rate is explained if $\delta > 0.01$.

[18] "Solar Models of low neutrino-counting rate: the central Black Hole" Donald D. Clayton, Michael J. Newman and Raymond J. Talbot, Jr, *Astrophysical Journal*, **201**, 489 (1975). Stothers and Ezer, *Astrophysical Journal Letters*, **13**, 45, had already published a short paper in 1973 concerning structural effects on the sun of a black hole at its center, but they had neglected the accretion luminosity onto the black hole, so they missed the key effect.

[19] A photo of Fowler and Clayton at this time is <1975 Fowler and Clayton in Fowler's home>. Fowler's home was but a block east on San Pasqual from Caltech's Athenaeum, by which he walked daily on his stroll to his office in Kellogg Lab.

Chapter 15
[1] "Grains of anomalous isotopic composition from novae", Donald D. Clayton and Fred Hoyle, *Astrophysical Journal* **203**, 490 (1976).

We had submitted this paper in April 1975 following Hoyle's second visit to Rice University. The idea had been mine, and it was intended to bring Hoyle, a great pioneer of stardust, into the group of advocates of its isotopic signatures. I wanted Fred on my team. In it we pointed to data favoring the rapid condensation of dust in novae from Nova Serpentis (1970). But the main thrust was to detail the unusual isotopic compositions expected to characterize the elements in that dust. About 15 years later, such nova dust was found.

[2] "On the development of infrared radiation from an expanding nove shell", Donald D. Clayton and N. C. Wickramasinghe, *Astrophysics and Space Science* **42**, 463 (1976). We submitted this paper in September 1975 at the end of my first summer in Cardiff. Owing to my appointment as Visiting Professor at University College, this paper bears that affiliation. One goal for me was to strengthen the credibility of nova stardust advanced in 1. A photo of DDC and NCW at work in Clayton's garden on this paper is <1975 Chandra Wickramasinghe and Clayton in Cardiff>.

[3] A photo of me delivering my talk before such a painting is <1975 Clayton speaking at Hoyle 60th in Venice>.

[4] See [17] and [18] of Chapter 14.

[5] My photo of Kirsten and Raymond Davis on that occasion can be seen at <1979 Kirsten and Davis at second GALLEX meeting>. We were at that moment strolling in the gardens at Schwetzingen.

[6] "Cosmoradiogenic ghosts and the origin of Ca-Al-rich inclusions", Donald D. Clayton, *Earth and Planetary Science Letters* **35**, 398 (1977). This paper listed Max-Planck Institute as my affiliation. This somewhat reflected my strong motivation to carve out a new scientific identity. This paper was submitted during my Heidelberg residence, and thanked the *Alexander von Humboldt Foundation* for their Senior Scientist Award to me.

It was at this time that I recognized the need for a position paper laying a new framework for the original ideas that I was introducing. It needed to address for cosmochemists and for astronomers the highly fractionated state of interstellar chemistry and its isotopes. In this position paper I laid out with broad strokes the differing types of interstellar dust, which I named SUNOCONs, STARDUST, and NEBCONs (initially capitalizing to make the word seem more specific in scientific meaning. These were original scientific words that I coined because they would be needed for the discoveries that would come. I was very sure of this, as reading that position paper (my manifesto) will show: "Precondensed Matter: Key to the early Solar System", *The Moon and Planets* **19**, 109 (1978). I presented that picture verbally in a talk to the first *Protostars and Planets* conference, January 3-7 (1978) in Tucson AZ. Astonishingly to me, this paper has had negligible citation rate. It is essentially ignored, and will probably remain so until a historian of science takes interest in it. I think, frankly, that the science world just does not know of it, and that it is an original masterpiece. I beg the reader's indulgence for my own high opinion of it. In a scientific autobiography, such as this book is, my opinion of that paper certainly shares with the reader some salient characteristic about me.

[7] "Interstellar potassium and argon", Donald D. Clayton, *Earth and Planetary Science Letters* **36**, 381 (1977). This paper was submitted from Heidelberg two months after my previous submission, reflecting my feverish work pace on stardust. I thanked four scientists from Heidelberg and Mainz for their encouragement and help.

[8] "Solar System Isotopic Anomalies: Supernova Neighbor or Presolar Carriers?" Donald D. Clayton, *Icarus* **32**, 255 (1977). I submitted this paper to Icarus in November 1976, shortly before leaving for our first residence in Germany.

[9] My photograph of the Darmstadt group, with me in it, is <1977 European Renaissance in Darmstadt>. Friedl Thielemann, who became one of the european leaders of nuclear astrophysics,

snapped this picture with my camera so that I, the guest, could be in it.

[10] The publication that had presented the idea of injection of supernova gas into a molecular cloud was by Al Cameron and Jim Truran, *Icarus* **30**, 447 (1977). This paper had a large preprint circulation and gained a huge following because no researcher save me was constructing alternative explanations for the ^{16}O-rich aluminum oxides and the putative presence of radioactive ^{26}Al within the meteorites. It was essentially the Cameron-Truran picture that Clayton and Schramm had supported in their article "Did a Supernova trigger the formation of the solar system?", David N. Schramm and Robert N. Clayton, *Scientific American*, **259**, Number 4 (October 1977), p. 124. Interestingly, they took part of my picture[8] by suggesting that grains from a near supernova had been injected into the solar system carrying the excess ^{16}O and the live ^{26}Al. They stated (p. 130) "It is more likely that the anomalous oxygen entered the solar cloud in chemically combined form within solid grains" One of many reasons that I found to totally discount their variation of that idea was the ratio of ^{26}Al atoms to pure extra ^{16}O atoms in grains from the carbon shell, where they (and I[6,8]) expected these grains to condense. I had already done the math on this for my papers[6,8], and I had found that those grains carry too much ^{26}Al. This is because the ^{26}Al atoms and the extra ^{16}O atoms from the carbon shell would have thermally condensed within Al_2O_3 crystal units, requiring that an extra 750,000 ^{16}O atoms per million silicon atoms (needed to make it 5% rich) would be accompanied by about 500 live ^{26}Al atoms from the carbon shell. This would have amounted to an initial ratio $^{26}Al/^{27}Al = 60 \times 10^{-5}$ owing to the ^{26}Al atoms in that dust. This is about twelve times as much ^{26}Al as is actually found in the Al-rich minerals. Saving their injection picture would have required a wait of 3.5 million years after the supernova before the meteorite minerals formed, a time long enough to allow most of the ^{26}Al to decay. After such a long time there would be no ^{41}Ca left; so it could not work. I had other equally convincing reasons

why this model using grains injected by the supernova trigger could not work, and therefore gave it no heed.

I preferred instead to utilize the interstellar expectation of Al_2O_3 crystal units from the full history of past supernovae, using the fraction of supernova dust that exisys throughout the interstellar matter. In these the excess ^{26}Mg was locked within the Al_2O_3 crystal units, within which they had long ago been created by the decay of the initial ^{26}Al atoms. Therefore, my works[6,8] required that the Al_2O_3 crystal units be able to assemble into larger Al_2O_3 minerals in a hot solar disk without loss of all of their carried ^{26}Mg atoms. They would lose *most* of the ^{26}Mg carried within aluminum, but not *all*. I called this *charmed ^{26}Mg*.

[11] My paper with my student Richard Ward, "*s*-Process Studies: Xenon and Krypton Isotopic Abundances" *Astrophysical Journal*, **224**, 1000 (1978) was a resubmission of the same paper rejected in 1975 by *Geochimica et Cosmochimica Acta*. My correspondence on this is in my papers; but it can also be found in the *GCA* office. It had not been rejected as wrong or uninteresting, but rather as too speculative for a journal devoted to cosmochemists. All doubted that the *s*-process pattern could actually be found, undiluted by *r*-process nuclei. I resubmitted it after the phone call from Edward Anders.

[12] "Cosmic Chemical Memory: A New Astronomy", *Quarterly Journal of the Royal Astronomical Society* **23**, 174 (1982). This was the written version of my honorary invited lecture, The George Darwin Lecture of the *RAS*.

[13] A good candid photograph of me with may baby daughter, Alia, in the Bartlett house is <1979 Clayton and daughter>

Chapter 16
[1]Donald D. Clayton, "Extinct Radioactivities: a three-phase mixing model", *Astrophysical Journal*, **268**, 381 (1983). This paper was submitted in July 1982 from Heidelberg, where I prepared it. It provided a pathbreaking new approach to the issue of the

mean expectations for the abundances of short-lived radioactive nuclei in molecular clouds, where new stars form. In the decades since this idea was introduced astronomers have observed many details of these interstellar gas phases and their relationships to the life of the ISM and to the birth of stars. But even with this added data the mixing history of the ISM is still not clear. What is clear is that my ideas changed the interpretation of the level of radioactive abundances that would be expected on average within molecular clouds. This paper was immediately accepted by astrophysics reviewers, unlike the rejections voiced by the referees of my papers on the relevance of nuclear astrophysics to meteoritics.

[2] Donald D. Clayton, "Discovery of s-process Nd in Allende residue", *Astrophysical Journal,* **271**, L107 (1983). At the March 1983 Lunar & Planetary Science meeting in Houston, Lugmair *et al.* had suggested that the excess ^{142}Nd and ^{143}Nd had resulted from radioactive decays of ^{146}Sm and ^{147}Sm in the meteorite residues left behind after dissolving the bulk of the meteorite in strong acids. This is a poetic argument, in a way, in that the ^{146}Sm nucleus has a halflife short enough (103 million years) that it is today extinct on earth whereas the ^{147}Sm nucleus has a halflife long enough (106 billion years) that it is hardly diminished at all by the 4.6 billion year age of the earth. It remains at almost its original abundance, although the 4% of it that has decayed may have augmented the daughter ^{143}Nd abundance if the chemical ratio Sm/Nd is great enough. Poetic though it is, I found that the argument did not convince me.

My argument, which I demonstrated was workable by renormalizing their data, was that the excess ^{142}Nd and ^{143}Nd had instead been carried in by stardust from red giant stars that had been surfaced-enriched by the *s* process within those stars. In those stars the Nd abundances have nothing to do with radioactive decay. About five years later it was demonstrated that silicon carbide grains from AGB giants contain almost pure *s*-process isotopes for many heavy elements. As frequently happened, these great discoveries remained almost unknown to astronomers, and even to most meteoriticists, owing to the technical expertise required.

[3] Photos of Kirsten and Clayton in later musical adventures can be seen on my Photo Archive: <2008 Kirsten and Claytons at Mannheim Opera>; <2004 Kirsten and Clayton in Charleston> during Spoleto music festival; <2001 Kirsten and Clayton at Schwetzingen Palace> during Schwetzingen music festival.

[4] A good summary of my gamma-ray line predictions is found in a book published in 1983 to celebrate the 70th birthday of Wm. A. Fowler, *Essays in Nuclear Astrophysics* (C.A. Barnes, D.D. Clayton, D. N. Schramm, eds.),Cambridge University Press. See chapter 18, "Cosmic Radioactivity: a gamma ray search for the origins of atomic nuclei, pp. 401-426.

[5] "*s*-process studies in the light of new experimental cross sections: distribution of neutron fluences and *r*-process residuals", F. Käppeler, H. Beer, K. Wisshak, D. D. Clayton, R.L. Macklin and Richard A. Ward, *Astrophysical Journal* **257**, 821 (1982) This updated the decomposition of heavy isotopes into their *s*-, r-, and p-process contributions from nucleosynthesis. It continued the approach pioneered by this writer. The decomposition became increasingly important for astronomy when it was later confirmed that the *r*-process began earlier in galactic history than did the *s*-process.

[6] "The Oriented Scintillation Spectrometer Experiment for the *Gamma Ray Observatory*", J.D. Kurfess, W.N. Johnson, R.L. Kinzer, G.H. Share, M.S. Strickman, M.P. Ulmer, D. D. Clayton and C.S. Dyer, *Advances in Space Research* **3**, No. 4, 109 (1983). This paper presented the scientific goals and the details of the gamma-ray detector to the astrophysics world. Jim Kurfess and Neil Johnson, who were Principal Investigator and Project Scientist, respectively, had been trained in Haymes' balloon-borne experiment at Rice University. They work at Naval Research Lab in Washington, where the detector was built.

[7] Mahoney *et al., Astrophysical Journal,* **262,** 742 (1982) presented their data from the *HEAO 3* spacecraft. This discovery paper reported how much ^{26}Al exists today in the interstellar medium. Lee *et al., Astrophysical Journal Letters,* **211,** L107 (1977) had shown data arguing that when the solar system formed the fraction of the aluminum that was radioactive was ^{26}Al/^{26}Al = 4.5 x 10^{-5}. Such a huge ratio could not exist in the ISM without creating gamma flux much greater than Mahoney *et al.* observed. My own paper, "^{26}Al in the interstellar medium", *Astrophysical Journal,* **280,** 144 (1984) presented confounding reasons why supernovae could not maintain as much ^{26}Al as Mahoney *et al.* observed. This impasse remained for several years until its incorrect assumptions could be pinpointed. I had used only the ^{26}Al /^{27}Al abundance ratio from the carbon burning core, which makes most of the stable aluminum. But more ^{26}Al now seems to be ejected from hydrogen burning in the outer layers of supernovae, which creates no new stable aluminum but does change ^{25}Mg into ^{26}Al. My 1971 paper that addressed observing the collected yields of supernovae rather than individual supernovae, using ^{60}Fe radioactivity as the prototype, was "New prospect for gamma-ray line astronomy", Donald D. Clayton, *Nature* **234,** 231 (1971). The situation was the same for ^{26}Al radioactivity, which has a comparable value of halflife. Both nuclei arise primarily from Type II supernovae.

[8] This photo <1983 Fowler Nobel Prize> was snapped at breakfast on the very day when Fowler learned of sharing the 1983 Nobel Prize in Physics. I looked older and more gaunt at this time, a bare three weeks after my father's death, whose emotional distress caused me to lose weight. My 1996 obituary for Fowler is "William Alfred Fowler (1911-1995)", Donald D. Clayton, *Publications of the Astronomical Society of the Pacific* **108,** 1 (1996)

[9] My paper creating a new tool for studies of the radioactive abundances in the interstellar medium was "Galactic Chemical Evolution and Nucleocosmochronology: A Standard Model", Donald D. Clayton, in *Nucleosynthesis: New Challenges and Developments,* W.D. Arnett and J. W. Truran, eds, University

of Chicago Press (1985). p. 65. What was novel about this work was that it presented exact solutions for the abundances of both radioactive and stable nuclei within the interstellar medium, albeit with an idealized family of models of galactic chemical evolution. These exact solutions illuminated many misunderstandings about interstellar radioactivity. I published in *Astrophysical Journal*, **285**, 411 (1984) a first sequel in which the rate of infall of metal-poor gas onto the galactic disk terminated at some time prior to the birth of the sun. That paper was submitted on February 24, 1984, and was published prior to the first paper on this standard model because the journal was published more expeditiously than the book. Three more sequels were published using these ideas: *Astrophysical Journal*, **288**, 569 (1985); *Astrophysical Journal*, **290**, 428 (1985); *Monthly Notices of the Royal Astronomical Society*, **234**, 1 (1988). Each of these publications bore my name as the single author, for I had fallen into that groove and did not easily escape from it because additional authors would simply slow me down!

A decade later the 1993 paper the paper that resolved why it was that interstellar ^{26}Al was thirty times more abundant than I could anticipate in my 1984 arguments was published with faculty coauthors: "On ^{26}Al and other short-lived interstellar radioactivity", Donald D. Clayton, Dieter H. Hartmann and Mark D. Leising, *Astrophysical Journal*, **415**, L25 (1993).

Chapter 17
[1] Our initial prediction of observable gamma rays having energies of 847 keV and 1238 keV was "Gamma Ray Lines from Young Supernova Remnants", Donald D. Clayton, S. A. Colgate and G.J Fishman, *Astrophysical Journal* **155**, 75, (1969). That paper provided the plan, the hope, that gamma ray lines could reveal fundamentals of nucleosynthesis in supernovae. Research groups designed instruments to be able to detect them. It was later selected by editors of *Astrophysical Journal* as one of the seminal works of the 20th century and was reprinted by them in the centennial volume.

My paper five years later, "Line ^{57}Co gamma rays:

new diagnostic of supernova structure", Donald D. Clayton, *Astrophysical Journal* **188**, 155 (1974), enlarged the consequences of a measurement. It was that opportunity that I felt to be slipping away by the launch delay. An opaque overlayer of 10 gm/cm^2 would be an absorption column for gamma rays about equal to a four-inch depth of water, which, though quite transparent to light in human experience, cuts the gamma-ray escape roughly in half. The overlayer above any point in the interior decreases rapidly with time owing to the fast expansion of the explosion. It had initially been much greater than this amount. It is when that thinning has fallen to about 10 gm/cm^2 that the gamma rays from the radioactivity begin to emerge from the surface and can be detected. If GRO had been up when SN1987A exploded, we could have measured the time required for that overlayer to fall low enough for the gamma rays to emerge. I also noted that the overlayer can be made smaller not only owing to general expansion but also owing to gaseous hot bubbles containing radioactive matter being stirred outward to thinner overlayers.

[2] "On presolar meteoritic sulfides ",Clayton, D. D. and S. Ramadurai, *Nature*, **265**, 427-428, 1977. Ramadurai and I met in Cardiff during summer 1976 when we first discussed the idea of this paper. My published Abstract for the 1987 meeting of the Meteoritical Society is " Oxidation and Sulfidation of the Isotopes of Ca, Ti, Cr and Fe", *Meteoritics*, **22**, 359 (1987). That public presentation attempted to elaborate of the idea that some isotopes should be more oxidized than others in the ISM. The presentation by Edward Anders that implicated silicon carbide (SiC) as carrier of Xe-s and other noble-gas components at the same meeting and proceedings is *Meteoritics*, **22**, 462 (1987) . My prediction of *s*-process xenon contained in grains from red giant stars had occurred a decade earlier: Clayton, D. D. and R. A. Ward, "s-process studies: xenon and krypton isotopic abundances", *Astrophys. J.*, **224**, 1000-1006, 1978.

[3] "Nuclear cosmochronology within analytic models of the chemical evolution of the solar neighborhood", Donald D. Clayton, *Monthly Notices of the Royal Astronomical Society*, **234**,

1 (1988). The section 3.1 of that paper presenting my explanation of contemporary growth of s- and r-process abundances was called "Simulating a primary s process", on page 21-23. I decided to use this paper rather than further correspondence with Fowler expose the weakness of Fowler's Milne Lecture *QJRAS* **28**, 87 (1987). My calculated models showed how intimately the data on radioactive chronology requires a sound and realistic model of the chemical evolution of the galaxy. In his misguided hope to reach an important conclusion, Fowler had used a simple and naïve prescription to represent galactic chemical evolution. And he did not address the uncertainty in the two cosmochronologies that he considered, namely $^{235}U/^{238}U$ and $^{238}U/^{232}Th$. My submitted paper demonstrated clearly that those two isotopic abundance ratios are woefully inadequate for determining the galactic age. This created a low in my long friendship with Fowler.

[4] "Th/Nd abundance ratio in the surfaces of G-dwarfs", Clayton, D. D., *Nature* **329**, 397, 1987. I elaborated somewhat more on this topic in section 3 of my full paper, p. 17[3].

[5] A photo of Hoyle opening the gate to his farmhouse is <1974 Hoyle at Cockley Moor home>. Another of us conversing in his farmhouse office is <1974 Hoyle and Clayton in Hoyle's Cockley Moor home>.

[6] "Stochastic models of refractory interstellar dust", Liffman, K. and D. D. Clayton, *Lunar and Planet. Sci.*, **XVIII**, Lunar and Planetary Institute (Houston), 552-553, (1987). A digest of this work also appeared in the Book *Dust in the Universe*; but the proceedings paper for the Houston NASA meeting was much more complete. A photograph with Kurt Liffman and Mark Leising is <1991 Leising and Liffman at dinner with Clayton>.

[7] My photo of Harry Kroto deep in thought was taken three years later in November 1990 while he visited Clemson University. See <1990 Workshop Kroto >. Kroto did share the Nobel Prize for discovery of Buckminsterfullerene.

Chapter 18

[1] The Lunar and Planetary Science Conferences, held each March in Clear Lake near Johnson Space Center, were important reporting venues for research supported by the Cosmochemistry Program. In 1987 and 1988 I was careful to make stimulating presentations in the effort to maintain NASA funding following my impending move to Clemson University. My plan was for cosmochemistry to remain an equal partner to astronomy with radioactivity and nucleosynthesis theory. Papers presented at those meetings were: Clayton, D. D., "New cosmic-chemical-memory mechanism for isotopic anomalies", *Lunar and Planetary Sciences*, **XIX**, Lunar and Planetary Institute (Houston), 195-196, 1988; Liffman, K. and D. D. Clayton, "Stochastic models of refractory interstellar dust", *Lunar and Planet. Sci.*, **XVIII**, Lunar and Planetary Institute (Houston), 552-553, 1987; Liffman, K. and D. D. Clayton, "Stochastic models of refractory interstellar dust", *Proc. Lunar Planet. Sci. Conf.*, **18**, Cambridge University Press (Cambridge), 637-658, 1987.

[2] The Leonard Medal Lectures are published in the journal *Meteoritics and Planetary Science*. My 1991 lecture is **27**, 5-17 (1992). I here quote its Abstract:

The science of nucleosynthesis was substantially inspired by chemical analyses of meteorites. As if in repayment, that theory now imbues meteoritics with enlarged meaning. I recount the emergence of four great issues for nucleosynthesis--issues that received decades of my own attention; and I describe unexpected abundance patterns within meteorites that were suggested by the resolution of those issues. The latter have altered the information content of meteoritic science. The issues are:

1. a quantitative s-process theory

2. cosmoradiogenic chronology

3. explosive nucleosynthesis and gamma-ray astronomy

4. cosmic chemical memory

Starting from historical origins for each issue, I comment upon both the broad cultural canvas in which they lie and my own work in their establishment. Examples of predicted (or rationalized) meteoritic measurements illustrate our surprised delight at the expansion of the range and power of meteoritic science.

[3] The first of two publications on the gamma-ray measurements was "OSSE Observations of ^{57}Co in SN1987A", Kurfess, J. D., Johnson, W.N., Kinzer, R. L., Kroeger, R. A., Strickman, M. S., Grove, J. E., Leising, M. D., Clayton, D. D., Grabelsky, D. A., Purcell, W. R., Ulmer, M. P., Cameron, R. A. and Jung, G. V.,. *Astrophys. J. (Lett.)*, **399**, L137-L140, 1992. The second on the interpretations of the value of our measured ^{57}Co abundance was "The ^{57}Co Abundance in Supernova 1987A", Clayton, Donald D., Leising, M. D., The, L.-S., Johnson, W. N. and Kurfess, J. D., *Astrophys. J. (Lett.)*, **399**, L141-L144, 1992. Both papers were submitted in June 1992, about eleven months after we began our long series of observations in space with OSSE.

[4] "New Ideas for SiC: Mg Burning in AGB Shell Flashes", Clayton, Donald D. and Brown, L. E., *Lunar Planet. Sci.* **23**, 229 (1992). Herein I was beginning my long struggle to attempt to understand how stars that were born, evolved and died before the sun was even born could have greater ^{29}Si/^{28}Si and ^{30}Si/^{28}Si abundance ratios than does our solar system. My discussion of galactic chemical evolution [Clayton, D. D., "Isotopic anomalies: Chemical memory of galactic evolution", *Astrophys. J.*, **334**, 191 (1988)] had shown that presolar stars should instead carry isotopic ratios *smaller* than those in the sun. Larry Brown and I submitted a fuller discussion for publication in a journal:"SiC particles from asymptotic giant-branch stars: Mg burning and the s process", *Astrophys. J.*, **392**, L79, (1988).

[5] "Weak s-process irradiations", J. G., Peters, W. A. Fowler, and D. D. Clayton, *Astrophys. J.* **173**, 637 (1972). We argued that the galaxy must provide an s-process stellar site that differes greatly from the stars that yiled the s-process nuclei heavier than A=88. There exist zones in supernovae that do not become heated enough to destroy s-process abundances that were created by the ^{22}Ne neutron source in the presupernova massive star.

[6] "New prospect for gamma-ray-line astronomy", Clayton, D. D., *Nature*, **234**, 291, 1971. In that paper I proposed that individual supernovae might not create enough ^{60}Fe to be observable at earth,

but that the roughly 100,000 supernovae that have exploded in the Milky Way during the last few million years may establish an observable steady-state concentration of ^{60}Fe in the ISM.

[7] The number of excess neutrons in a single nucleus is defined as N-Z, the difference between the number of neutrons (N) and the number of protons (Z). For ^{48}Ca that excess is eight, since N=28 and Z=20. Thus the fractional excess of neutrons in ^{48}Ca is 8/48 = 1/6. This is a huge number within the spectrum of natural, stable nuclei.

[8] "^{48}Ca Production in Matter Expanding from High Temperature and Density", Meyer, B.S., Krishnan, T. D. & Clayton, D. D., *Astrophys. J.* **462**, 825 (1996) presented the basic astrophysics conclusions of this study. The more fundamental theory of quasiequilibrium appeared later in "Theory of quasiequilibrium nucleosynthesis and applications to matter expanding from high temperature and density", Clayton, D. D., Krishnan, T. D. & Meyer, B. S., *Astrophys. J* **498**, 808 (1998). I personally judged these to be among the best nucleosynthesis papers of the decade, so it satisfied my scientific soul to have been able to formulate them in my sixties. But they never could have happened without Brad Meyer.

Chapter 19
[1] My paper that first pointed to cosmic-chemical-memory effects based on the galactic chemical evolution of silicon isotopes was "Isotopic anomalies: Chemical memory of galactic evolution", Clayton, D. D., *Astrophys. J.* **334**, 191-195, 1988. After Frank Timmes joined me at Clemson University as a CGRO-supported postdoc, we published more detailed studies based on his numerical modeling of the evolution of the galaxy: "Galactic Evolution of Silicon Isotopes: Application to Meteoritic Grains", Timmes, F. X. & Clayton, D. D., *Astrophys. J.* **472,** 723 (1996); "Placing the Sun in Galactic Chemical Evolution: Mainstream SiC Particles", Clayton, D. D. & Timmes, F. X., *Astrophys. J.* **483**, 220 (1997). The bottom line was our confirmation with the use of computer models of the

slope-unity mainstream line that would be produced by galactic chemical evolution. The mainstream line is defined as the plotted points for each SiC grain on a graph showing the $^{29}Si/^{28}Si$ and $^{30}Si/^{28}Si$ abundance ratios. Each grain is a separate point. As it develops, the mainstream line has a slope somewhat greater than unity, so something else is going on, perhaps coupling unit-slope galactic evolution to an additional effect.

[2] "Placing the Sun and Mainstream SiC Particles in Galactic Chemodynamic Evolution", Clayton, D. D., *Astrophys. J. Letters* **484**, L67 (1997). An article for a more general audience was written for publication on the web: "Moving Stars and Shifting Sands of Presolar History", Clayton, D. D., *PSR Discoveries*, www. soest.hawaii.edu/PSRdiscoveries (U. Hawaii: 1997)

[3] "A Presolar Galactic Merger Spawned the SiC-grain Mainstream", Clayton, D.D. *Astrophys. J.* **598**, 313 (2003). This paper bids strongly to provide the correct explanation for the mainstream line.

[4] My prediction paper was "^{22}Na, Ne-E, Extinct radioactive anomalies and unsupported ^{40}Ar", Clayton, D. D., *Nature* **257**, 36, (1975). This paper was scoffed at repeatedly during the next two decades. Then Nittler *et al.* in *Astrophys. J.* **462**, L31 (1997) demonstrated with secondary-ion mass spectrometry that excess ^{44}Ca existed in a certain class of SiC stardust owing to the *in situ* decay within them of radioactive ^{44}Ti that is created only in exploding supernovae. This subgroup of SiC particles is unrelated to the mainline, which originates in AGB stars, but is instead a SiC class of supernova SUNOCON.

[5] The first published journal paper on condensing graphite in oxygen-rich gas was "Condensation of Carbon in Radioactive Supernova Gas", Clayton, D. D., Liu, W. and Dalgarno, A.,, *Science* **283**, 1290-1292 (1999). Preliminary versions of our research strategy were presented in March 1999 in Houston: Clayton, D. D. and Liu, W., "Condensation of Carbon in Supernovae", *Lunar Planet. Sci.* **30** (1999) CDROM, Lunar & Planetary Institute (Houston); Liu, W. and Clayton, D. D., "Condensation

of Carbon in Radioactive Supernova Gas", *Lunar Planet. Sci.* **30** (1999) CDROM, Lunar & Planetary Institute (Houston). Our approach was to kinetically model the growth of linear carbon-chain molecules in the presence of destructive oxygen until they became large enough that isomerization to ringed-form carbon molecules occurred, after which time they become more resistant to oxidation.

[6]   Our first paper with my Ph.D. student Ethan Deneault was "Condensation of Carbon in Radioactive Supernova Gas", Clayton, D.D., Deneault, E. & Meyer, B. S., *Astrophysical Journal* **562**, 480-493 (2001). Trying to account for the existence of isotopes from zones overlying the condensation zone for SiC, I next examined possible effects of the reverse shocks that exist in young supernovae on the grain chemistry in "Supernova Reverse Shocks and Presolar SiC Grains", Deneault, E., Clayton, D. D. & Heger, A., *Astrophys. J.* **594**, 312-25 (2003). Then one more final paper containing several sophistications from Ethan's Ph. D. thesis: "Growth of Carbon Grains in Supernova Ejecta", E. A. Deneault, Meyer, B.S. & Clayton, D.D., *Astrophys. J.* **638**, 234-40 (2006). I personally believe these papers will have a large future influence on astrophysics; but I think it will take another decade. I do not think it will be possible to understand SUNOCON isotopes without our condensation picture. It will take time for astronomers to become aware of the vast data bank within SUNOCONs and equally long for most geochemists to grasp that these papers discovered a new road for kinetic carbon chemistry. Of course, these papers could be flawed in some way that I can not see.

[7] "Astrophysics with Presolar STARDUST", Clayton, D.D. & Nittler, L.R., *Ann. Rev. Astron. Astrophys* **42**, 39-78 (2004). A photograph of us during the writing period can be seen at <2003 Nittler and Clayton at Carnegie Symposium IV>.

[8] I took that chapter-by-chapter view of the varied characteristics of the isotopes of the chemical elements and their abundances in my recent book *Handbook of Isotopes in the Cosmos* (Cambridge University Press, Cambridge UK 2003). For a general reader,

whatever his background, I recommend its introductory chapter, which I wrote for laymen as a stand-alone essay on nature's intricate canvas of abundances.

INDEX

abundance scale   136

acceleration of gravity   36, 81, 95, 123

Adair County, Iowa   2, 6, 531

Adams, Gerry   262

AGB stars   417, 458, 518, 569

airplanes   17-19, 21-4, 30-1, 34-5, 66, 119

Aluminum-26   (^{26}Al)   421-24, 427, 429, 452, 510, 560, 564

Al_2O_3 sapphires   390, 392, 551, 560

Alley Theater (Houston)   233

alpha-rich freezeout   238

Amari, Sachiko   517

Allende meteorite   310, 389, 417

ancestors   6-9

Anders, Edward   173, 331, 377, 380, 400, 445, 449, 490, 554, 555, 557, 561

Anderson, Carl   111, 183, 312

antiques   282

Apollo 11   254, 290, 297, 488

apprentice (Caltech)   127

argon-40 (^{40}Ar)   394

argon-36 (^{36}Ar)   395

ARPA Defense Science (UCLA)   189-94

Arnett, W. David   235, 244, 250, 289, 545, 546

Arnould, Marcel   395, 463

Atlantis (shuttle)   494

Audouze, Jean   413, 416

Bahcall, John   157, 209, 359

barium   135, 353, 457

Barnes, Charles 131, 372
Beer, Hermann 462
Bethe, Hans 126, 321, 354
B^2FH 121, 147, 150, 187, 202, 241, 537, 539
Big Bang 123, 157, 210, 364, 531
births
  mine 27-9
  Keith 33
  Donald 140-1
  Devon 144
  Alia 397
  Andrew 437
Black, David 352, 556
black hole, solar 358, 557
Blairbuie House (Scotland) 407
Bodansky, David 203, 224, 239, 352, 543, 550
Bomar house (Houston) 407, 430, 431
Buckyballs 479-80
Burbidge, Geoffrey 219, 339, 372, 525, 544
Burrows, Adam 491
Butcher, Harvey 454
calcium-40,42 ($^{40,42}Ca$) 444
calcium-41 (^{41}Ca) 353, 395, 560
calcium-48 (^{48}Ca) 513, 570
calcium oxidation/sulfidation 443, 445
CalciumAluminum inclusions (CAI) 307, 391, 551, 558
californium hypothesis 167-9, 227, 544
Calhoun, John C. 484
Caltech (general) 101, 104, 106
Cambridge 178-9, 216, 327
Cameron, Alastair G. W. 198, 201, 271, 288, 309, 344, 378, 543, 549, 552, 560
carbon, condensation of 416, 522-4, 373, 400, 518, 522, 524, 571, 572
carbon, combustion of 523, 524
carbon monoxide 522, 523
carbon-13 neutron source 125-6, 147, 458
carbon-13 isotope (^{13}C) 137
Cardiff 362, 369-74, 558, 566
CasA supernova 511
Challenger 433
Chandrasekhar, S. 165
charmed ^{26}Mg 393
chess 95, 419

Clayton, Alia (daughter)	381, 395, 398, 408, 412, 431, 439
Clayton, Andrew Adair (son)	438, 471-72, 500-03
Clayton, Annette Hildebrand	300, 315, 318-20, 322-25, 360, 364, 374, 381, (second wife) 395, 401
Clayton, Avis Irene Kembery	7, 10-16, 19-20, 24-9, 27-9, 41-2, 44-6, 55, 66, 68, (mother) 82, 534
Clayton, Barbara Anne (sister)	72
Clayton, Carolyn Ruth (sister)	43-4, 47, 64
Clayton, Delbert Homer (father)	10-16, 37-8, 40, 44-6, 58, 66, 75-9, 80, 105, 106, 118-9, 161, 407, 412, 423-4, 532
Clayton, Devon Charles (son)	144, 214, 322, 364, 371, 375, 399, 409, 412, 438, 472-73, 514-16
Clayton, Donald Douglas (son)	144, 214, 274, 323, 363, 399, 409, 412, 414-15, 438
Clayton, Fowler & Hoyle (book)	179-83, 186-9, 221
Clayton, Keith Eugene (brother)	42, 44, 48, 52, 70, 72, 80, 99, 360
Clayton, Mary Lou	98-9, 103-5, 106-9, 142-4, 146, 163, 171, 173, 196, (first wife) 214, 234, 256, 259-67, 273, 276, 294, 323, 398
Clayton, Nancy McBride (wife)	404-08, 412, 418, 430-32, 438, 441, 446, 471
Clayton, Paul V (grandfather)	17, 67
Clayton, Robert M.(gr. grandfather)	7
Clayton, Robert N.	305, 333, 398, 524, 551, 555, 560
Clemson, Thomas Green	484
Clemson University	451, 484
CN cycle	125-6, 244
cobalt-56 (^{56}Co)	228, 435, 497, 499, 545, 550
cobalt-57 (^{57}Co)	252, 436, 495, 497, 499, 547
Cockroft, Sir John	257
Colgate, Stirling	230, 543
Compton Gamma Ray Observatory	see Gamma Ray Observatory
COMPTEL	508, 510
computers arrive	157
contract bridge	96
cosmic chemical memory	387, 400, 559
cosmochronology, nuclear	154-7, 164, 427, 452, 460, 465, 539, 540, 565
Cosmos Club	491
Coulomb deexcitation	175, 541
Crab Nebula	167-8, 494, 540
Craddock, Wade	168, 540
cricket	279

Curl, Robert	479
Dalgarno, Alex	522, 569
Dallas News delivery boy	61, 73
The Dark Night Sky	249, 320, 330, 360, 535, 551
George Darwin Lecturer (RAS)	400, 559
Davenport, Iowa	30-8, 40-3
Davis, Raymond Jr	206, 357, 383, 551, 556
delayed radioactive power	498
Deneault, Ethan	523, 570
Department of Space Science	159, 176
Dessler, Alexander	159, 162, 289
Diehl, Roland	510
Disneyland APS meeting	359
divorce	295, 322, 402, 406
Domingo, John	129, 212
Donald's schooling	195-6, 223, 278
drive to California	106, 190-1
Durham Cathedral	439, 440
Durham University	440
Dust in the Universe (book, conf.)	478
Dwek, Eli	358, 381, 557
Dyson, Freeman	376
Earthrise behind the Moon	298
Eberhardt, Peter	349, 397
Einstein, Albert	397
El Eid, Mounib	395, 508
El Goresy, Ahmed	388, 490
electric charge, quantized	102
energy levels	112, 140, 538, 539
europium	457
Explorer 1	123, 537
extinct radioactivity	349-51, 364, 388, 394, 410, 429
farm machinery	1-2, 81
Feynman, Richard	141-44, 157, 208, 213, 438
Fisher, John	97, 536
Fishman, Gerald J.	222, 226-8, 476, 497, 545
fission	226, 334, 555
Fitzsimmons, Robert	433
Fontanelle, Iowa	4, 79-82
Fowler, William A.	112-4, 120-2, 124, 147, 174, 179-83, 186-9, 191-2, 207, 219-20, 231, 237, 270-72, 281-4, 310, 362-5, 373-4, 448, 459-62, 467, 525, 530, 537, 541, 542, 555, 561, 564

Fowler Nobel Prize	165, 320-21, 366, 369, 426, 467-70, 564
Foucault's pendulum	88, 95, 100, 537
free-decay interval	410
Friday evenings in Kellogg	130
galactic chemical evolution	421, 427, 453, 540, 562
galactic merger, cannibalism	521
GALLEX	382-5
gamma-ray astronomy	166, 227-8, 231, 287
Gamma Ray Observatory	385, 420, 434, 440, 494, 509
Gell-Mann, Murray	114, 354
Germania Haus (Würzburg)	421
ghost, isotopic	392
George W. Gignilliat House	485, 498
Gignillat, Thomas	485
globular cluster	164
Goddard Space Flight Center	489
golf	72, 85
Gordon Research Conference	165, 172-3, 296
The Grange (Cambridge)	276, 287, 474
grief of loss	48-9, 51-54, 173, 295, 324, 402, 413, 424, 514-16
Greenstein, Jesse	133
Gregynog (conference on isotopes) 373	
Grossman, Lawrence	305, 308, 549
Hartmann, Dieter	493-94
Harvard scholarship	85
Hawking, Stephen	234, 359
Haxel, Otto	200
Haymes, Robert C.	166-9, 222, 241, 420, 497, 546
Hebbard, Dale	128
Heger, Alexander	525
Heidelberg	178-9, 369
Heymann, Dieter	310
Highland Park High School	55, 64-5, 71, 74, 83, 89-90, 536
Hillebrandt, Wolfgang	397
Howard, William Michael	243, 250, 352, 416, 546
Hoyle, Barbara	215, 467
Hoyle, Fred	112, 123, 179, 186-9, 199, 201-2, 210, 220, 270-72, 284, 298, 303, 310-16, 320-21, 337-40, 348, 370, 374, 448, 465-70, 480, 525, 530, 536, 537, 541, 543, 545, 555, 557
Hoyle's equation	235, 545
Iben, Icko	157, 452

Icarus	270
Inst. Theoretical Astronomy	179, 212, 219, 232
interdisciplinary insight	166, 522
iodine-129 (^{129}I)	332, 369, 379, 409, 552
iron abundance peak	199, 237
iron-60 (^{60}Fe)	242, 424, 511, 567
isotopes, silicon	518-20
Isotopes in Cosmos, Handbook	532, 550
Jacoby, Oswald	97, 536
Johnson, W. Neil	241-42, 385, 546, 563
Johnson Space Center	177
Joshua Factor	317, 407, 551
journeyman rating	152
Kaeppeler, Franz	131, 462, 539, 563
Kellogg Radiation Lab	128, 147, 557
Kembery, Frank and Martha	8, 51-2, 79-81, 173, 402, 533
Kennedy, John F.	250
Kirsten, Till	369, 382-5, 410, 419, 556, 561
Korschinek, Guenter	512
Krishnan, Tracy	513
Kroto, Harry	479-80, 565
Kurfess, Jim	242, 385, 438, 492, 561
Lauritsen, Thomas	111, 140, 146, 537
Leising, Mark	420, 474-77, 493
Liffman, Kurt	488, 565
Leonard Medal	489, 490, 505, 566
Liffman, Kurt	478
Liu, Weihong	523, 571
Lugmair, Guenter	417
Lunar Science Conference	297, 369, 566
Lunatic Asylum (Caltech)	364, 380
Macklin, Richard L.	152, 171, 539, 540
Magellanic Clouds	433
magic numbers (nuclear)	152, 200, 512, 537, 549
mainstream SiC	518, 569
manifesto (stardust)	393, 559
mathematics prize	83, 536
Mathews, Grant	459
Max Planck Inst. for Nuclear Phys	369, 382, 556
MPI for Extraterrestrial Physics	396
May, Robert M.	205, 542
McAuliffe, Crista	433
McDonald, Frank C	92-95, 101-04,
McGraw-Hill	208
McMurtry, Larry	224

McNulty, Peter	492
de Menil, John and Dominique	245
metal-poor star	134, 456-7
meteorites	138, 302, 331, 355
Meteoritical Society meeting	
Albuquerque	416
Bethlehem	378
Cambridge	389, 391,397
Durham	442-45,
Meyer, Bradley S.	415, 493, 513, 523, 568
merger, galactic	521
Michel, F. Curtis	159, 175, 292, 541
Millikan oil-drop experiment	102
mixing, galaxies	521
mixing, stars	458, 548
mixing, supernova	437
mixing, molecular cloud	411, 562
Mt. Humphries	118
Naval Research Laboratory	419, 492
NASA Achievement Medal	508
NASA Cosmochemistry Program	378, 488, 568
NASA Origins of Solar Systems	479, 488, 507
National Academy of Science	449-51,
National Science Foundation Fellow	104
neodymium, s-process	417, 443, 455, 562
neon-E, ^{22}Ne	352, 397, 548, 556, 571
neutrinos, SN1987A	434, 440
neutrinos, solar	157, 357, 382, 553, 557
neutron-capture cross sections	152, 171, 563
Newman, Michael J.	253, 358, 548, 557
Newton's milkpail	81, 123
Nickel-56 (^{56}Ni)	200, 227, 235
Nickel-57 (^{57}Ni)	238, 495
Nickel-58 (^{58}Ni)	236
Nittler, Larry	517, 525, 572
nitrogen-13 (^{13}N)	312, 553
nitrogen-14 (^{14}N)	126, 546
^{15}N+^{3}He	127
nova explosions	311-3, 339, 555
nova positron annihilation	312, 553
nova stardust	558
nuclear shell model	200,551
nucleosynthesis watershed	197
Nuth, Joseph	507
Oak Ridge National Lab	152, 171, 175

O'Brian, Patrick	299
"OK then, you fly it"	jacket, 37, 38
Opik, Ernst	262
OSSE	386, 419, 437, 509, 511, 561
overabundance in stars	135
oxygen-16 (^{16}O) excess	305, 310, 333, 560
Paisley, Ian	263-69
Paris honeymoon	414
Pasadena rentals	108, 195
Pearson, John	130
Penn State University	447
Phi Beta Kappa	99
photo-induced beta decay	175, 543
physics in childhood	15-16, 36, 55, 70, 78, 81, 84
Pillinger, Colin	446
Pippard, Brian	234
plutonium (^{244}Pu)	332, 555
polio	16-17, 82
positron, positronium	241, 312, 546, 553
potassium-40 (^{40}K)	394
potassium-39 (^{39}K)	394
potassium-41 (^{41}K)	353
Potter, Beatrix	432
primary nucleosynthesis	456-58
Principles of Stellar Evolution &	
Nucleosynthesis	162, 179-83, 186-9, 197, 213, 222,
	224, 255
Proust, Marcel	279
psychological fear	140
Quaide, William (NASA)	378
quasiequilibrium	203-4, 226, 514
quasiequilibrium bridge	204
The Queen Elizabeth (Cunard)	214
r process	131, 153, 167, 178, 541
r-process abundances	175, 542
radiogenic iron	543
Ramaty, Reuven	386-7, 420
Read, Clark	245
Rees, Martin	372
referees (SUNOCON)	341-44
Renken, James	362
Reynolds, John H.	376, 554
rhenium-osmium chronology	154-7, 428, 455, 463, 541
Rice University	159, 250, 310-14, 412, 415, 487
Ringberg Castle (Bavaria)	511

Robertson, Howard P.	113
Rocky Mountains	530
Rose Parade	118
Rosselini, Roberto	245, 330, 547
running roads	281, 430
Ryle, Martin	338, 366
s process	125, 147-52, 175, 253, 421, 540, 548
s process in stardust	330-34, 354, 400, 443, 554
s process xenon	330-34, 354, 400, 417, 446, 549, 556, 559, 564
s/r polarization	444
sabbatical leave	178, 300, 437
Salpeter, Edwin	123, 188, 262, 538
samarium-146,147	417
Sands, Matthew	141
Sargent, Wallace	220, 272
scarlet fever	40, 534
Schmidt, Maarten	188
school report cards	62-3, 71
Schramm, David	364, 372, 459, 560
Schwarzschild, Martin	165
science truth	138, 236, 240, 303, 312, 368, 393, 481, 491, 552
Scotland mountains	219, 233, 271-6, 326, 407, 532, 549
Scowen, Paul	488
Sears, Richard	157
secondary nucleosynthesis	456-60
Seeger, Phillip A.	131, 153, 157, 167, 178, 331, 362, 542
Shaw, Peter B.	175, 543
Sherry Lane (Dallas)	57, 82
silicon carbide (SiC)	446, 504, 518, 569
$^{29}Si/^{28}Si$ isotope ratio (graph)	518-20
silicon melting	198
Silk, Joseph	286, 550
slide rule	207
Sloan Foundation Fellow	189
Smalley, Richard	479
sodium-22 (^{22}Na)	244, 352, 360, 479, 553, 556
Solar Maximum Mission	475
soot	522
South Carolina Watermedia Society	529
Southern Methodist Univ. (SMU)	56, 88
Space Science Laboratory (Rice)	176

Sputnik 1	122
stardust	303, 310, 354, 389, 394, 552, 559, 572
steady-state cosmology	124
St. Clements Gardens, Camb.	300, 474
St. Mary's College, Durham	440
story time	408, 432
Suess, Hans	187, 201, 539, 551
SUNOCON	303, 333-4, 389, 444, 522, 559, 572
supernova 1987A	362, 434, 436, 439, 496, 548
gamma-ray lines from	362, 475, 496
Takahashi, Koji	397
Talbot, Raymond J.	251, 357, 398, 557
Tangley house	223,
Texas Mafia in Cambridge	250
The, Lih-Sin	491
Thielemann, Friedl	397, 559
Thorium chronology	428, 455, 467
Thorne, Kip	209
time dependence	
s process	149-51, 540
r process	153, 174
quasiequilibrium	198, 203
Timmes, Frank	519, 520
Titanium-44 (^{44}Ti)	239, 353, 511, 522
Tongue Hotel (Scotland)	405
transformation, personal	302-04, 345-55, 363, 368
Truran, James W.	204, 550, 560
University Park (Park Cities)	55, 57, 534, 535
Van de Graaff accelerator	127-8, 171, 538
Vance, Nina	234
Venice	371-72, 558
da Vinci, Leonardo	418
von Humboldt Award	370, 381, 396, 409, 558
Wagoner, Robert V.	210, 219
Walker, Doak	88-90
Walker, Robert M.	173, 506
Wasserburg, G. J.	364, 367, 374, 380, 393
Ward, Richard A.	250, 355, 400, 548, 549, 556
wedding	
first	98
second	301
third	412
Robert A. Welch Foundation	378
West Bridge Lab. (Caltech)	116

West University Place   (Houston)   161
Wetherill, George   375
White Cottage   (Cambridge)   216-18, 232, 474
White, Edward H.   177
white dwarf star   311-13, 339
Wickramasinghe, Chandra   341, 370, 372, 558
Wiess College (Rice)   301, 325
windbagging   191-2
Wind River Range   529
Wolf, Richard A.   203
Wolfendale, Arnold   439
Woosley, Stanford E.   224, 235, 250, 352, 372, 375, 494, 519, 545, 550

World War II   65
Wutöschingen   (Schwarzwald)   318-20, 381, 395
xenology   331-2, 376, 554
Yorkshire Moors   442
Zimmerman, Barbara   207
Zinner, Ernst   505, 506, 507, 517, 522, 525
Ziurys, Lucy   522
zur Forstquelle, Heidelberg   382, 409